Research Methods for
BSW Students
—eighth edition—

Research Methods for BSW Students

Richard M. Grinnell, Jr.
Western Michigan University

Margaret Williams
University of Calgary

Yvonne A. Unrau
Western Michigan University

PairBondPublications.com

Check Out Our Website

PairBondPublications.com

√ *Student Workbook Exercises*
√ *Chapter Power Point Slides*
√ *On-line Glossaries*
√ *General Links*
√ *Specific Chapter Links*

Copyright © 2010 by Pair Bond Publications.

ALL RIGHTS RESERVED. No part of this work covered by the copyright hereon may be reproduced or used in any form or by any means—graphic, electronic, or mechanical, including photocopying, recording, taping, Web distribution, or information storage and retrieval systems—without the written permission from: permissions@pairbondpublications.com

Printed in the United States of America

Credits appear on page 380 and represents an extension of the copyright page

Pair Bond Publications
6985 Oak Highlands Drive
Kalamazoo, Michigan 49009-7508
(269) 353-7100

Contents

PART I   AN INTRODUCTION TO INQUIRY
    1 Scientific Inquiry and Social Work   2
    2 Research Questions and Problems   26
    3 Research Ethics   48

PART II   APPROACHES TO KNOWLEDGE DEVELOPMENT
    4 The Quantitative Research Approach   66
    5 The Qualitative Research Approach   98

PART III   MEASURING VARIABLES
    6 Measuring Variables   116
    7 Measuring Instruments   134

PART IV   SAMPLING AND THE LOGIC OF RESEARCH DESIGNS
    8 Selecting Research Participants   152
    9 Case-Level Research Designs   166
    10 Group-Level Research Designs   186

PART V   COLLECTING DATA
    11 Collecting Quantitative Data   230
    12 Collecting Qualitative Data   250
    13 Selecting a Data Collection Method   266

PART VI   ANALYZING DATA
    14 Analyzing Quantitative Data   282
    15 Analyzing Qualitative Data   310

PART VII   RESEARCH PROPOSALS AND REPORTS
    16 Quantitative Proposals and Reports   332
    17 Qualitative Proposals and Reports   354
    Credits   380 / Index   381

Preface

THIS INTRODUCTORY RESEARCH METHODS TEXT is intended for BSW students as their first introduction to social work research methodology and statistics. We have selected and arranged the contents so that it can be used in a one-semester (or quarter) beginning social work research methods course.

GOAL OF BOOK

As in the previous editions, our goal is to produce a highly accessible inexpensive "student-friendly," straightforward introduction to social work research methods couched within the quantitative and qualitative traditions—the two approaches most commonly used to generate relevant social work knowledge.

To accomplish our goal, we have strived to meet the following simple objectives:

- Our book complies with the Council on Social Work Education's (Council) research requirements. In fact, this book meets all of the Council's requirement for the research content to be taught in all accredited social work programs.
- We include only the *core* material that is realistically needed in order for BSW students to appreciate and understand the role of research in social work; that is, our guiding philosophy is to include only material that they realistically need to know to function as entry-level generalist practitioners; information overload is avoided at all costs.

    After teaching social work research for decades we asked ourselves a simple question, "What research methodology and statistical content can *realistically* be taught in a one-semester course given the student population?" The answer to our question is contained in the Content's section of our book (Page v). However, more advanced additional content is easily accessible from the book's Website, such as: evidence-based practice, meta-analyzes, program evaluation, evaluating quantitative and qualitative research reports, systematic reviews, and multivariate statistics, to name just a few.

In addition to the access of additional advanced content, the book's Website also contains relevant Web-based links to tutorials and advanced readings for each chapter in the book.

- We provide students with a solid foundation for more advanced social work research courses and texts.

- Our book prepares students to become beginning critical consumers of the professional research literature. It also provides them with an opportunity to see how social work research is actually carried out.

- We explain terms with social work examples that students will appreciate. Many of our examples center around women and minorities, in recognition of the need for social workers to be knowledgeable of their special needs and problems.

- This book is written in a crisp style and uses direct language. It is easy to teach *from* and *with*.

- Our book has an extensive companion student-orientated Website that contains numerous student (and instructor) resources.

- A free 396-page *Student Study Guide* is now available (from the book's Website). Students can easily download the homework exercises, complete them in Word, and then e-mail their completed exercises to their instructors. The Instructors Resources section (password protected) on the Website contains the answers to the exercises.

- Our book sells for $40.00 new and $15.00-$30.00 used. Hands down, this book is the least expensive research methods text on the market today. We feel that requiring undergraduate students to purchase a far more expensive book (for two or three times the price) is asking way too much of them considering their tuition and textbook costs have been increasing faster than their income levels.

- On the environmentally-side of things, we continually strive for our book to be 100% green; that is, we comply, whenever possible, on meeting all the requirements of the Green Press Initiative (www.greenpressinitiative.org).

- On the social/professional responsibility-side of things, we donate 10% of the yearly net profits to the National Association of Social Workers.

Like all introductory social work research books, ours had to include *relevant* and *basic* research content. Our problem here was not so much what content to include as what to leave out. As previously mentioned, the research methodology and statistical content that we touch on in passing is treated in-depth elsewhere. But our elementary book is a primer, an introduction, a beginning. Our aim is to skim the surface of the research enterprise—to put a toe in the water, so to speak, and to give beginning social work students a taste of what it might be like to swim in the research arena.

ORGANIZATION

With the above goal and objectives in mind, our book is organized to follow the basic phases of the generic research process—from a quantitative perspective and from a qualitative perspective. The book begins where every researcher begins; that is, with finding a meaningful problem area to study and developing research questions and hypotheses. It then proceeds from measuring variables, selecting samples, constructing research designs, collecting and analyzing data, through report writing.

Our book is organized in a way that makes good sense in teaching research methods. Many other sequences that could be followed would make just as much sense, however. The chapters in our book are consciously planned to be independent of one another. They can be read out of the order in which they are presented, or they can be selectively omitted.

For example, some research instructors require their students to complete a research proposal for one of the course requirements. In this case, an instructor can require the students to read Chapters 16 and 17 (on quantitative [Chapter 16] and qualitative [Chapter 17] proposal writing) somewhere at the beginning of the course.

The creative use of our book is highly encouraged. We have only provided the minimal course content that we feel should be taught to BSW students. It is up to the instructor to add to our foundation through the use of the book's Website and/or their own personal material, additional readings, videos, group assignments, guest lectures, and so forth.

WHAT'S NEW IN THIS EDITION?

We solicited extensive feedback from instructors who adopted the previous editions. From their comments we made the following changes in this edition.

- Many instructors felt that our book would be more useful to students if it was physically larger so students could write notes in the margins. So, we expanded its size to 8½ X 11.

- Some instructors thought it would be useful if the chapters were broken down into specific concrete sections, or parts, that followed the research process. In his spirit, we now have seven parts that contain relevant chapters that reflect the research process (see Contents on Page v)
- Some instructors thought it would be useful to have every chapter preceded by a chapter outline. Presto, their wish was our command. Also, there is a PowerPoint presentation for each chapter on the book's Website. They can be used by students and instructors alike.
- Many instructors thought a new chapter would be useful that focused on how to formulate research questions and problems in addition on how to use the literature in formulating them. A new chapter (Chapter 2) now addresses these issues.
- Most of the instructors thought the accompanying 396-page *Student Study Guide* (that previously sold for $10.00) should be available to students free of charge. So, it is now free and can be downloaded from the book's Website.
- And, most importantly, there was an unanimous agreement that the cost of the book should remain as low as possible. So, we have kept the cost to $40.00.

In short, students can receive this book, a 396-page downloadable student study guide, and an accompanying Website for $40.00—an unheard of price in today's textbook market.

A FINAL WORD

The field of research in our profession is continuing to grow and develop. We believe this edition will contribute to that growth. A ninth edition is anticipated, and suggestions for it are more than welcome. E-mail your comments directly to: rick.grinnell@wmich.edu.

We hope that the apparent levity with which we have treated the social work research enterprise will be accepted in the same spirit as it was intended. Our goal was not to diminish research; it was to present the research process with warmth and humanness so that the student's first experience with it will be a positive one. After all, if wetting their big toes scares them, they will never learn to swim.

Richard M. Grinnell, Jr.
Margaret Williams
Yvonne A. Unrau

Check Out Our Website

PairBondPublications.com

√ *Student Workbook Exercises*
√ *Chapter Power Point Slides*
√ *On-line Glossaries*
√ *General Links*
√ *Specific Chapter Links*

Part I

An Introduction to Inquiry

Chapter 1: Scientific Inquiry and Social Work   2

Chapter 2: Research Questions and Problems   26

Chapter 3: Research Ethics   48

Chapter 1

Scientific Inquiry and Social Work

RESEARCH AND ACCOUNTABILITY
    The Council on Social Work Education
    The National Association of Social Workers
HOW DO WE ACQUIRE KNOWLEDGE?
    Authority
        Questioning the Accuracy of Authority Figures
        Sources of "Evidence"
    Tradition
        Knowledge vs. Beliefs
    Experience
    Intuition
        Intuition and Professional Judgment
    The Research Method
        Value Awareness
        Skeptical Curiosity
        Sharing
        Honesty
PHASES OF SCIENTIFIC INQUIRY
    Example of the Scientific Inquiry Process
RESEARCH AND PRACTICE: BOTH PROBLEM-SOLVING PROCESSES
RESEARCH ROLES
    The Research Consumer
    The Creator and Disseminator of Knowledge
    Integrating the Two Research Roles
SUMMARY
REFERENCES AND FURTHER READING

1

MADAME CLEO IS AN ASTROLOGICAL CONSULTANT. She advertises widely on television promising that her astounding insights into love, business, health, and relationships will help her viewers to achieve more fulfilling and gratifying lives. Hah! you think. I bet she can't do this for me! I bet she's just out for the money! But if she could, but if she could only tell me ... !

How do I know if she's for real or I'm just getting taken for a ride? Perhaps the unscrupulous banks that dolled out thousands upon thousands of high risk (subprime) home loans to folks who couldn't afford them could have used her services. What about the Enron Corporation as well?

There is a parallel here between the people who receive social services—sometimes called clients—and you, the future social worker. Most people who we help—in common with all those people who are never seen by social workers—would like more fulfilling and rewarding lives.

Like Madame Cleo's naive clientele who get suckered into calling her, many of our clients also have personal issues, money issues, relationships issues, or health issues. Unlike Madame Cleo, however, who only has to be accountable to her checkbook, we, as a profession, are required to be accountable to society and must be able to provide answers to three basic accountability questions:

- How do our *clients* know that we can help them?
- How does our *profession* know that we have helped our clients?
- How do the *funding bodies* that fund the programs* (which employ us) know how effectively their dollars are being spent?

RESEARCH AND ACCOUNTABILITY

What is the role that research plays in answering the above three accountability questions? In one word *significant!* That is the position of both the Council on Social Work Education (CSWE) and the National Association of Social Workers (NASW). These two prestigious national accountability organizations have a tremendous amount of jurisdiction on what curriculum content is required to be taught to all social work students (CSWE) and how the students, after they graduate, practice their trade (NASW).

The Council on Social Work Education

The CSWE is the official "educational organization" that sets minimum curriculum standards for BSW and MSW programs throughout the United States. This accreditation organization firmly believes that all social work students should know the basic principles of research. For example, the latest version of CSWE's *Educational Policy and Education Standards* (2008) contains the following two "research-type" policy statements (i.e., 2.1.3, 2.1.6) that all professional social workers must adhere to:

> *Educational Policy 2.1.3*—Apply critical thinking to inform and communicate professional judgments. Social workers are knowledgeable about the principles of logic, scientific inquiry, and reasoned discernment. They use critical thinking augmented by creativity and curiosity. Critical thinking also requires the synthesis and communication of relevant information. Social workers...
>
> - distinguish, appraise, and integrate multiple sources of knowledge, including research-based knowledge, and practice wisdom
> - analyze models of assessment, prevention, intervention, and evaluation
> - demonstrate effective oral and written communication in working with individuals, families, groups, organizations, communities, and colleagues
>
> *Educational Policy 2.1.6*—Engage in research-informed practice and practice-informed research. Social workers use practice experience to inform research, employ evidence-based interventions, evaluate their own practice, and use research findings to improve practice, policy, and social service delivery. Social workers comprehend quantitative and qualitative research and understand scientific and ethical approaches to building knowledge. Social workers...
>
> - use practice experience to inform scientific inquiry
> - use research evidence to inform practice

With the preceding two policy statements by CSWE in mind, we need to know basic research methodology so we can be in compliance with the two policy statements; that is, it is extremely difficult for social workers to meet the two policy statements without an understanding of the contents in this book.

The National Association of Social Workers

Just like CSWE, NASW is a parallel "practice organization" that works to enhance the professional growth and development of practicing social workers. Like social work students with CSWE, NASW believes that social work practitioners should also know the basics of research:

(a) Social workers should monitor and evaluate policies, the implementation of programs, and practice interventions.

(b) Social workers should promote and facilitate evaluation and research to contribute to the development of knowledge.

(c) Social workers should critically examine and keep current with emerging knowledge relevant to social work and fully use evaluation and research evidence in their professional practice.

(d) Social workers engaged in evaluation or research should carefully consider possible consequences and should follow guidelines developed for the protection of evaluation and research participants. Appropriate institutional review boards should be consulted.

(e) Social workers engaged in evaluation or research should obtain voluntary and written informed consent from participants, when appropriate, without any implied or actual deprivation or penalty for refusal to participate; without undue inducement to participate; and with due regard for participants' well-being, privacy, and dignity. Informed consent should include information about the nature, extent, and duration of the participation requested and disclosure of the risks and benefits of participation in the research.

(f) When evaluation or research participants are incapable of giving informed consent, social workers should provide an appropriate explanation to the participants, obtain the participants' assent to the extent they are able, and obtain written consent from an appropriate proxy.

(g) Social workers should never design or conduct evaluation or research that does not use consent procedures, such as certain forms of naturalistic observation and archival research, unless rigorous and responsible review of the research has found it to be justified because of its prospective scientific, educational, or applied value and unless equally effective alternative procedures that do not involve waiver of consent are not feasible.

(h) Social workers should inform participants of their right to withdraw from evaluation and research at any time without penalty.

(i) Social workers should take appropriate steps to ensure that participants in evaluation and research have access to appropriate supportive services.

(j) Social workers engaged in evaluation or research should protect participants from unwarranted physical or mental distress, harm, danger, or deprivation.

(k) Social workers engaged in the evaluation of services should discuss collected information only for professional purposes and only with people professionally concerned with this information.

(l) Social workers engaged in evaluation or research should ensure the anonymity or confidentiality of participants and of the data obtained from them. Social workers should inform participants of any limits of confidentiality, the measures that will be taken to ensure confidentiality, and when any records containing research data will be destroyed.

(m) Social workers who report evaluation and research results should protect participants' confidentiality by omitting identifying information unless proper consent has been obtained authorizing disclosure.

(n) Social workers should report evaluation and research findings accurately. They should not fabricate or falsify results and should take steps to correct any errors later found in published data using standard publication methods.

(o) Social workers engaged in evaluation or research should be alert to and avoid conflicts of interest and dual relationships with participants, should inform participants when a real or potential conflict of interest arises, and should take steps to resolve the issue in a manner that makes participants' interests primary.

(p) Social workers should educate themselves, their students, and their colleagues about responsible research practices.

This book provides the beginning research content to comply with the research standards set out by CSWE and NASW. Unlike Madame Cleo, however, social work students and practitioners are expected to have a substantial research knowledge base to guide and support their interventions. This knowledge base is generally derived from your social work education. Of course, we, as a profession, tend to have more credibility than astrological consultants like Madame Cleo.

We have graduated from accredited social work programs (CSWE) and have recognized practice qualifications (NASW, 1996). You are expected to have not only good intentions but the skills and knowledge to convert your good intentions into desired practical results that will help your clients. It all boils down to the fact that we have to be accountable to society and to do so means that we need to acquire the knowledge and skills to help our clients in an effective and efficient manner.

HOW DO WE ACQUIRE KNOWLEDGE?

Our discussion so far automatically leads us to the question of, "where do we acquire the necessary knowledge base to help our clients?" As can be seen in Figure 1.1, you will acquire your knowledge base to help others through five highly interrelated sources: (1) authority, (2) tradition, (3) experience, (4) intuition, and (5) the research method. All of these "ways of knowing" overlap to some degree but it greatly simplifies things to discuss them separately.

Authority

Some things you "know" because someone in authority told you they were true. Had you lived in Galileo's time, for example, you would have "known" that there were seven heavenly bodies: the sun, the moon, and five planets. Since "seven" was a sacred number in the seventeenth century, the correctness of this belief was "self-evident" and was proclaimed by professors of philosophy.

But Galileo peeked through his telescope in 1610 and saw four satellites circling Jupiter. Nevertheless, it was clear to those in authority that Galileo was wrong. Not only was he wrong, he had blasphemed against the accepted order. They denounced Galileo and his telescope and continued to comfortably believe in the sacredness of the number "seven."

But the "authorities" could have looked through Galileo's telescope! They could have seen for themselves that the number of heavenly bodies had risen to eleven! In fact, they refused to look because it wasn't worth their while to look because they "knew" that they were right. They had to be right because, in Galileo's time, the primary source of "how you knew something" was by authority—not by reason, and certainly not by observation.

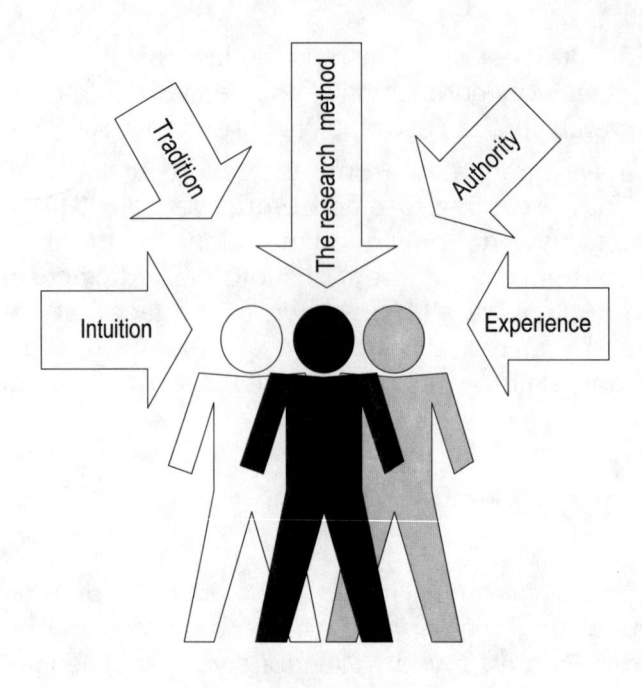

FIGURE 1.1
How We Know Something

Today, this may seem a bit strange, and we may feel a trifle smug about the fact that, in *our* time, we also rely on our own observations *in addition to* authority. Even today, entrenched opinions are extremely difficult to change and facts are more often than not disregarded if they conflict with cherished beliefs.

Questioning the Accuracy of Authority Figures

Fortunately and unfortunately you have little choice but to believe authority figures. You wouldn't progress very fast in your social work program if you felt it necessary to personally verify everything your professors said. Similarly, practicing social workers lack the time to evaluate the practice recommendations that were derived from research studies; they have no choice but to trust statements made by researchers—the authority figures—who conducted the research studies and claimed their findings "to be true."

Experts can be wrong, however, and the consequences can sometimes be disastrous. A few decades ago, for example, authority figures in family therapy believed that children who were schizophrenic came from parents who had poor parenting skills. Researchers emphasized such causative factors as parental discord, excessive familial interdependency, and mothers whose overprotective and domineering behaviors did not allow their children to develop individual identities.

In accordance with these "research findings," many social workers assumed that all families were dysfunctional who had a child with schizophrenia. Because the social workers focused their interventions on changing the family system, they often inadvertently instilled guilt into the parents and increased tensions rather than helping the parents to cope with their child's situation.

However, recent research studies now show that schizophrenia is caused largely by genetic and other biological factors, not by bad parenting. According to *these* findings, the most effective social work intervention is to support the family system by providing a nonstressful environment. This is what social workers *currently* do, again relying on *current* authority figures. More than likely the authorities are correct this time. However, not as quite as exact *today* as they will be *tomorrow* when our knowledge of schizophrenia has progressed a little bit more.

So what are we to do when we need to trust the experts but the experts might be wrong? Put simply, we need to evaluate the quality of the research studies and the "kind of evidence" the studies generate. This means that we must be able to distinguish good research studies from bad, and from, well quite frankly, the awful ones.

One of the purposes of this book is to enable you, the future social worker, to evaluate research articles, which were written from research studies, with a more discerning eye. You need to decide for yourself which research findings you will clutch to your heart and use to help your clients, and which research findings you will disregard until more information is forthcoming.

Sources of "Evidence"

The "kind of evidence" on which a practice statement is based must always be evaluated on the source(s) of the evidence that was used to make the practice statement in the first place. And, the media as a source of evidence must always be questioned. For example, we obtain knowledge by watching television shows and movies in addition to reading newspapers, journals, and magazine articles. These forms of communication provide rich information (right and wrong) about the social life of individuals and society in general.

Most people, for example, who have had absolutely no contact with criminals learn about crime by these forms of communications. However, as we all know too well, the media can easily perpetuate the myths of any given culture (Neuman, 2009):

> The media show that most people who receive welfare are African American (most are actually non-African American), that most people who are mentally ill are violent and dangerous

(only a small percentage actually are), and that most people who are elderly are senile and in nursing homes (a tiny minority are).

Also, a selected emphasis on an issue by the media can change public thinking about it. For example, television repeatedly shows low-income, inner-city African American youth using illegal drugs. Eventually, most people "know" that urban African Americans use illegal drugs at a much higher rate than other groups in the United States, even though this notion is false.

Tradition

The second way of adding to your social work knowledge base is through tradition. Authority and tradition are highly related to one another. For example, some things you "know" because your mother "knew" them and her mother before her, and they are a part of your cultural tradition. Your mother was also an authority figure who learned her bits and pieces through tradition and authority.

More often than not, people tend to accept cultural beliefs without much question. They may doubt some of them and test others for themselves, but, for the most part, they behave and believe as tradition dictates. To be sure, such conformity is useful, as our society could not function if each custom and belief were reexamined by each individual in every generation.

On the other hand, unquestioning acceptance of "traditional dictates" easily leads to stagnation and to the perpetuation of wrongs. It would be unfortunate, for example, if women were never allowed to vote because women had never traditionally voted, or if racial segregation were perpetuated because traditionally that's just the way it was. Some traditional beliefs are based on the dictates of authority, carried on through time. The origins of other beliefs are lost in history. Even in social service programs, where history is relatively brief, things tend to be done in certain ways because they have always been done in these ways.

When you first enter a social service program as a practicum student your colleagues will show you how the program runs. You may be given a manual detailing program policies and procedures which contains everything from staff holidays, to locking up client files at night, to standard interviewing techniques with children who have been physically and emotionally abused. Informally, you will be told other things such as, how much it costs to join the coffee club, whom to ask when you want a favor, whom to phone for certain kinds of information, and what form to complete to be put on the waiting list for a parking space.

In addition to this practical information, you may also receive advice about how to help your future clients. Colleagues may offer you a few of their opinions about the most effective treatment intervention strategies that are used within your practicum setting. If your practicum is a child sexual abuse treatment program, for example, it may be suggested to you that the nonoffending mother of a child who has been sexually abused does not need to address her own sexual abuse history in therapy in order to empathize with and protect her daughter.

Such a view would support the belief that the best interventive approach is a behavioral/learning one, perhaps helping the mother learn better communication skills in her relationship with her daughter. Conversely, the suggestion may be that the mother's personal exploration into her psyche (whatever that is) is essential and therefore, the intervention should be of a psychodynamic nature. Whatever the suggestion, it is likely that you, as a beginning social work student, will accept it, along with the information about the coffee club.

To be sure, you will want to fit in and become a valued member of the team. If the nonoffending mother is the first client for whom you have really been responsible for, you may also be privately relieved that the intervention decision has been made for you. You may believe that your colleagues, after all, have more professional experience than you and they should surely know best. In all likelihood, they probably do know best.

At the same time, they also were once beginning social work students like yourself and they probably formed their opinions in the same way as you are presently forming yours. They too once trusted their supervisors' knowledge bases and experiences. In other words, much of what you will initially be told is based upon the way that your practicum site has traditionally worked. This might be a good moment to use your new-found skills to evaluate the research literature on the best way to intervene with children who have been sexually abused.

But if you do happen to find a different and more effective way, you may discover that your colleagues are unreceptive or even hostile. They "know" what they do already works with clients—they "know it works" because it has worked for years.

Thus, on one hand, tradition is useful. It allows you to learn from the achievements and mistakes of those who have tried to do your job before you. You don't have to reinvent the wheel as you've been given a head start. On the other hand, tradition can become way too comfortable. It can blind you to better ways of doing things.

Knowledge vs. Beliefs

At this point it is useful to differentiate between knowledge and beliefs (or faith). Knowledge is an accepted body of facts or ideas acquired through the use of the senses or reason. Beliefs are similarly a body of facts or ideas that are acquired through the reliance on tradition and/or authority. We now have *knowledge* that the earth is round because we have been into space and observed it from above. A few centuries ago, we would have *believed* that it was flat, because someone "in authority" said it was or because tradition had always held it to be flat.

Knowledge is never final or certain. It is always changing as new facts come to our attention and new theories explaining the facts are developed, tested, and accepted or rejected. Belief systems, on the other hand, have remarkable staying power. Various beliefs about life after death, for example, have been held since the beginning of time by large numbers of people and will doubtless continue to be held, without much change, because there is nothing to change them.

More recently, the belief that one acquires worth through work is strongly held in North American society. The harder you work, the more virtue you acquire by doing the work. At the same time, it is believed that people will avoid work if at all possible—presumably they value ease over virtue—so the social service programs we have in place are designed to punish our clients' "idleness" and reward their "productivity."

Experience

The third way of acquiring knowledge is through experience. You "know" that buttered bread falls when you drop it—buttered side down of course. You "know" that knives cut and fire burns. You "know," as you gain experience in social work, that certain interventive approaches tend to work better than others with certain types of clients in particular situations. Such experience is of enormous benefit to clients, and its unfortunate that the knowledge gained by individual social workers over the years is rarely documented and evaluated in a way that would make it available to others.

However, as with anything else, experience has it advantages and disadvantages. Experience in one area, for example, can blind you to the issues in another. Health planners from mental health backgrounds, for example, may see mental illness as the most compelling community health problem because of their experiences with the mentally ill. Mental health issues may therefore command more dollars and attention than other public health issues, equally deserving. Awareness of your own biases will allow you to make the most of your own experience while taking due account of the experiences of others.

Intuition

Intuition is fourth on our countdown to the ways of knowing. It can be described in a number of ways: revelation through insight, conviction without reason, and immediate apprehension without rational thought. In short, you "know" something without having a clue of how you "know" it. It has been suggested that intuition springs from a rational process at the subconscious level.

Intuition and Professional Judgment

Perhaps intuition works that way. Perhaps it doesn't. Some of us trust it. Some of us don't. Whatever it is, intuition should not be confused with an experienced social worker's professional judgment. Professional judgment is a *conscious* process whereby facts, as far as they are known, are supplemented with the knowledge derived from experience to form the basis for rational decisions.

In this eminently reasonable process, you know what facts you have and how reliable they are, you know what facts are missing, and you know what experiences you're using to fill in the gaps. You are thus in a position to gauge whether your judgment is almost certainly right (you have all the facts), probably right (you have most of the facts) or possibly out to lunch (you know you are almost entirely guessing). A reasoned professional judgment on your part, no matter how uncertain you may be, is far more beneficial to your client than your intuition.

The Research Method

We have now come to the fifth and last way of knowing. This way of acquiring knowledge is through the use of the research method—the main focus of this book. It is sometimes called the problem-solving method, the scientific method, or the research process. The research method is a relatively new invention. For example, Aristotle was of the opinion that women had fewer teeth than men.

Although he was twice married and the number of teeth possessed by women and men was a contentious issue in his day, it never occurred to him to ask both of his wives to open their mouths so he could observe and count the number of teeth each had. This is a solution that would occur to anyone born in the twentieth century because we are accustomed to evaluate our assumptions in the light of our observations. The social work profession—and modern society—is enamored with knowledge development through the use of the research method.

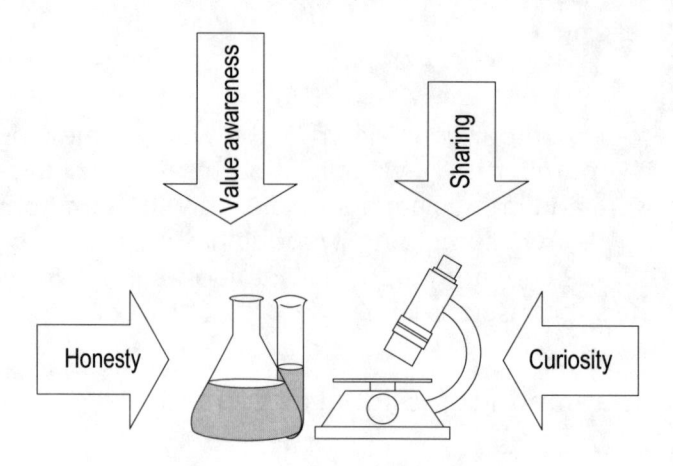

FIGURE 1.2
Characteristics of the Research Method

Acquiring knowledge through the use of research findings that were derived from the research method is the most objective way of "knowing something." For example, as can be seen in Figure 1.2, when researchers do research studies they must: (1) be value aware (2) be skeptics, (3) share their findings with others, and (4) be honest.

Value Awareness

You must be aware of and be able to set aside your values when you do a research study—you must be unbiased and impartial to the degree it is possible—like a judge. This means that you, as a social work researcher, should be able to put aside your personal values both when you are conducting research studies and when you are evaluating research results obtained by other researchers.

If your personal value system dictates, for example, that health care should be publicly funded and equally available to everyone, you should still be able to use the research method to acquire knowledge about the advantages and disadvantages of a privatized system.

If the evidence from your own or someone else's study shows that privatized health care is superior in some respects to your own beliefs, you should be able to weigh this evidence objectively, even though it may conflict with your personal value system.

Skeptical Curiosity

Now that you are valueless, you must become insatiably curious. We now know that knowledge acquired using the research method is never certain. Scientific "truth" remains true only until new evidence comes along to show that it is not true, or only partly true. Skeptical curiosity means that all findings derived from the research method should be—and most importantly, must be—questioned. Wherever possible, new studies should be conducted by different researchers to see if the same results are obtained again. In other words, research studies, whenever possible, should be replicated.

Replication of the same study, with the same results, by another researcher makes it less likely that the results of the first study were affected by bias, dishonesty, or just plain old error. Thus, the findings are more likely to be "true" in the sense that they are more likely to reflect a reality external to the researchers.

We will come back to this business of "external reality" later on. For now, it is enough to say that the continual replication of research studies is a routine practice in the physical sciences but is far more rare in the social sciences, especially in the social work profession, for two main reasons. First, it is much more difficult to replicate a study of people than a study of physical objects. Second, researchers in the social sciences have a harder time of finding money to do research studies than researchers in the physical sciences.

Sharing

Like your mother said, "you must share your stuff with others." The results of a research study and the methods used to conduct it should be available to everyone so that the study's findings can be critiqued and the study replicated. It is worth noting that sharing findings from a research study is a modern value. It is not so long ago when illiteracy among peasants and women were valued by those who were neither. Knowledge has always been a weapon as well as a tool. Those who know little may be less likely to question the wisdom and authority of those who are above them in the social hierarchy. Public education is thus an enormously powerful social force that allows people to access and question the evidence upon which their leaders make decisions on their behalf.

Honesty

Not only must you share your research findings with others, you must be honest in what you share. Honesty means, of course, that you are not supposed to fiddle with the results obtained from your study. This may sound fairly straightforward but, in fact, the results of research studies are rarely as clear cut as we would like them to be. Quite often, and in the most respectable of scientific laboratories, theories are formulated on the basis of whether one wiggle on a graph is slightly longer than the corresponding woggle. If "dishonesty" means a deliberate intention to deceive, then probably very few researchers are dishonest. If it means that researchers allow their value systems and their preconceived ideas to influence their methods of data collection, analysis, and interpretation, then there are probably a few guilty ones. In this sense, the term "honesty" includes an obligation, on the part of researchers, to be explicit about what their values and ideas are.

They need to be sufficiently self-aware to both identify their value systems and perceive the effects of these upon their work; and then, they need to be sufficiently honest to make an explicit statement about where they stand so that others can evaluate the conclusions drawn from the research studies.

PHASES OF SCIENTIFIC INQUIRY

On a very general level, and in the simplest of terms, the scientific inquiry process is illustrated in Figure 1.3. As can be seen, it begins with Phase 1—some kind of an observation and/or measurement. Suppose, for example, we find in the garage an unidentified bag of seeds and we don't know what kind of seeds they are.

We plant a random seed from the bag into the ground and it grows into a petunia. This might be a coincidence but, if we plant thirty-seven more seeds from the same bag and all of them grow into petunias, we might assume that all the seeds in our bag have something to do with the petunias. We have now reached the second phase in the research method; we have made an assumption based on our observations.

The third phase is to test our assumption. This is done by planting yet another seed (the 38th) in the same way as before. If the 38th seed, too, becomes a petunia, we will be more certain that all the seeds in our bag will grow into petunias. On the other hand, if the 38th seed grows into a cabbage, we will begin to wonder if our original assumption—the bag contains all petunia seeds—was wrong, the fourth phase of the research method.

It is possible, of course, that we are quite mad and we only imagined those petunias in the first place. We would be more certain of the real existence of those petunias if someone else had seen them as well. The more people who had observed them, the surer we would become.

The scientific inquiry process holds that, in most cases, something exists if we can observe *and* measure it. To guard against objects that are seen without existing, such as cool pools of water observed by people dying of thirst in deserts, the research method has taken the premise one step further. A thing exists if, and only if, we can measure it. The cool pools of water that we observed, for example, probably

```
        Phase 1
Observing and/or measuring something
  about a person, an object, or an event

   Phase 4                    Phase 2
Revising the assumption → Making an assumption on the basis of the
on the basis of the test    observations and/or measurements

          Phase 3
     Testing the assumption
   to see what extent it is true
```

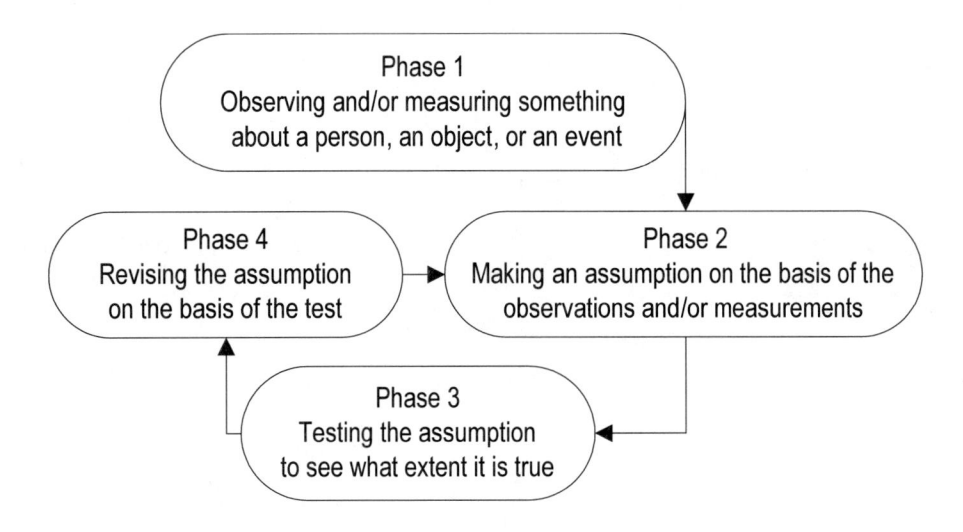

FIGURE 1.3
Phases of Scientific Inquiry

could not be measured by a thermometer and a depth gauge. Things that have always occurred in sequence, such as summer and fall, probably will continue to occur in sequence. In all likelihood, rivers will flow downhill, water will freeze at zero degrees centigrade, and crops will grow if planted in the spring.

But nothing is certain, nothing is absolute. It is a matter of slowly acquiring knowledge by making observations and measurements, deriving assumptions from those observations, and testing the assumptions by making more observations and measurements. Even the best-tested assumption is held to be true only until another observation comes along to disprove it. Nothing is forever. It is all a matter of probabilities.

Let's say you have lived your whole life all alone in a log cabin in the middle of a large forest. You have never ventured as much as a hundred yards from your cabin and have had no access to the outside world. You have observed for your entire life that all of the ducks that flew over your land were white. You have never seen a different-colored duck. Thus, you assume, and rightfully so, that all ducks are white. You would only have to see one nonwhite duck to disprove your white-duck assumption. Nothing is certain no matter how long you "objectively observed" it.

Example of the Scientific Inquiry Process

Suppose, for a moment, you are interested in determining whether the strength of a child's attachment to his/her mother affects the social skills of the child. In order to test your assumption (hypothesis, if you

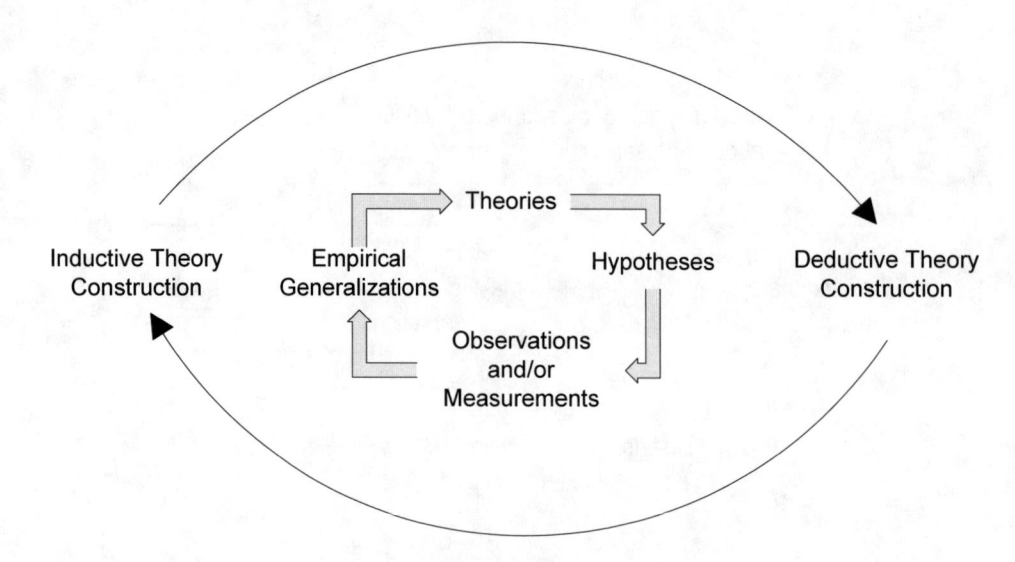

FIGURE 1.4
Inductive/Deductive Cycle of Theory Construction

will), you must now decide on what you mean by "child" (say, under 6), and you need to find some young children and their respective mothers.

Next, you need to decide what you mean by "attachment" and you need to observe how attached the children are to their mothers. Because you need to measure your observations, you will also need to come up with some system whereby certain observed behaviors mean "strong attachment," other behaviors mean "medium attachment" and still other behaviors mean "weak attachment." Then you need to decide what you mean by "social skills" and you now need to observe and measure the children's social skills. All of these definitions, observations, and measurements constitute Phase 1 of the research study.

On the basis of your Phase 1 data, you might formulate an assumption, hunch, or hypothesis, to the effect (say) that the stronger a child's attachment to his/her mother, the higher the child's social skills. Or, to put it another way, children who have higher attachments with their mothers will have higher social skills than children who have lower attachments to their mothers. This is Phase 2 of the scientific inquiry process and involves *inductive* logic (see Figure 1.4). In short, you begin with detailed observations and/or measurements of the world obtained in Phase 1 and move toward more abstract generalizations and ideas.

If your assumption is correct, you can use it to predict that a particular child with a strong attachment to his/her mother will also demonstrate strong social skills. This is an example of *deductive* logic where you are deducing from the general to the particular. In Phase 3, you set about testing your assumption, observing and measuring the attachment levels and social skills of as many other children as you can manage.

Data from this phase may confirm or cast doubt upon your assumption. The data might also cause you to realize that "attachment" is not so simple of a concept as you had imagined. It is not just a matter of the *strength* of the attachment, but the *type* of the attachment is also a factor (e.g., secure, insecure, disorganized).

If you have tested enough children from diverse cultural backgrounds, you might also wonder if your assumption holds up better in some cultures than it does in others. Is it more relevant, say, for children raised in nuclear families than for children raised in a more communal environment such as a First Nations reserve or an Israeli kibbutz?

These considerations will lead you to Phase 4 where you revise your conjecture in the light of your observations (inductive logic) and begin to test your revised hunch all over again (deductive logic). Hopefully, this will not be a lonely effort on your part. Other researchers interested in attachment will also examine your assumption and the evidence you formulated it from, and conduct their own studies to see how right you really were. This combined work, conducted with honesty, skepticism, sharing, and freedom from entrenched beliefs, allows our knowledge base in the area of attachment to increase.

RESEARCH AND PRACTICE: BOTH PROBLEM-SOLVING PROCESSES

Believe it or not, social work research and practice have much in common. They are both problem solving processes (Grinnell, Unrau, & Williams, 2008, Duehn, 1985). As can be seen in Figure 1.5, there are parallels between social work research and social work practice.

All social work activities, both practice and research, are organized around one central assumption: There is a preferred order of thinking and action which, when rigorously and consciously followed, will increase the likelihood of achieving our objectives

This way of looking at our practice and research activities is not new. Social work practitioners and researchers base their conclusions on careful observation, systematic trial, and intelligent analysis. Both observe, reflect, conclude, try, monitor results, and continuously reapply the same problem-solving process until the problem at hand is addressed.

RESEARCH ROLES

We have looked at the reasons why social workers need to engage in research and how we go about acquiring the knowledge we need to help our clients help themselves—the purpose of social work. There are two main research-related roles that social workers engage in. These

Figure 1.5

Phase 1: Defining the problem
  General problem solving
    Recording that a problem exists
  Social work research
    Identifying the research problem
  Social work practice
    Diagnosis and assessment

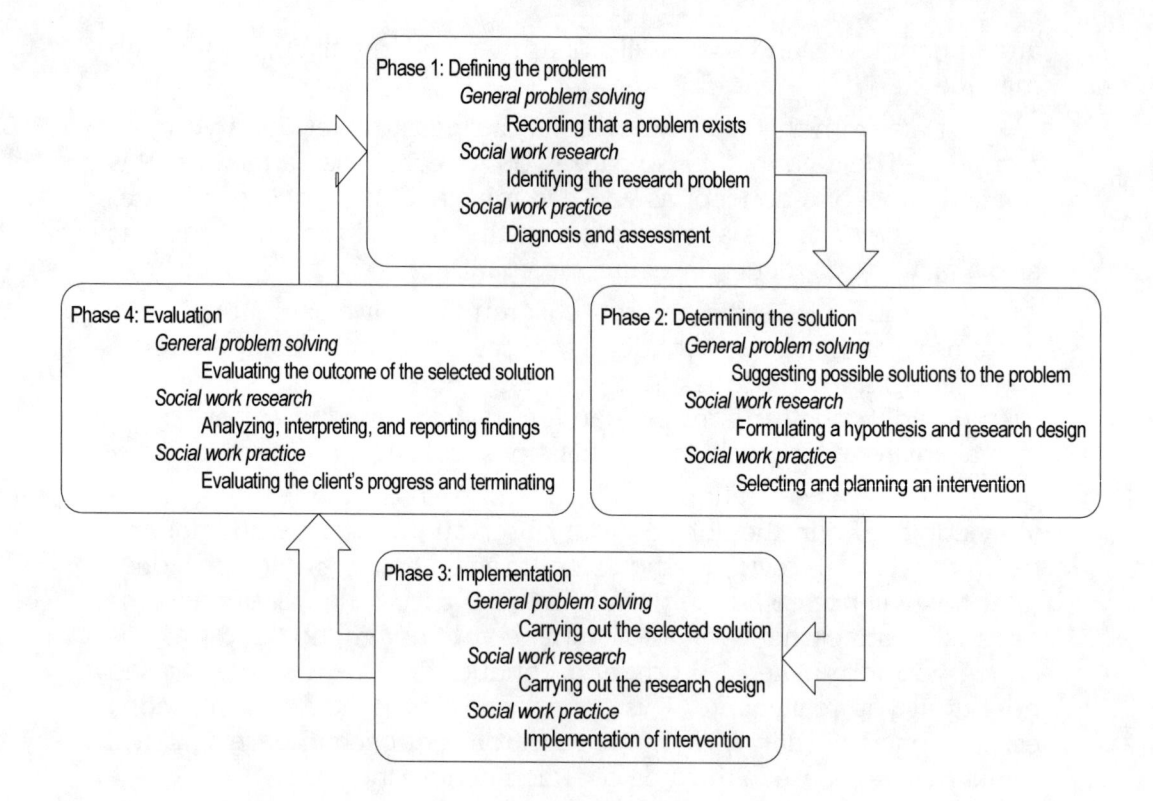

Phase 2: Determining the solution
  General problem solving
    Suggesting possible solutions to the problem
  Social work research
    Formulating a hypothesis and research design
  Social work practice
    Selecting and planning an intervention

Phase 3: Implementation
  General problem solving
    Carrying out the selected solution
  Social work research
    Carrying out the research design
  Social work practice
    Implementation of intervention

Phase 4: Evaluation
  General problem solving
    Evaluating the outcome of the selected solution
  Social work research
    Analyzing, interpreting, and reporting findings
  Social work practice
    Evaluating the client's progress and terminating

FIGURE 1.5
Parallels among General Problem Solving, Research, and Social Work Practice

are: (1) the research consumer, and (2) the creator and disseminator of knowledge.

The Research Consumer

If you go to your doctor to discuss your arthritis, you expect the doc to be aware of the most recent advances in the management and treatment of arthritis. All professionals, in all disciplines, are expected by their clients to keep up with the latest developments in their fields. They do this by attending conferences, reading books and journals, and paying attention to the results derived from research studies.

In other words, these professionals—which include you as a social worker—are *research consumers* and, as previously noted, they need to know enough about the research method to consume it wisely, separating the nutritious wheat from the dubious chaff.

The Creator and Disseminator of Knowledge

You may be quite determined that you will never yourself conduct a research study. Never ever! you say, and then you find that you are the only staff person in a small voluntary social service program that desperately requires a needs assessment if the program is to serve its clients and keep its funding base.

You look up "needs assessment" in forgotten research texts, and sweat, and stumble through them anyway because someone has to do the study and there is no one there but you. This may seem like an unlikely scenario, but in fact many social service programs are very small and are run on a wing and a prayer by a few paid staff and a large volunteer contingent. They rise and flourish for a time and die; and death is often hastened along by their inability to demonstrate, in research terms, how much good they are doing on their clients' behalf, how little it is costing, and what the dreadful social consequences would be if they weren't there to do it.

You may escape being the sole social worker in a program that needs research know-how. But even if you are a mere cog in a immense machine of interlocking social workers, the time might come when you want to try something new. Most social workers do. Most of them, however, don't try the something in any structured and systematic way. They don't write down exactly what the something was (perhaps a new intervention for raising Jody's self-esteem), they don't say why they needed it (nothing else was working), how they tested it (they measured Jody's self-esteem before and after doing it), or how effective it was (Jody's self-esteem score rose triumphantly, from X to Y, and was still at its higher level three months later).

Worse, they don't tell anyone else they did it, except for a few murmurs, rapidly forgotten, to a colleague over coffee. One consequence of this is that other Jody-types, who might benefit from the same intervention, never have the opportunity to do so because their social workers don't know that it exists. In reality, many social service programs conduct some kind of research studies from time to time, particularly evaluative studies.

Many more agree to host studies conducted by researchers external to the program, such as university professors and graduate students. Unlike studies conducted by psychologists, social work research rarely takes place in a laboratory but, instead, is usually conducted in field settings. Data may be drawn from program clients or their records and may be collected in the program or in the clients' homes.

Since social workers are mostly employed by social service programs, they are often drawn into the program's research activities by default. Such activities are normally conducted by a team, consisting of researchers and program staff members. Today, the solitary social work researcher, like the solitary mad scientist, is very much a figment of the past. Staff members contributing to research inquiry may have specific skills to offer which they never imagined were research-related.

One may be particularly acute and accurate when it comes to observing client behaviors. A second worker on the other hand, may work well as a liaison between clients and researcher, or between one program and another. Some social workers are cooperative in research endeavors, and others are less so, depending on their attitudes toward knowledge development through the use of the research method.

Those of us who know most about research methods tend to be the most cooperative and also the most useful. Hence, the greater the number of social workers who understand research principles, the more likely it is that relevant studies will be successfully completed and social work knowledge will be increased.

Integrating the Two Research Roles

Just about everything in life is interdependent on everything else. Chaos theory comes readily to mind on the idea of interdependence. Presto, the same holds true with the two research roles noted above—they are not independent of one another. They must be integrated if research is to accomplish its goals of increasing our profession's knowledge base and improving the effectiveness of our interventions.

The issue is not whether we should consume research findings and/or produce and disseminate research results. Rather it is whether we can engage the full spectrum of available knowledge and skills in the continual improvement of our practices. Social workers who adopt only one research role are shortchanging themselves and their clients (Reid & Smith, 1989):

> ... If research is to be used to full advantage to advance the goals of social work, the profession needs to develop a climate in which both doing and consuming research are normal professional activities. By this we do not mean that all social workers should necessarily do research or that all practice should be based on the results of research, but rather that an ability to carry out studies at some level and the facility in using scientifically based knowledge should be an integral part of the skills that social workers have and use. (p. xi)

A research base within our profession will not guarantee its public acceptance, but there is no doubt that the absence of such a base and the lack of vigorous research efforts to expand it will, in the long run, erode our credibility and be harmful to our clients.

SUMMARY

Knowledge is essential to human survival. Over the course of history, there have been many ways of knowing, from divine revelation to tradition and the authority of elders. By the beginning of the seventeenth century, people began to rely on a different way of knowing—the research method. Social workers derive their knowledge from authority, tradition, professional experience, personal intuition, as well as from findings derived from studies.

Social workers engage in two main research roles. They can consume research findings by using the findings of others in their day-to-day practices, and they can produce and disseminate research results for others to use. Now that we have briefly explored the place of research in social work, the following chapter discusses how social workers formulate research problems to study.

REFERENCES AND FURTHER READING

Council on Social Work Education (2008). *Baccalaureate and masters curriculum policy statements*. Alexandria, VA. Author.

Duehn, W.D. (1985). Practice and research. In R.M. Grinnell, Jr. (Ed.), *Social work research and evaluation* (2nd ed., pp. 19–48). Itasca, IL: F.E. Peacock.

Epstein, I. (1988). Quantitative and qualitative methods. In R.M. Grinnell, Jr. (Ed.), *Social work research and evaluation* (3rd ed., pp. 185–198). Itasca, IL: F.E. Peacock.

Garvin, C.D. (1981). Research-related roles for social workers. In R.M. Grinnell, Jr. (Ed.), *Social work research and evaluation* (pp. 547–552). Itasca, IL: F.E. Peacock.

Grinnell, R.M., Jr., Rothery, M., & Thomlison, R.J. (1993). Research in social work. In R.M. Grinnell, Jr. (Ed.), *Social work research and evaluation: Quantitative and qualitative approaches* (4th ed., pp. 2–16). Itasca, IL: F.E. Peacock.

Grinnell, R.M., Jr., & Siegel, D.H. (1988). The place of research in social work. In R.M. Grinnell, Jr. (Ed.), *Social work research and evaluation* (3rd ed., pp. 9–24). Itasca, IL: F.E. Peacock.

Grinnell, R.M., Jr., Unrau, Y.A., & Williams, M. (2008). Introduction. In R.M. Grinnell, Jr. & Y.A. Unrau (Eds.), *Social work research and evaluation: Foundations of Evidence-Based Practice* (7th ed., pp. 4–21). New York: Oxford University Press.

National Association of Social Workers (1996). *Code of ethics*. Silver Spring, MD: Author.

Neuman, W.L. (2009). *Understanding research*. Boston: Allyn & Bacon.

Reid, W.J., & Smith, A.D. (1989). *Research in social work* (2nd ed.). New York: Columbia University Press.

- Advanced current social work research methods texts:

Drake, B., & Johnson-Reid, M. (2008). *Social work research methods: From conceptualization to dissemination.* Boston: Allyn & Bacon.

Engel, R.J., & Schutt, R.K. (2008). *The practice of research in social work* (2nd ed.). Thousand Oaks, CA: Sage.

Grinnell, R.M., Jr., & Unrau, Y.A., (Eds.). (2008). *Social work research and evaluation: Foundations of evidence-based practice* (8th ed.). New York: Oxford University Press.

Krueger, L.M., & Neuman, W.L. (2006). *Social work research methods.* Boston: Allyn & Bacon.

Rubin, A., & Babbie, E. (2008). *Research methods for social work* (6th ed.). Pacific Grove, CA: Thomson.

Thyer, B.A. (Ed.). (2008). *The handbook of social work research methods* (2nd ed.). Thousand Oaks, CA: Sage.

Yegidis, B.L, & Weinbach, R.W. (2009). *Research methods for social workers* (6th ed.). Boston: Allyn & Bacon.

- Current social work statistics texts:

Faherty, V. (2007). *Compassionate statistics: Applied quantitative analysis for social services.* Thousand Oaks, CA: Sage.

Weinbach, R.W., & Grinnell, R.M., Jr. (2010). *Statistics for social workers* (8th ed.). Boston: Allyn & Bacon.

Check out our Website for useful links and chapters (PDF) on:
- introductory research concepts
- evidence-based social work practice
- tutorials on the use of the scientific method

Go to: www.pairbondpublications.com
Click on: Student Resources
　　　　　 Chapter-by-Chapter Resources
　　　　　 Chapter 1

Check Out Our Website

PairBondPublications.com

√ *Student Workbook Exercises*
√ *Chapter Power Point Slides*
√ *On-line Glossaries*
√ *General Links*
√ *Specific Chapter Links*

Chapter 2

Research Problems and Questions

WHAT IS SOCIAL WORK RESEARCH?
    Pure and Applied Research
    The Research Attitude
    Approaches to the Research Method
THE KNOWLEDGE-LEVEL CONTINUUM
    Exploratory Research Studies
    Descriptive Research Studies
    Explanatory Research Studies
CLASSIFICATION OF RESEARCH QUESTIONS
    Existence Questions
    Composition Questions
    Relationship Questions
    Descriptive-Comparative Questions
    Causality Questions
    Causality-Comparative Questions
    Causality-Comparative Interaction Questions
FORMULATING RESEARCH QUESTIONS
    Factors Contributing to Good Research Questions
REVIEWING THE LITERATURE
    Connecting the Research Question to Theory
    Identifying Previous Research Studies
ACCESSING INFORMATION
    Libraries and the Web
SUMMARY
REFERENCES AND FURTHER READING

2

IN THE LAST CHAPTER WE DISCUSSED how we go about knowing something and placed a considerable emphasis on the fact that knowing something derived from the research method is more objective, if you will, than knowing that same thing through authority, tradition, experience, and intuition. In this chapter, we will begin to look at how research questions are formulated that can be answered through the "research way" of knowing.

WHAT IS SOCIAL WORK RESEARCH?

So far, we have discussed the various ways of obtaining knowledge and briefly looked at the characteristics and phases of the research method. Armed with this knowledge, we now need a definition of research, which is composed of two syllables, *re* and *search*.

Dictionaries define the former syllable as a prefix meaning again, anew, or over again, and the latter as a verb meaning to examine closely and carefully, to test and try, or to probe (Duehn, 1985). Together, these syllables form a noun describing a careful and systematic study in some field of knowledge, undertaken to establish facts or principles. Social work research therefore can be defined as:

> a systematic and objective inquiry that utilizes the research method to solve human problems and creates new knowledge that is generally applicable.

We obtain much of our knowledge base from the findings derived from research studies that utilize the research method. However, all research studies have built-in biases and limitations that create errors and keep us from being absolutely certain about the studies' outcomes.

This book helps you to understand these limitations and to take them into account in the interpretation of research findings, and helps you to avoid making errors or obtaining wrong answers. One of the principal products of a research study is obtaining "objective and systematic" data—via the research method—about reality as it is, "unbiased" and "error-free."

Pure and Applied Research

Social work research studies can be described as pure or applied. The goal of pure research studies is to develop theory and expand our profession's knowledge base. The goal of applied studies is to develop solutions for problems and applications in practice. The distinction between theoretical results and practical results marks the principal difference between pure and applied research studies.

The Research Attitude

The research method, or "scientific method" if you will, refers to the many ideas, rules, techniques, and approaches that we—the research community—use. The research attitude, on the other hand is simply a way that we view the world. It is an attitude that highly values craftsmanship, with pride in creativity, high-quality standards, and hard work. As Grinnell (1987) states:

> Most people learn about the "scientific method" rather than about the scientific attitude. While the "scientific method" is an ideal construct, the scientific attitude is the way people have of looking at the world. Doing science includes many methods: what makes them scientific is their acceptance by the scientific collective. (125)

Approaches to the Research Method

The research method of knowing contains two complementary research approaches—the quantitative approach and the qualitative approach.

Simply put, the quantitative portion of a research study relies on quantification in collecting and analyzing data and uses descriptive and inferential statistical analyses. If data obtained within a research study are represented in the form of numbers, then this portion of the study is considered "quantitative."

On the other hand, a qualitative portion of a research study relies on qualitative and descriptive methods of data collection. If data are presented in the form of words, diagrams, or drawings—not numbers as in the quantitative approach—then this portion of the study is considered "qualitative."

It should be pointed out, however, that a research study can be solely quantitative in nature. It can also be exclusively qualitative. As we will see throughout this book, a good research study uses both approaches in an effort to generate useful knowledge for our profession.

The unique characteristics and contributions of the quantitative approach and qualitative approach to knowledge building are examined in Chapters 4 and 5 respectively. The quantitative and qualitative research approaches complement each other and are equally important in the generation of social work knowledge.

THE KNOWLEDGE-LEVEL CONTINUUM

We will now turn our attention to how the quantitative and qualitative research approaches answer various types of research questions. Any research study falls anywhere along the knowledge-level continuum depending on how much is already known about the topic (see Figure 2.1). How much is known about the research topic determines the purpose of the study. If you didn't know anything, for example, you will merely want to explore the topic area, gathering basic data.

Studies like this, conducted for the purpose of exploration are known, logically enough, as *exploratory* studies and fall at the bottom of the knowledge-level continuum as can be seen in Figure 2.1. Usually exploratory studies adopt a qualitative research approach.

When you have gained some knowledge of the research topic area through exploratory studies, the next task is to describe a specific aspect of the topic area in greater detail, using words and/or numbers. These studies, whose purpose is description, are known as *descriptive* studies and fall in the middle of the knowledge-level continuum as presented in Figure 2.1. As can be seen, they can adopt a quantitative and/or qualitative research approach.

After descriptive studies have provided a substantial knowledge base in the research topic area, you will be in a position to ask very specific and more complex research questions—causality questions. These kinds of studies are known as *explanatory* studies.

The division of the knowledge continuum into three parts—exploratory, descriptive, and explanatory—is a useful way of categorizing research studies in terms of their purpose, the kinds of questions they can answer, and the research approach(s) they can take in answering the questions. However, as in all categorization systems, the three divisions are totally arbitrary and some social work research studies defy categorization, falling nastily somewhere between exploratory and descriptive, or between descriptive and explanatory.

This defiance is only to be expected since the knowledge-level continuum is essentially that—a *continuum*, not a neat collection of categories. Let's take a moment here to look at exploratory, descriptive and explanatory studies in more detail and the kinds of questions each type of study can answer.

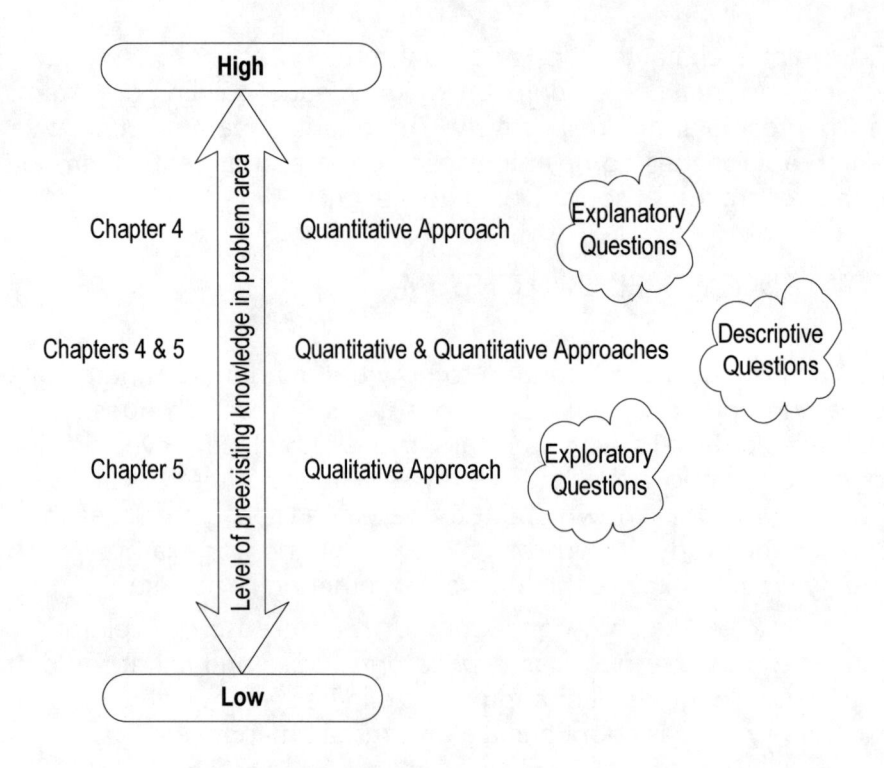

FIGURE 2.1
The Knowledge-Level Continuum and
Approaches to the Research Method

Exploratory Research Studies

Exploratory studies are most useful when the research topic area is relatively new. In the United States during the 1970s, for example, the development of new drugs to control the symptoms of mental illness, together with new federal funding for small, community-based mental health centers, resulted in a massive discharge of people from large state-based mental health institutions.

Some folks applauded this move as restoring the civil liberties of the mentally ill. Others were concerned that current community facilities would prove inadequate to meet the needs of the people being discharged and their families. Social workers active in the 1970s were anxious to explore the situation, both with an eye on influencing social policy and in order to develop programs to meet the perceived needs of this group of people. The topic area here is very broad. What are the consequences of a massive discharge of people who are psychiatrically challenged and were recently institutionalized? Many different questions pertaining to the topic can be asked.

Where are these people living now? Alone? In halfway houses? With their families? On the street? Are they receiving proper medication and nutrition? What income do they have? How do they spend their time? What stresses do they suffer? What impact have they had on their family members and the communities in which they now reside? What services are available to them? How do they feel about being discharged?

No single study can answer all these questions. It is a matter of devising a sieve-like procedure where the first sieve with the biggest holes identifies general themes. Each general theme is then put through successively finer sieves until more specific research questions can be asked (Figure 2.2).

For example, you might begin to explore the consequences of the massive discharge by gathering together a group of these discharged people and asking them a basic exploratory question: What have been your experiences since you were discharged?—what are the components that make up the discharge experience?—and it will be answered using qualitative data. Individual answers, via words—not numbers—will generate common themes.

You may find, for example, that they have had a number of different living arrangements since they were discharged, vary greatly in their ability to manage their budget, leisure, food and medication, feel rejected or supported, suffer less or more stress, and so forth.

You might then take one of these major themes and try to refine it, leaving the other major themes to be explored by someone else at a latter date. Suppose you choose living arrangements (Figure 2.2) and mount a second exploratory study to ask respondents what living arrangements they have experienced since discharge. You may find now that those who were institutionalized for a long time have tended to move between halfway houses, shelters, and the street, while those who were institutionalized for a shorter time moved in with family first, and stayed regardless on how accepting the family was.

At this point, you might feel a need for numbers. How many of them are living where? How many times have they moved on average? What percentage of those who moved in with their families stayed there? These are *descriptive* questions, aimed at describing, or providing an accurate profile, of this group of people.

You are now moving up the knowledge continuum from the exploratory to the descriptive category but, before we go there, let's summarize the general goals of exploratory research studies. These are to (Neuman, 2009):

- Become familiar with the basic facts, people, and concerns involved
- Develop a well-grounded mental picture of what is occurring

Composition question (Exploratory)
What are the consequences of a massive discharge of people who are psychiatrically challenged and were recently institutionalized?

Budget　Food　**Living arrangements**　Medications　Support　Stress

Composition question (Exploratory)
What living arrangements have you experienced since discharge?

Halfway house　Shelter　**Family**　Friends　Street

Comparative question (Descriptive)
Is there a real difference between **accepting** families and **rejecting** families?

Accepting families　　　　Rejecting families

Causality question (Explanatory)
Do accepting families lead to discharged families members to stay at home?

Causality question (Explanatory)
Do rejecting families prevent discharged family members from staying at home?

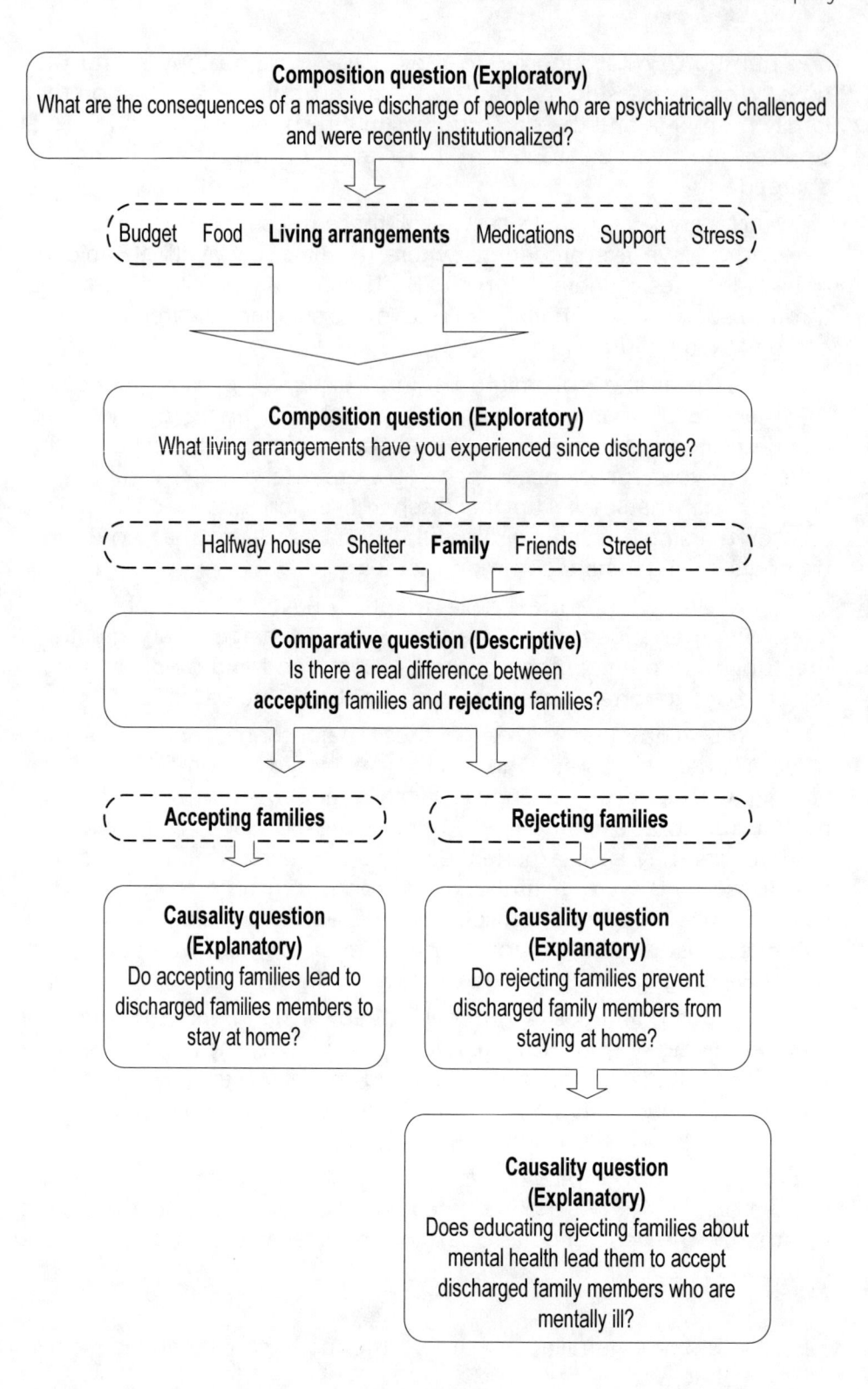

Causality question (Explanatory)
Does educating rejecting families about mental health lead them to accept discharged family members who are mentally ill?

FIGURE 2.2
Example of a Sieving Procedure

Chapter 2: Research Problems and Questions

- Generate many ideas and develop tentative theories and conjectures
- Determine the feasibility of doing additional research
- Formulate questions and refine issues for more systematic inquiry
- Develop techniques and a sense of direction for future research

Descriptive Research Studies

At the descriptive level, suppose you have decided to focus on those people who moved in with their families. You have an idea, based on your previous exploratory study that there might be a relationship between the length of time spent in the institution and whether or not this group of people moved in with their families after discharge. You would like to confirm or refute this relationship, using a much larger group of respondents than you used in your exploratory study.

Another tentative relationship that emerged at the exploratory level was the relationship between staying in the family home and the level of acceptance shown by the family. You would like to know if this relationship holds with a larger group. You would also like to know if there is a real difference between accepting and rejecting families: Is Group *A* different from Group *B*? and what factors contribute to acceptance or rejection of the discharged family member?

Eventually, you would like to know if there is anything social workers can do to facilitate acceptance, but you don't have enough data yet to be able to usefully ask that question. In general, the goals of descriptive research studies are to (Neuman, 2009):

- Provide an accurate profile of a group
- Describe a process, mechanism, or relationship
- Give a verbal or numerical picture (e.g., percentages)
- Find information to stimulate new explanations
- Create a set of categories or classify types
- Clarify a sequence, set of stages, or steps
- Document information that confirms or contradicts prior beliefs about a subject

Explanatory Research Studies

Suppose you have learned from your descriptive studies that there are real differences between accepting and rejecting families and that these differences seem to have a major impact on whether or not the discharged person stays at home. Now you would like to ask two related causality questions: "Does an accepting family lead to the discharged person staying at home?" and "Does a rejecting family prevent the discharged person from staying at home?"

In both cases, the answers will probably be yes, to some extent. Perhaps 30% of staying at home is explained by an accepting family and the other 70% remains to be explained by other factors: the severity of the discharged person's symptoms, for example, or the degree of acceptance shown by community members outside the family.

Now, you might want to know whether acceptance on the part of the family carries more weight in the staying-at-home decision than acceptance on the part of the community. The answer to this question will provide a direction for possible intervention strategies. You will know whether to focus your attention on individual families or on entire communities.

Suppose you decide, for example, on the basis of your own and other explanatory studies, to focus on families and the intervention you choose to increase their acceptance is education around mental illness. In order to evaluate the effectiveness of your intervention, you will eventually need to ask another explanatory (or causality) question: Does education around mental illness lead to increased acceptance by families of their discharged family members?

With the answer to this question, you have concluded your sieving procedures as outlined in Figure 1.2, moving from a broad exploratory question about discharge experiences to a tested intervention designed to serve the discharged people and their families. In general, the goals of explanatory research studies are to (Neuman, 2009):

- Determine the accuracy of a principle or theory
- Find out which competing explanation is better
- Link different issues or topics under a common general statement
- Build and elaborate a theory so it becomes more complete
- Extend a theory or principle into new areas or issues
- Provide evidence to support or refute an explanation

CLASSIFICATION OF RESEARCH QUESTIONS

Figure 2.3 shows how research questions can be placed on a continuum from simple (exploratory studies) to complex (explanatory studies). Not surprisingly we need to ask the simple questions first. When we have the answers to these simple questions, we then proceed to ask more complex ones. We are thus moving from "little knowledge about our research question" to "more knowledge about our research question."

Figure 2.3 presents the knowledge continuum (from high to low—middle arrow), seven general classifications that research questions can take (left side), and the appropriate research approach (right side) that is most appropriate to answer each question classification.

On a very general level, there are seven types of questions that research studies can answer: (1) existence questions, (2) composition questions, (3) relationship questions, (4) descriptive-comparative questions, (5) causality questions, (6) causality-comparative questions, and (7) causality-comparative interaction questions.

Existence Questions

Suppose for a moment you have an assumption that there is an association between low self-esteem in women and spousal abuse. You are going to study this topic—over a number of studies—starting at the beginning. The beginning, at the bottom of the knowledge continuum, is an existence question: in fact, two existence questions since your assumption involves two concepts: (1) self-esteem and (2) spousal abuse.

First, knowing nothing whatsoever about either self-esteem or spouse abuse, let alone whether there is an association between them, you want to know if self-esteem and spousal abuse actually exist in the first place. Self-esteem and spouse abuse are concepts—they are nothing more than ideas, human inventions if you like—that have become very familiar to social workers, and it is tempting just to say, "Of course they exist. I know they exist."

But there must have been a time when self-esteem was no more than a vague idea among students of human nature that some people seem to feel better about themselves than other people do. It would then have been just a matter of contacting of Ms. Smith—and Ms. Jones, and Ms. Tasmania—and asking them, "Do you feel good about yourself? What is it that makes you feel good about yourself?"

This interpretive study, more commonly referred to as qualitative—and many others like it, conducted by different researchers over time—would have provided an indication that yes indeed, some people do feel better about themselves than other people do. Self-esteem, if that is what you want to call (or label) the feeling, does in fact exist. The same process can be done to determine if spouse abuse exists. However, spouse abuse can be more easily observed and measured than self-esteem.

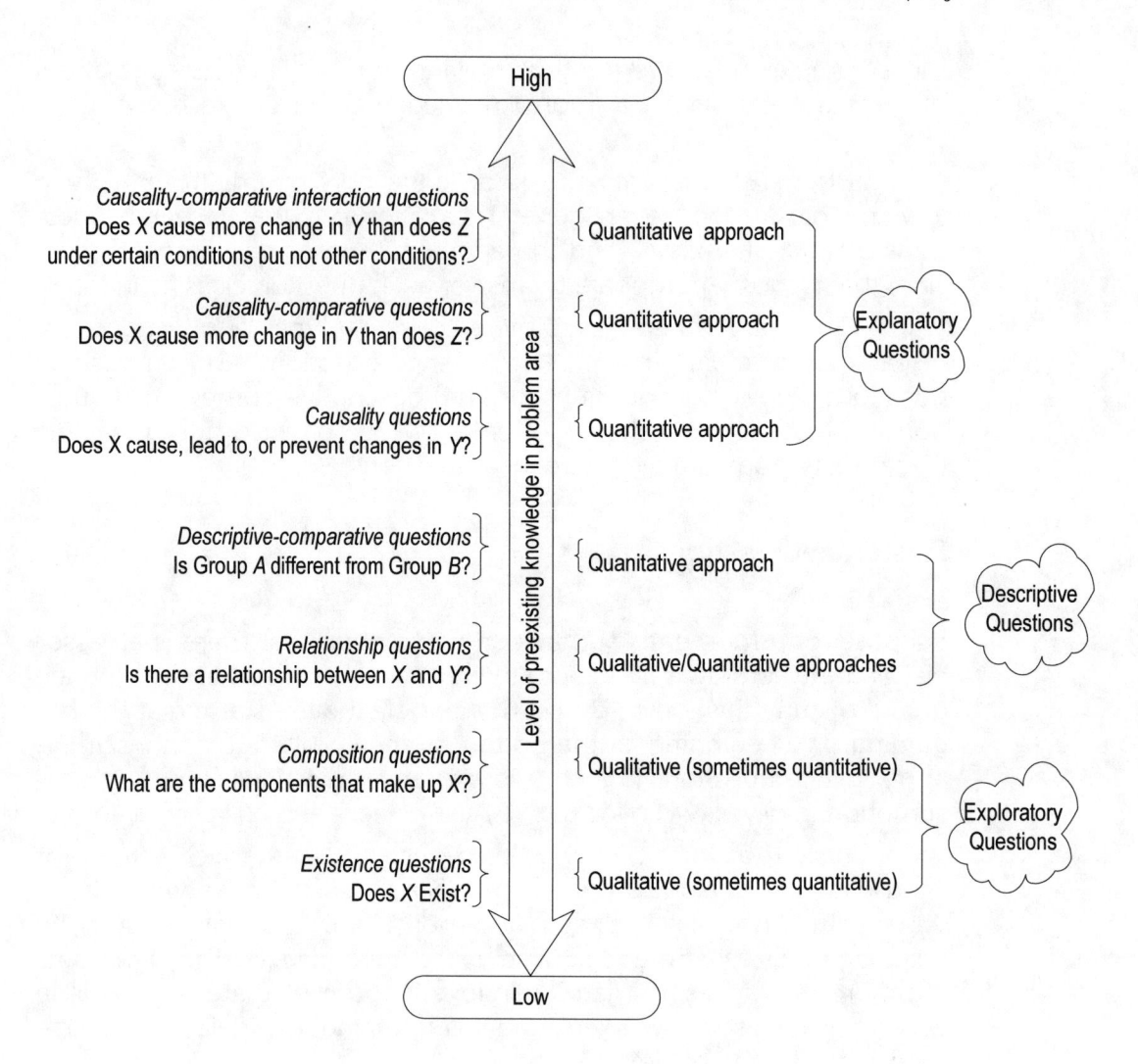

FIGURE 2.3
Types of Research Questions, the Knowledge-Level Continuum, and Approaches to the Research Method

Composition Questions

The second question "what is it that makes you feel good about yourself?" is an attempt to find out what particular personal attributes contribute to self-esteem. That is, it answers the *composition* question next on the knowledge continuum. Qualitative studies exploring this dimension may have discovered that people who feel good about themselves in general also specifically feel that they are liked by others, that they are competent, intelligent, caring, physically attractive, and have a host of other attributes that together make up the concept "self-esteem."

Thus, qualitative studies provide *descriptive*, or *qualitative data*, indicating that self-esteem exists and what it is. Similarly, they indicate that women are in fact sometimes abused by their partners and what particular forms such abuse can take.

Relationship Questions

Going up one more notch on the list, you next come to *relationship questions*. What, if any, is the relationship between women's self-esteem and spousal abuse? Here, you might begin with another qualitative study, trying to determine, on an individual basis, whether there seems to be any connection between having low self-esteem and being abused.

If there does seem to be enough evidence to theorize that such a relationship may exist, you might then use the quantitative approach to see if the relationship holds for a larger number of women rather than just the small number you interviewed in your qualitative study.

Descriptive-Comparative Questions

Next on the continuum you come to *descriptive-comparative questions:* Is Group *A* different from Group *B*? Here, you might go in a number of different directions. You might wonder, for example, whether there is any difference in self-esteem levels between women in heterosexual relationships and women in lesbian relationships. Does self-esteem differ between First Nations women and non-First Nations women? Do women with different sexual orientations or from different cultural groups also differ in how often they are abused, or how severely, or in what particular way?

Because you are asking about differences between social groups, which involve large numbers of people, you will need quantitative methods to address these questions. These will be discussed in Chapter 4.

Causality Questions

Third from the top of the continuum are the *causality* questions. Does low self-esteem actually cause women to be abused? Or, for that matter, does being abused cause low self-esteem? Most complex behaviors don't have single causes. Being abused might certainly contribute to having low self-esteem but it is highly unlikely to be the sole cause.

Similarly, it is unlikely that having low self-esteem leads inevitably to being abused. Quantitative studies here, with their use of impressive statistics, can tell us what percentage of the abuse can be explained by low self-esteem and what percentage remains to be explained by other factors.

Causality-Comparative Questions

If there are factors other than low self-esteem that cause abuse, it would be nice to know what they are and how much weight they have. Perhaps heavy drinking on the part of the abuser is a factor. Perhaps poverty is, or the battering suffered by the abuser as a child. Once these possible factors have been explored, using the same process you used in your exploration of self-esteem and spousal abuse, quantitative methods can again tell us what percentage of abuse is explained by each factor.

If low self-esteem accounts for only 2 percent, say, and heavy drinking accounts for 8 percent, you will have answered the *causality-comparative question*. You will know, on the average, that heavy drinking on the abuser's part has more effect on spousal abuse than does the woman's self-esteem level.

Causality-Comparative Interaction Questions

At the tip-top of the continuum are the *causality-comparative interaction questions*. They ask whether your research findings only hold up under certain conditions. For example, if it is true that heavy drinking contributes more to spousal abuse than does the woman's level of self-esteem, perhaps that is only true for couples who are living in poverty. Or, it is only true for couples with children, or it is only true if the abuser was himself abused.

This final type of question, again answered through quantitative methods, reflects the highest aim of social work research—explanation. If we are to give our clients the best possible service, we need to know about causes and effects. What action or attribute on the part of whom causes how much of what effect, and under what conditions? What type of treatment will cause most change in a particular type of client in a particular situation?

FORMULATING RESEARCH QUESTIONS

How do you go about formulating research questions anyway? We have already talked about what motivates social workers to do research studies in Chapter 1. Your supervisor wants you to, or you have to for the sake of your clients, or you are just interested in something. Sometimes the research question is mostly already there and it only remains for you to word it more precisely so it can become researchable.

Perhaps, for example, your social service program provides substance abuse treatment to First Nations adolescents and it wants to know whether the spirituality component it has recently introduced makes the treatment more effective. At other times, you begin more generally with a broad topic area—poverty, for example—and then you will need to decide what particular aspects of the topic you want to address before you can formulate the research question.

Social work research topics deal with social problems. This may not sound particularly profound but consider for a moment what it is that causes some social circumstances and not others to be defined as "problems." In our day, poverty is seen as a problem. In earlier times, poverty certainly existed but it was not considered a problem in the sense that we felt we ought to do something about it. The attitude was very much that the poor are always with us, they always have been, they always will be, and that is just the way life is.

The first essential factor, then, in defining a social circumstance as a social problem is that society has to believe the circumstance *ought to be changed*. The second factor is that society has to believe the circumstance *can be changed*. If we saw poverty as an evil but were convinced that our society had too few resources to change the situation, we would not be conducting research studies into how the situation might best be changed.

This is an encouraging thought as we look at the vast amount of social problems that still confront us. In short, we do not define a social problem in the first place unless we believed that change both ought to and probably can occur.

A social problem is something we wish we didn't have. Again, profundity may seem to be lacking here, but think about the average social worker's professional attempts to look on the shiny side. If you want to change John's behavior, you might focus on decreasing his negative behaviors or increasing his positive behaviors, and most social workers will hone-in on the positive behaviors if at all possible.

Social workers are trained to think of problems in terms of opportunities for change, and they can use this hopeful propensity when they come to formulating research questions. All else being equal, a good research question will ask how to increase the positives—things like empowerment and resiliency—rather than how to decrease the negatives.

Factors Contributing to Good Research Questions

While we are on the subject of what makes a good research question, let us consider four more factors. A good research question must be: (1) relevant, (2) researchable, (3) feasible, and (4) ethical.

Relevance

Relevance is one of those things, like efficiency, that is never absolute. The same research question may be less or more relevant depending on who is defining what is meant by "relevant" in what particular context. A research question about Chinese pottery, for example, might be relevant to researchers who are interested Chinese potters or to archeologists but may be less relevant to social workers.

To social workers, a relevant research question is one whose answers will have an impact on policies, theories, or practices related to the social work profession. Within this framework, the degree of relevance of any particular study is usually decided by the organization who funds it, the social service program who houses it and the research team who undertakes it.

Researchability

Some very interesting questions cannot be answered through the research process, either because of the nature of the question or because of the difficulties inherent in collecting the necessary data. For example, research studies cannot answer the question "Does God exist?" Neither can it answer the question, "Is abortion wrong?" With respect to the latter, we may be able to collect evidence related to the effects of experiencing or being denied an abortion, but we cannot answer the underlying moral question. In general, questions concerning faith or moral decisions are outside the province of research.

Then, there is the matter of technical difficulties inherent in the question. For example, Aristotle believed that children would be healthier if conceived when the wind was from the north. We are quite as interested in the health of children as Aristotle was, but, even supposing that we accepted the wind as a possible contributing factor, the question would not be researchable because of the practical difficulties associated with determining the direction of the wind at conception in a large enough number of cases.

Researchability, then, has to do with whether the question is appropriate for scientific enquiry and whether it is possible to collect the necessary data.

Feasibility

Feasibility carries the "collection of the necessary data" issue one step further. Perhaps it *is* possible to collect data about the direction of the wind at conception (wind socks attached to every home combined with careful record keeping) but it is not possible for *you* to do it, given your limited resources and your inability to influence home-building and protection of privacy standards. The question about the relationship between the wind and the health of children might therefore be *researchable* but it not *feasible* so far as you are concerned.

Your available resources have a profound effect on the scope of the research study you are able to mount and thus on the questions you are able to answer. In other words, given all these things, could you practically do what you have planned?

Ethical Acceptability

There is a potential for harm not just in the way the study is conducted but in the way the research question is phrased. Researchers are in a wonderful position to offend just about everyone if they fail to pay attention to the cultural, gender, and sexual orientation aspects of their research questions. Not only might they offend people but they might cause harm, not just to research participants but to entire groups of people. We will discuss ethics in-depth in the following chapter.

REVIEWING THE LITERATURE

We have said that social work concepts are usually broad and complex, and can only be completely described through the use of a large number of variables. Feasibility considerations (can you actually do what you have planned?) often mean that the researcher must select a few variables to be studied out of many possible variables. You make this selection based on what you read in the literature and on your knowledge of the context (often your agency or community) in which the study is to take place.

Reviewing the literature helps to develop the research question in two ways: by connecting the research question to existing theory; and by identifying previous research studies in your research area. Journal articles, which provide much valuable information, essentially fall into three categories:

- Theoretical articles (in which the author typically draws on research literature to advance theory in some area)
- Research articles (which report on research studies)
- Review articles (which draw conclusions from a number of published research studies, or other materials, exploring the same area)

Obviously, the theoretical journal articles, together with relevant books, are of most use when it comes to connecting the research question to theory.

Connecting the Research Question to Theory

Before you turn to the literature, you have a broad idea of the concepts your study will be concerned with. If you are looking at ethnicity in relation to accessing social work services, for example, you need to know what has been discovered already in this area and what theories have been proposed. This will help you to place your question correctly on the knowledge continuum presented in Figure 2.3. It will also help you to identify the variables that best describe your concepts.

For example, previous studies may have established that people from ethnic minority groups do indeed access social work services less often than people from the dominant culture. The descriptive-comparative question "Is there a difference between ethnic groups with respect to accessing social work services?" has already been answered. The next level on the knowledge continuum is the causality level, where research questions have to do with what factors cause the difference.

For instance, perhaps a theory has been proposed to the effect that social work services are generally delivered from an individualistic perspective: that is, intervention tends to focus on creating change in the individual, and this is acceptable to people from the dominant North American culture which may value individual achievement. However, people from other cultures may place more value on collective or community achievement and see individualistic social work service as unhelpful or irrelevant.

At this point in your reading, you stop to think. The question you originally had in mind—"Is there a relationship between accessing social work services and the client's ethnic background?"—is no longer appropriate because you have just read the answer.

Perhaps you should be asking instead whether the social work services provided in your unit are perceived to be helpful by ethnic minorities. Perhaps the word has spread and the reason you see so few ethnic minority clients is that they know you wouldn't be helpful anyway.

Reviewing theories in this way can help you to formulate research questions that address practical problems while taking into account existing knowledge. Even if the question is essentially formulated already, you will need to understand the theories underlying the question. For example, perhaps you are trying to improve John's behavior by using positive reinforcement and, in order to evaluate your own practice for the benefit of your client, you want to answer the question,

"Has positive reinforcement increased John's positive behaviors?" This is a relationship question, in which you are trying to relate one variable—*positive reinforcement*—to a second variable—*John's positive behaviors*. A review of behavior theory, which includes positive reinforcement, will allow you to assign values to positive reinforcement so that you can measure it more accurately and completely.

Similarly, you may be involved in a program evaluation for the benefit of your social service program where you work. Perhaps your program provides early intervention services to the families of children with developmental delays and the question is: "Do the program's services increase the ability of parents to cope with their children's developmental delays?"

Again, this is a relationship question with two variables: *the program's services* and *the ability of parents to cope with their children's developmental delays*. In order to assign values to, and thus measure, the coping ability of parents, you have to know enough about developmental delay to understand what constitutes *coping* in this context. Much more will be said about this in Chapter 4.

Identifying Previous Research Studies

We have said that identifying previous studies allows you to know what questions have already been answered so that you can place your question correctly on the knowledge continuum and select appropriate variables for consideration in your own study. Previous studies may or may not have been conducted in a context similar to your own.

For example, if you are studying the needs of the homeless in your community, you will find that the needs of the homeless have been studied before in other communities. Whether or not you need to do another assessment specific to your own community will depend on how similar the communities already studied are to yours. Or perhaps you will find that previous surveys of the homeless have already been conducted in your community.

If the latest survey is recent, you will not need to repeat it, but if it was conducted say 10 years ago, you might feel that new data are needed. If you do feel that, you can look at the way the previous survey was done to see how you might approach such practical aspects as finding people who are homeless to survey. Previous studies often provide valuable suggestions about things like sampling and data collection methods, as well as letting you know whether your study or a similar study has been done before.

ACCESSING INFORMATION

In this information age, most everything you need to know is available somewhere and it is largely a matter of navigating your way through the information jungle in order to find it. Libraries are still a primary source of information, and so is the World Wide Web.

Libraries and the Web

Academic libraries provide both printed materials and the catalogs to access them. Computerized catalogs show what is available in a particular library and also provide links to online journals, books, and other materials. Different libraries have different systems for searching their catalogs but the principles are similar. You can enter an author's name, a title, or subject words and you can usually tailor your search by setting various parameters such as a publication date.

When you have found a relevant item, its keywords will help you to conduct other searches. Searching "ethnicity" for example, may bring you to "racism," "discrimination," and "professional ethics" which, in combination with "social services," may bring you to materials dealing with the relationship between accessing social services and ethnic background.

You can also access databases and indexes related to your topic. Indexes will help you to locate literature reviews, which not only discuss current research in your area but provide references to dissertations, books, book chapters and other documents. These will not necessarily be held in your library but the library may be able to get them for you through an interlibrary loan.

Databases allow you to access the full text of some articles or documents, while providing abstracts of others or links to the full text. Another source of material is government documents, which include census data, government information, and statistics. Some libraries provide a separate online system for finding these since an increasing amount of government information is only available on the Web.

The Web is a growing source of information, providing access to international as well as local sites. Contrary to popular opinion, it is not all free. Many materials acquired by the library through subscriptions are still acquired through subscriptions even though they are now web-based rather than in print. These are made available to library users through license agreements via the library's own web-site.

The Web is in a state of constant flux. Web addresses (URLs) appear, disappear, and alter; Web sites are re-organized; links change. In an attempt to impose order on the chaos, some schools of social work and organizations of social workers have developed gateway sites that provide organized access to sites of interest to social workers.

With so much information available, it may sometimes be difficult to evaluate the accuracy of whatever texts you select. When you read a printed book or journal, for example, it maybe tempting to assume that the information given is accurate and current: otherwise, why would the material be in print? However, authors are human beings who may sometimes be mistaken, and researchers are responsible for confirming the accuracy of the material they use whether the information came from a printed source or was posted on the Web.

There is no fail-safe way of doing this, but you may feel more confident that a piece of information is correct if several authors say the same thing. Finding confirmation is a matter of precise information retrieval. Present search engines are not particularly precise but new engines are continually being developed which will make it easier to find what you are looking for.

Finally, remember that you are never alone in a library. Library staff are trained to find things. They can save you time in the short run and help you to learn in the long run, so that next time you will be able to find whatever you want yourself.

SUMMARY

This chapter started out with a definition of social work research. We then discussed the main three forms that research studies can take: exploratory, descriptive, and explanatory. Next, we talked about how research studies can be classified via the research questions they address. We then went on to discuss the various considerations that all research questions must contain: relevance, researchability, feasibility, and ethical acceptability. Finally, we presented a very brief discussion on how the professional literature can be used to formulate research questions. The next chapter discusses research ethics in-depth.

REFERENCES AND FURTHER READING

Cooper, H.M. (1998). *Synthesizing the literature: A guide for literature reviews* (3rd ed.). Thousand Oaks, CA: Sage.

Corcoran, J. (2008). *Systematic reviews and meta-analysis*. New York: Oxford University Press.

Fang, L., Manuel, J., Bledsoe, S.E., & Bellamy, J.L. (2008). Finding existing knowledge. In R.M. Grinnell, Jr., & Y. Unrau (Eds.). *Social work research and evaluation: Foundations of evidence-based practice* (8th ed., pp. 465-480). New York: Oxford University Press.

Fink. A. (2004). *Conducting research literature reviews: From the Internet to paper* (2nd ed.). Thousand Oaks, CA: Sage.

Grinnell, R.M., Jr., Unrau, Y.A., & Williams, M. (2008). Group-level designs. In R.M. Grinnell, Jr., & Y.A. Unrau (Eds.), *Social work research and evaluation: Foundations for evidence-based practice* (8th ed., pp. 179–204). New York: Oxford University Press.

Manual, J., Fang, L., Bellamy, J.L., & Bledsoe, S.E. (2008). Evaluating evidence. In R.M. Grinnell, Jr., & Y. Unrau (Eds.). *Social work research and evaluation: Foundations of evidence-based practice* (8th ed., pp. 481-495). New York: Oxford University Press.

Neuman, W.L. (2009). *Understanding research.* Boston: Allyn & Bacon.

Williams, M., Tutty, L.M., & Grinnell, R.M., Jr. (1995). *Research in social work: An introduction* (2nd ed.). Itasca, IL: F.E. Peacock.

Williams, M., Unrau, Y.A., & Grinnell, R.M., Jr. (1998). *Introduction to social work research.* Itasca, IL: F.E. Peacock.

Yegidis, B.L., & Weinbach, R.W. (2005). Using existing knowledge. In R.M. Grinnell, Jr., & Y. Unrau (Eds.) *Social work research and evaluation: Quantitative and qualitative approaches* (7th ed., pp. 45-57). New York: Oxford University Press.

Check out our Website for useful links and chapters (PDF) on:
- how to formulate research questions
- how to do literature reviews

Go to: www.pairbondpublications.com

Click on: Student Resources
              Chapter-by-Chapter Resources
              Chapter 2

Check Out Our Website

PairBondPublications.com

√ *Student Workbook Exercises*
√ *Chapter Power Point Slides*
√ *On-line Glossaries*
√ *General Links*
√ *Specific Chapter Links*

Chapter 3

Research Ethics

ETHICS AND THE USE OF CLIENTS AS RESEARCH PARTICIPANTS
    Obtaining Informed Consent
        Bribery, Deception, and Other Forms of Coercion
    Designing the Study In an Ethical Manner
    Informing Others about the Study's Findings
ETHICS AND OUR SOCIAL SERVICE PROGRAMS
    Misuses of Research Results
        Justifying Decisions Already Made
        Public Relations
        Performance Appraisals
        Fulfilling Funding Requirements
SUMMARY
REFERENCES AND FURTHER READINGS

3

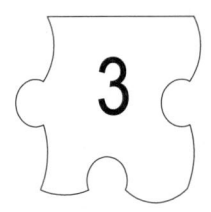

IN THE LAST TWO CHAPTERS, we discussed the different ways of acquiring knowledge by focusing on the research method and how to formulate research questions. The rest of the book looks at the actual process of undertaking research studies, exploring both the quantitative and qualitative approaches. But before we can discuss in-depth the actual doing of research studies, however, we need to know something about research ethics and what you need to take into account before you start.

ETHICS AND THE USE OF CLIENTS AS RESEARCH PARTICIPANTS

Many social work research studies use clients as research participants. When doing so, we need to be extremely careful not to violate any of their ethical rights. Sounds simple you say. Well read on. Consider Aristotle, for example. Aristotle believed that the bite of the female shrew mouse was dangerous to horses, particularly if the mouse was pregnant.

Since it would not have occurred to him to test this belief by subjecting horses to the bites of pregnant shrew mice—and non-pregnant shrew mice too, of course—he would not have had to solve the resulting ethical dilemma. Are we justified in subjecting some horses to discomfort and possible danger in order to establish a definite danger against which we can protect a greater number of horses in the future? Supposing the horses were people?

To social workers, the answer is apparent. No, we are not justified. Increasing the sum of knowledge is a worthy aim, but the client must not be harmed in the process. We are not in the business of committing lesser evils for the sake of greater goods.

Not harming the client, by commission or omission, is a cardinal rule of research and imposes a major limitation. There are a number of bodies devoted to ensuring that harm does not occur. In the United States, for example, there is a committee known as the National Commission for the Protection of Human Subjects of Biomedical and Behavioral Research. All universities have ethics committees, and many large social service programs do as well. There are also various professional associations and lay groups which focus on protecting research participants.

However, it is likely that the participants in your research study will never have heard of any of these bodies. They will do what you ask them to do either because they trust you or because they think they have no choice. The responsibility if they are harmed is yours. There are, therefore, three precautions that must be taken when undertaking any research study. These are: (1) obtaining the participant's informed consent, (2) designing the study in an ethical manner, and (3) properly informing others about the findings.

Obtaining Informed Consent

Before you involve any human being in a research study, you must obtain the person's informed consent. The key word here is "informed." You must make sure that the would-be participant understands what is going to happen during the study, why it is going to happen, how long it will take, and what the risks and benefits will be for him/her. For example, if she is not capable of full understanding—if she is senile, perhaps, or handicapped, or a minor—then you must explain to someone else, such as a parent, adult child, guardian, social worker, or spouse.

All written communications must be couched in a simple language that the potential participant will understand. Some researchers, particularly academics, have a tendency to confuse obscurity with profundity. They use technical terms so firmly embedded in convoluted sentence structures that the meaning falters and disappears altogether at the second comma.

Consent letters or forms that you are expecting the participant to sign should be no longer than two pages of single-spaced type. Box 3.1 contains all the necessary information that has to go into a consent letter. Using the contents of Box 3.1, Box 3.2 provides an example of a consent letter written to parents who were to participate in a study designed to explore non-traditional gender roles in families. Note that confidentiality issues are fully discussed and anonymity is not guaranteed.

There is a distinct difference between confidentiality and anonymity. A participant is anonymous when no-one knows who he is, not even the researcher. For example, mailed questionnaires might be sent out with instructions to the respondents to return the completed form without a name, address, or any other identifying information. The questionnaires are then anonymous because the researcher has no idea who responded, who didn't, and who sent back which response. This is good from the point of view of concealing the identity of the respondent. It is less good from the point of view of the researcher, who has now lost the opportunity to send out polite reminders to those anonymous souls who failed to respond.

BOX 3.1
Obtaining Informed Consent In a Nutshell

The most important consideration in any research study is to obtain the participants' *informed* consent. The word "informed" means that each participant fully understands what is going to happen in the course of the study, why it is going to happen, and what its effect will be on him or her. If the participant is psychiatrically challenged, mentally delayed, or in any other way incapable of full understanding, our study must be fully and adequately explained to someone else—perhaps a parent, guardian, social worker, or spouse, or someone to whom the participant's welfare is important.

It is clear that no research participant may be bribed, threatened, deceived, or in any way coerced into participating. Questions must be encouraged, both initially and throughout the course of the study. People who believe they understand may have misinterpreted our explanation or understood it only in part. They may say they understand, when they do not, in an effort to avoid appearing foolish. They may even sign documents they do not understand to confirm their supposed understanding, and it is our responsibility to ensure that their understanding is real and complete.

It is particularly important for participants to know that they are not signing away their rights when they sign a consent form. They may decide at any time to withdraw from the study *without penalty*, without so much as a reproachful glance. The results of the study will be made available to them as soon as the study has been completed. No promise will be made to them that cannot be fulfilled.

A promise that is of particular concern to many research participants is that of anonymity. A drug offender, for example, may be very afraid of being identified; a person on welfare may be concerned whether anyone else might learn that he or she is on welfare. Also, there is often some confusion between the terms "anonymity" and "confidentiality."

Some studies are designed so that no one, not even the person doing the study, knows which research participant gave what response. An example is a mailed survey form, bearing no identifying mark and asking the respondent not to give a name. In a study like this, the respondent is *anonymous*. It is more often the case, however, that we do know how a particular participant responded and have agreed not to divulge the information to anyone else. In such cases, the information is *confidential*. Part of our explanation to a potential research participant must include a clear statement of what information will be shared with whom.

All this seems reasonable in theory, but ethical obligations are often difficult to fulfill in practice. There are times when it is very difficult to remove coercive influences because these influences are inherent in the situation. A woman awaiting an abortion may agree to provide private data about herself and her partner because she believes that, if she does not, she will be denied the abortion. It is of no use to tell her that this is not true: She feels she is not in a position to take any chances.

There are captive populations of people in prisons, schools, or institutions who may agree out of sheer boredom to take part in a research study. Or, they may participate in return for certain privileges, or because they fear some penalty or reprisal. There may be people who agree because they are pressured into it by family members, or they want to please the social worker, or they need some service or payment that they believe depends on their cooperation. Often, situations like this cannot be changed, but at least we can be aware of them and try to deal with them in an ethical manner.

A written consent form should be only part of the process of informing research participants of their roles in the study and their rights as volunteers. It should give participants a basic description of the purpose of the study, the study's procedures, and their rights as voluntary participants. All information should be provided in plain and simple language.

A consent form should be no longer than two pages of single-spaced copy, and it should be given to all research participants. Survey questionnaires may have a simple introductory letter containing the required information, with the written statement that the completion of the questionnaire is the person's agreement to participate. In telephone surveys, the information will need to be given verbally and must be standardized across all calls. A written consent form should contain the following items, recognizing that the relevancy of this information and the amount required will vary with each research project:

1. A brief description of the purpose of the research study, as well as the value of the study to the general/scientific social work community (probability and nature of direct and indirect benefits) and to the participants and/or others.

2. An explanation as to how and/or why participants were selected and a statement that participation is completely voluntary.

3. A description of experimental conditions and/or procedures. Some points that should be covered are: (1) The frequency with which the participants will be contacted; (2) The time commitment required by the participants; (3) The physical effort required and/or protection from overexertion; (4) Emotionally sensitive issues that might be exposed and/or follow-up resources that are available if required; (5) Location of participation (e.g., need for travel/commuting); and (6) Information that will be recorded and how it will be recorded (e.g., on paper, by photographs, by videotape, by audiotape).

4. Description of the likelihood of any discomforts and inconveniences associated with participation, and of known or suspected short- and long-term risks.

5. Explanation of who will have access to the collected data and to the identity of the participants (i.e., level of anonymity or confidentiality of each person's participation and information) and how long the data will be stored.

6. Description of how the data will be made public (e.g., scholarly presentation, printed publication). An additional consent is required for publication of photographs, audiotapes, and/or videotapes.

Chapter 3: Research Ethics

7. Explanation of the participants' rights: (1) That they may terminate or withdraw from the study at any point; (2) That they may ask for clarification or more information throughout the study; and (3) That they may contact the appropriate administrative body if they have any questions about the conduct of the people doing the study or the study's procedures.
8. Description of other projects or other people who may use the data.

BOX 3.2
EXAMPLE OF A CONSENT LETTER
[AGENCY LETTERHEAD]

Date

Ms. Smith
Smith Street
Smithtown, USA

Dear Ms. Smith:

As we discussed on the telephone, I am interested in the different ways that couples organize their family lives. In many families where both partners are earning income, women still tend to do most of the childcare and housework, while men are more involved in their jobs or careers. My focus is on couples like you who are doing things differently. For example, some couples are trying to divide family and work responsibilities more equally and some have switched roles due to illness or job loss or for some other reason. I am interested in developing a picture of how things work in these non-traditional families.

If you agree to participate, I would like to interview you for about an hour and a half in your home or my office or wherever you feel most comfortable. I will first ask you some background questions about your work and family history. Then I will ask you how responsibilities are divided within your family and how you came to do it that way. Finally, I will ask how it is working out for family members, what you like about it particularly, what you wish you could change, and so forth.

You do not have to answer any question you would prefer not to and you are free to end our interview at any time. If you change your mind about participating once you have begun, any information you have provided up to that time will be destroyed.

I would like to tape-record the interview so that I will have a more accurate record to work with later. The tape-recording will be transcribed by someone paid to help me, only she and I will listen to it, and she will be asked to sign a confidentiality agreement. Your name will not appear on the tape so only I will know

> which tape belongs to you. If any part of my interview with you is included in a research report, conference presentation, book or journal article, care will be taken to change identifying details. Real names will not be used.
>
> However, although every care will be taken to conceal your identity, confidentiality cannot be absolutely guaranteed. As you know, your partner is taking part in this study as well and may be able to identify you because of the experiences you are reporting or the way you express yourself in interview excerpts. In the same way, it is possible that someone in your social circle will recognize your style of speech. The chance of this happening is remote but you should be aware of the possibility.
>
> This interview will take up some of your time. As well, after the interview is transcribed, I will ask you to read through the typed document to make sure that it accurately reflects what you said. Your participation will not benefit you personally but I hope that the finished research will give people a more complete picture of what modern family life looks like, and perhaps show some new ways to help make families work. I plan to provide every participant with a summary of the research findings at the end of the project.
>
> Your signature on this form indicates that you have understood to your satisfaction the information I have given you about the project and have agreed to participate. In no way does your signature waive your legal rights or release the researchers or sponsors from their legal and professional responsibilities. You should feel free to ask for clarification or new information at any time. If you have further questions, please contact::
>
> Researcher's name, Address, Phone number, E-mail:
>
> YES: I AM WILLING TO PARTICIPATE IN THE PROJECT
>
> _____   _____
> Participant's signature   Date

A promise of anonymity is uncommon in social work research studies. More usually, participants are promised confidentiality, which means that no-one *except* the researcher, and people named by her, know who they are. The first step in the process for ensuring confidentiality is often to assign a code number to each participant.

The researcher and her assistants alone know that Ms. Smith, for example, is number 132. All data concerning Ms. Smith are then combined with data from all the other participants to produce results that do not identify Ms. Smith in any way. No-one reading the final research report or any publication stemming from it will know that Ms. Smith took part in the study at all.

Sometimes, however, complete confidentiality cannot be guaranteed. In a study undertaken in a small community, for example, direct quotes from an interview with "a" social worker may narrow the field to three because there are only three social workers there. Then, the flavor of the quote may narrow it again to Mr. Jones, who said the same thing in church last Sunday.

If there is any risk that Mr. Jones might be recognized as the author of the quote, then this possibility must be clearly acknowledged in the letter of consent that Mr. Jones is asked to sign. Such an acknowledgement is included in the sample letter contained in Box 3.2.

Although the ideal is to obtain written consent from the potential participant before the study begins, it is not always possible to obtain the consent in writing. In a telephone interview, for example, the information that would have been contained in a letter of consent is usually read to the participant and oral consent is obtained.

If a number of participants are to be interviewed by telephone, the same information, set out in exactly the same way, must be read to them all. Mailed questionnaires are sent out with an accompanying introductory letter which contains the required information and states that filling out the questionnaire and sending it back constitutes consent.

Bribery, Deception, and Other Forms of Coercion

It goes without saying that consent must never be obtained through bribery, threats, deception or any form of coercion. You may feel insulted that such a possibility should even be mentioned in a text addressed to social workers, but consider bribery for example. If you offer $1,000 to the chief executive officer of an agency to persuade him to take part in your study, this is bribery. Your study will not go forward and your sanity as well as your ethics will be questioned. On the other hand, if you offer $10 as an incentive to a homeless person, you might well be hailed as a sensitive caring soul who has a proper respect for the contributions made to research by the homeless.

The differences lie in the amount of money involved and the person to whom it is offered. It is unfortunately true that in our society—as in all societies everywhere, including barnyards—there is a pecking order. The higher up the social scale you go, the more power people have and the less ethically acceptable it is to offer them any form of inducement. On a general level, no professional person should ever be offered anything. Other research participants, particularly the more deprived, are commonly offered a small amount in cash or voluntary services of some kind. This small sum and/or in-kind services acknowledges their contribution and is viewed as a reward—or honorarium—rather than a bribe.

Next on the list we come to deception. You should never deceive anyone when trying to get them to participate in a research study. Unfortunately, there are still situations where deception can be very tempting. People's behavior, for example, often changes when they know they are being watched.

If you want to know how they *really* behave when no-one else is looking, you will have to deceive them into believing that they are *not* being watched. You might think you can do this using an interview room with a one-way mirror, or you might pretend to be an ordinary member of a group when you are, in fact, a glint-eyed observer. Neither of these behaviors is ethically acceptable.

The only conditions under which deception might be countenanced—and it is a very large "*might*"—are when the data to be obtained is vitally important and there is no other way to get it. If you can persuade the various ethics committees which review your proposal that both these conditions exist, you *might* be given permission to carry out the study. Even then, you would have to be sure that the deception was thoroughly explained to all the participants when the study was over and that arrangements had been made—free counseling, for example—to counter any harm they might have suffered.

Last, but not least, there are threats. No researcher would ever persuade potential participants to cooperate by threatening that, if they do not, worse things will befall. But a *perceived* threat, even if not intended, can have the same effect. For example, a woman awaiting an abortion may agree to provide private information about herself and her partner because she believes that, if she does not, she will be denied the abortion. It is of no use to tell her that this is not true. She feels she is not in a position to take any chances.

There are captive populations in prisons, schools, and institutions who may agree out of sheer boredom to take part in a research study. Or they may participate in return for certain privileges, or because they fear some reprisal. There may be people who agree because they are pressured into it by family members, or they want to please their social workers, or they need some service or payment that they believe depends on their cooperation. Often, situations like this cannot be changed, but at least you can be aware of them and do your best to deal with them in an ethical manner.

Designing the Study In an Ethical Manner

The second precaution that must be taken to protect research participants is to design the study in an ethical manner. One of the more useful research designs, presented in Chapter 9, involves separating research participants into experimental and comparison groups. The experimental group receives a treatment while the comparison group does not. The idea is to determine how effective the treatment is and there is usually an underlying assumption that it will prove to be effective, at least to some extent.

The dilemma here is whether it is ethically acceptable to withhold treatment, assumed to be beneficial, from the members of the comparison group. Even if they are on a waiting list and will receive the treatment at a later date, is it right to delay service in order to conduct the study?

Proponents of this model of research argue that people on a waiting list will not receive service any faster whether they are involved in the study or not. Furthermore, the treatment is only *assumed* to be beneficial. If its effects were known for sure, there would be no need to do the study. Our ethical responsibility to our clients is to test such assumptions through research studies before we continue with treatments that may be ineffective or even harmful.

But if it may be harmful, counter the critics, you have no right to give it to the experimental group either. But, say the proponents, then we will never know if it is beneficial. Both these arguments have valid points and it is up to the researcher—and the various ethics committees—to determine which of them is most convincing. Ethics, like politics, hinge on points of view, ideologies, cultural beliefs, and values, which change over time.

The same kind of controversy arises around the random assignment of research participants to groups receiving different treatments. The researchers will argue that nobody knows which treatment is best and that is why they are doing the study. Ms. Smith's social worker, on the other hand, might argue that she does know (or, at least she thinks she does) which treatment is best for Ms. Smith, and she is taking no chances that Ms. Smith might be randomly assigned to the group receiving existential therapy when the behavioral therapy being received by the other group would be more beneficial.

Since both sides know that they are right, the dispute will have to be settled by an unhappy program administrator who will try to satisfy the social worker without alienating the researchers. Politics as well as ethics will come into play here, and we will explore the political context further when we come to talk about the social service program itself as yet another factor that affects research studies.

Informing Others about the Study's Findings

The third precaution that must be taken to protect research participants is to report the study's findings in an ethical manner. It may be very tempting, for example, to give great weight to the positive findings while playing down or ignoring altogether the negative or disappointing findings. There is no doubt that positive findings tend to be more enthusiastically received—often by journal editors who should know better—but it is obviously just as important to know that a thing did not work well as to know that it did.

Perhaps the disappointing findings were due to the inherent limitations within the research design. All studies have limitations because practical and ethical considerations make it difficult to use the complex and costly designs that yield the most certain results. Since studies with more limitations yield less trustworthy findings, it is important to be honest about the limitations, so that readers can weigh the practice implications and the recommendations in light of the trustworthiness of the results.

For example, the current debate in the field of children's services about the effectiveness of residential care is largely based on the results of research studies carried out in the 1980s and 1990s. The general conclusion drawn from these studies is that there is little evidence for the effectiveness of residential care. However, proponents of residential care argue that such a conclusion is unjustified because it fails to take into account the limitations of the studies on which they were based. This squabble among the "authoritarians" may sound uninteresting but it will eventually have a major impact on the lives of children who are being cared for by the state. Ethical reporting of the study's findings also includes giving appropriate credit to colleagues.

With the exception of case-level designs, discussed in Chapter 9, where one social worker might do all the work, research studies are normally conducted by teams. The principal researcher, whose name is usually listed first on the report, must make sure that all team members receive recognition. It is also the responsibility of the principal researcher to ensure that the report is circulated to everyone entitled to receive it, including the study's participants, and that meetings are held as necessary to discuss the findings.

Often, the sharing of findings will be a delicate matter. Agency staff may be reluctant to hear, for example, that their program is less effective than they thought. If they were not engaged in the research process in the first place and they know little about research methods, they may be tempted to dismiss the findings and block any attempt on the part of the researcher to discuss recommendations for improvement. Findings must be presented carefully, therefore, to the right people, in the right order, at the right time.

Practitioners wrestle every day with a similar problem. Mr. Yen might not want to be told that his daughter is still threatening to run away despite all those parenting classes and family therapy sessions he attended. His daughter might not want him to know. His wife might not want him to know either in case this bit of data spurs him to inappropriate disciplinary steps.

The social worker must decide whom to tell, as well as how, when, and how much. The same hold true when doing research studies. All this has been summarized in the *Code of Ethics* published by the National Association of Social Workers (1996) in which scholarship and research are addressed (see Chapter 1).

ETHICS AND OUR SOCIAL SERVICE PROGRAMS

We noted in Chapter 1 that one thing all agency administrators have in common is worry. They worry about how effectively they are serving their clients and how they can demonstrate their effectiveness to funders. They worry about offending people. They worry about whether they will be seen to be doing the right thing, with the right people, in the right place, at the right time.

All these worries have ethical and political overtones. Take effective service to clients, for example. There is nothing absolute about effectiveness. The same service may be deemed to be effective or ineffective depending on how effectiveness is defined and who does the defining. For instance, the aim of a drug rehabilitation program might be to ensure that every client who comes through its doors will abstain from using any type of drug for evermore. This aim, though worthy, is impossible.

The program's actual achievement—in terms of the percentage of clients who completely abstained, partially abstained, or did not abstain from using a particular drug over a particular period—will be seen as successful or not depending on the political climate. If marijuana is currently viewed as a gateway to hell, the program's lack of success in curtailing its use will be judged more harshly than in a more tolerant climate. If a more immediate concern is gas sniffing among First Nations children, the program is likely to be judged on how well it is dealing with solvent abuse. Effectiveness thus tends to be defined and evaluated with respect to external political and social issues.

There are also the internal politics of the program to be considered. Staff members will each have their individual views about which of the many issues facing the program should be given priority, how scarce resources should be used, which treatment modality is best, what the reporting structure ought to be, and, in general, who ought to do what, where, how and when. These inevitable tensions affect the research climate within the program.

Most social service programs will be less or more receptive to research endeavors at different times, depending on their internal political situations and their current level of staff morale. When staff are feeling overworked, underpaid, and unappreciated, they will be less inclined to co-operate with research efforts, but, at the same time, this is the moment when they might benefit most from an evaluative study designed to improve client service delivery or to optimize the organizational structure.

Conversely, when morale is low due to conflicts with administrators, an evaluative study might just serve to exacerbate the conflicts. Researchers, like all social workers, have the potential to help or to harm. If they are to help, they must be sensitive to internal and external political considerations so that they can conduct a useful study in the right way, in the right place at the right time.

It might be worthwhile here to consider the misuses of research results since a study that sets out to be useful can turn out to be harmful if the results are used in an unethical way.

Misuses of Research Results

There are four primary ways in which research results may be misused within social service programs: (1) to justify decisions already made, (2) to safeguard public relations, (3) to appraise the performance of staff, (4) and to fulfill funding requirements.

Justifying Decisions Already Made

We said earlier that a common aim among agency administrators is to avoid offending people. Indeed, in the present political climate where services to clients are increasingly community based, programs make it a priority to establish and maintain good relationships with community stakeholders. Since there are usually a large number of stakeholder groups, each with its own agenda, maintaining good relationships with all of them requires a judicious juggling of political and ethical balls.

For example, mindful of the needs of its client group, the agency might propose to build a hospice for AIDS patients in a nearby residential area. The homeowners in the area object. In company with other public-spirited citizens, they approve of hospices for AIDS patients—and rehabilitation programs for young offenders, and homes for pregnant adolescents—but they don't want them in their own backyards.

The agency, trapped between its duty to its clients and its duty to its neighbors, will go through the normal processes of convening meetings, establishing committees, collecting comments and issuing reports. It is also quite likely that a research study will be commissioned into the advisability of building the hospice as proposed.

Again, the researchers need to be sensitive to the political currents both within and outside the agency setting. Probably some stakeholders have already decided that they want the hospice built and are looking to the research report to confirm this opinion. Other stakeholders will not want the hospice built and also expect that the research report will support their position. Trapped, like the agency, be-

Chapter 3: Research Ethics

tween a rock and a hard place, the researchers will need to consider very carefully the uses that are likely to be made of their research results.

If the results will be used solely to justify a decision that has already been made on other grounds, then it is not ethical to undertake the study. On the other hand, if it is possible that a disinterested appraisal of the advantages and disadvantages of building the hospice will actually sway the decision, then the study should be done. The researchers must decide whether or not to undertake the study, based on their judgment of the ethical and political factors involved.

Public Relations

Another of the worries shared by agency administrators is negative publicity. Perhaps a worker in a group home has been indicted for the sexual abuse of residents or a child has been abused by a foster parent. These kinds of incidents inevitably attract intense media scrutiny, and it is tempting for administrators to immediately commission an evaluative study to investigate the problem, declining to comment until the research results become available. Now, there is nothing wrong with an evaluative study. It may, indeed, be the best way to determine why the problem occurred and what can be done to prevent it from occurring again in the future.

But if the sole purpose of the study is to delay comment until the furor has died down, then it is not ethical for a researcher to undertake the study. Neither is it ethical if there is an unspoken expectation that the researcher will discover what the foster care program wants.

If parts of the report are taken out of context in order to show the program in a good light, or if the results are distorted to avoid public embarrassment, then the integrity of the study will obviously be compromised, and the researcher will be at fault as well. It used to be the case that researchers were not held responsible for the use made of study results and recommendations. Their jobs were merely to produce them. Now, however, it is considered essential for researchers to do the best they can to ensure that results are used in an ethical way.

Performance Appraisals

The third possible misuse of research results is for performance appraisals. Again, there is nothing wrong with a performance appraisal. Most workers are required to undergo an annual evaluation of how well they have performed during the previous year. However, a performance appraisal and a research study are two separate things. For example, consider a research study designed to document overall client progress within a program by aggregating (or adding up) the progress made by the clients of each individual social worker.

Before they are aggregated, the study results have the potential to demonstrate that the clients of one social worker made more progress overall than the clients of another. And this, in turn, may be taken to indicate that one social worker is more competent than the other. In a study of this kind, it is not ethical for researchers to release results before they have been aggregated; that is, before they are in a form that will protect the confidentiality of the social workers who participated in the study.

In the event that they are prematurely released, it is certainly not ethical for administrators to use the information to appraise a particular social worker. Nevertheless, there have been instances when research results have been used for political purposes, to promote or undermine a specific worker or program, and researchers need to be wary of this.

Fulfilling Funding Requirements

Most evaluation studies are used, at least in part, to demonstrate accountability to funders. Indeed, almost all social service programs, particularly new or pilot projects, are funded with the stipulation that they should be evaluated. It is not unethical, therefore, to use evaluation research to fulfill funding requirements. It is unethical, however, to use evaluation research solely to fulfill funding requirements without any real intention of using the information gathered to improve the program.

SUMMARY

This chapter briefly looked at a few of the ethical factors that affect the social work research enterprise. Now that we know how research is used in social work (Chapter 1), how to formulate research questions (Chapter 2), and how to behave in an ethical manner when answering these questions (this chapter), the following chapter discusses the quantitative research approach.

REFERENCES AND FURTHER READINGS

Congress, E.P., Black, P.N., & Strom-Gottfried, K. (2007). *Teaching social work values and ethics: A curriculum resource* (2nd ed.). Alexandria, VA. Council on Social Work Education.

Council on Social Work Education (2007). National Statement on Research Integrity in Social Work. Alexandria, VA: Author.

Ford, G.A. (2006). *Ethical reasoning for mental health professionals.* Thousand Oaks, CA: Sage.

Israel, M., & Hay, I. (2006). *Research ethics for the social scientists.* Thousand Oaks, CA: Sage.

Ivanhoff, A., Blythe, B., & Walters, B. (2008). The ethical conduct of research. In R.M. Grinnell, Jr., & Y.A. Unrau (Eds.), *Social work research and evaluation: Foundations of evidence-based practice* (8th ed., pp. 29–59). New York: Oxford University Press.

National Association of Social Workers (1996). *Code of ethics.* Silver Spring, MD: Author.

Reamer, F.G. (1987). Informed consent in social work. *Social Work, 32,* 425–429.

Reamer, F.G. (1998). *Ethical standards in social work: A critical review of the NASW Code of Ethics.* Washington, DC: NASW Press.

Reamer, F.G. (1998). *Social work research and evaluation.* New York: Columbia University Press.

Reamer, F.G. (1999). *Social work values and ethics* (2nd ed.). New York: Columbia University Press.

Reamer, F.G. (2005). Research ethics. In R.M. Grinnell, Jr., & Y.A. Unrau (Eds.), *Social work research and evaluation: Quantitative and qualitative approaches* (7th ed., pp. 34–45). New York: Oxford University Press.

Reichert, E. (2006). *Understanding human rights: An exercise book.* Thousand Oaks, CA: Sage.

Weinbach, R.W. (2005). Research contexts. In R.M. Grinnell, Jr., & Y.A. Unrau (Eds.), *Social work research and evaluation: Quantitative and qualitative approaches* (7th ed., pp. 24–32). New York: Oxford University Press.

Wronka, J. (2008). *Human rights and social justice: social action and service for the helping and health professions.* Thousand Oaks, CA: Sage.

Check out our Website for useful links and chapters (PDF) on:
- ethics and social work research

    Go to: www.pairbondpublications.com
    Click on: Student Resources
              Chapter-by-Chapter Resources
              Chapter 3

Check Out Our Website

PairBondPublications.com

√ *Student Workbook Exercises*
√ *Chapter Power Point Slides*
√ *On-line Glossaries*
√ *General Links*
√ *Specific Chapter Links*

Part II

Approaches to Knowledge Development

Chapter 4: The Quantitative Research Approach   66

Chapter 5: The Qualitative Research Approach   98

Chapter 4

The Quantitative Research Approach

WHAT IS THE POSITIVIST WAY OF THINKING?
    Striving Toward Measurability
    Striving Toward Objectivity
    Striving Toward Reducing Uncertainty
    Striving Toward Duplication
    Striving Toward the Use of Standardized Procedures
STEPS WITHIN THE QUANTITATIVE RESEARCH APPROACH
    Steps 1 and 2: Developing the Research Question
    Step 3: Designing the Research Study
    Step 4: Collecting the Data
    Steps 5 and 6: Analyzing and Interpreting the Data
    Steps 7 and 8: Presentation and Dissemination of Findings
EXAMPLE OF THE QUANTITATIVE RESEARCH APPROACH
    Step 1: What's the Problem?
    Step 2: Formulating Initial Impressions
    Step 3: Determining What Others Have Found?
    Step 4: Refining the General Problem Area
    Step 5: Measuring the Variables
    Step 6: Deciding on a Sample
    Step 7: Obeying Ethical Principles
    Step 8: Collecting the Data
    Step 9: Gathering and Analyzing the Data
    Step 10: Interpreting the Data
    Step 11: Comparing Results with Others Findings
    Step 12: Specifying the Study's Limitations
    Step 13: Writing and Disseminating the Study's Results
SUMMARY / REFERENCES AND FURTHER READINGS

4

CHAPTER 1 PRESENTED A DISCUSSION of why the generation of knowledge is best acquired through the use of the research method. As was delineated in the chapter, the research method contains two complimentary approaches—the quantitative approach, which is the topic of this chapter—and the qualitative approach—the topic of the following chapter.

No matter which approach we use to obtain our professional knowledge base, "knowing" something that resulted from either approach is much more objective than "knowing" that exact same something that was derived from the other ways of knowing. Before we discuss the quantitative approach to knowledge development, however, you need to understand how this approach is embedded within the "positivist way of thinking."

WHAT IS THE POSITIVIST WAY OF THINKING?

The positivist way of thinking strives toward: measurability, objectivity, reducing uncertainty, duplication, and the use of standardized procedures (see, for example, the left side of Figure 4.1). Our discussion on how positivists think has be adapted and modified from Grinnell and Williams (1990), Williams, Tutty, and Grinnell (1995), Williams, Unrau, and Grinnell (1998), and Williams, Unrau, and Grinnell (2003)

Striving Toward Measurability

The positivist way of thinking tries to study only those things that can be objectively measured. That is, knowledge gained through this belief is based on "objective measurements" of the real world, not on someone's opinions, beliefs, or past experiences.

Conversely, and as you know from Chapter 1, knowledge gained through tradition or authority depends on people's opinions and beliefs and not on "objective" measurements of some kind. Entities that cannot be measured, or even seen, such as id, ego, or superego, are not amenable to a positivistic-orientated research study but rather rely on tradition and authority. In short, a positivist principle would be that the things you believe to exist must be able to be measured.

However, at this point in our discussion, it is useful to remember that researchers doing studies within a positivistic framework believe that practically everything in life is measurable.

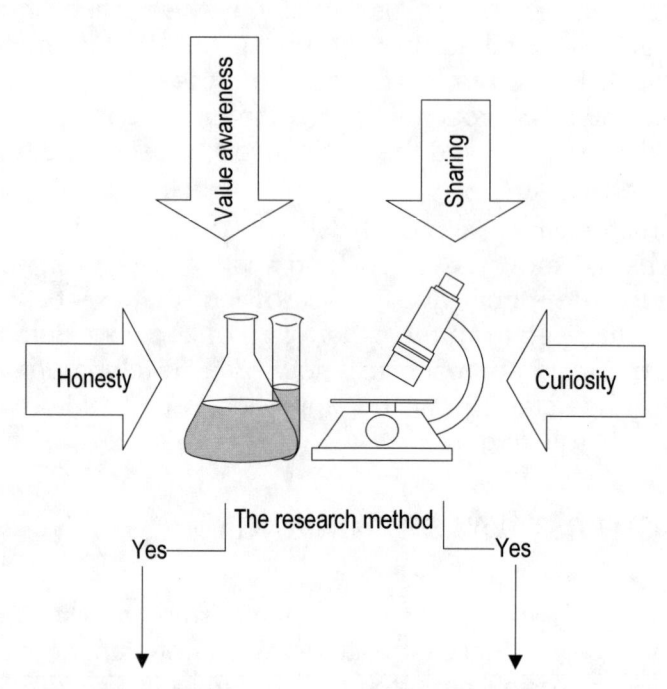

The Positivistic Research Approach

One objective reality
Seeks to be objective
Reality unchanged by observations & measurements
Researcher puts aside own values
Social and physical sciences are a unity
Passive roles for research subjects
Many research subjects involved
Data obtained through observations & measurements
Data are quantitative in nature
Deductive logic applied
Causal information obtained
Seeks to explain or predict
Tests hypotheses
Reliance on standardized measuring instruments
High generalizability of findings

The Interpretive Research Approach

Many subjective realities
Admittedly subjective
Reality changed by observations & measurements
Researcher recognizes own values
Social and physical sciences are different
Active roles for research participants
Few research participants involved
Data obtained by asking questions
Data are qualitative in nature
Inductive logic applied
Descriptive information obtained
Seeks to understand
Produces hypotheses
Researcher is the measuring instrument
Limited generalizability of findings.

FIGURE 2.1
The Positivistic (Quantitative) and Interpretive (Qualitative) Research Approaches within the Research Method

Chapter 4: The Quantitative Research Approach

Striving Toward Objectivity

The second ideal of the positivist belief is that research studies must be as "objective" as possible. The things that are being observed and/or measured must not be affected in any way by the person doing the observing or measuring. Physical scientists have observed inanimate matter for centuries, confident in the belief that objects do not change as a result of being observed.

In the sub-world of the atom, however, physicists are beginning to learn what social workers have always known. Things do change when they are observed. People think, feel, and behave very differently as a result of being observed. Not only do they change, they change in different ways depending on who is doing the observing and/or measuring.

There is yet another problem. Observed behavior is open to interpretation by the observer. To illustrate this point, let's take a simple example of a client you are seeing, named Ron, who is severely withdrawn. He may behave in one way in your office in individual treatment sessions, and in quite another way when his mother joins the interviews. You may think that Ron is unduly silent, while his mother remarks on how much he is talking. If his mother wants him to talk, perhaps as a sign that he is emerging from his withdrawal, she may perceive him to be talking more than he really is.

All folks doing research studies with the positivistic framework go to great lengths to ensure that their own hopes, fears, beliefs, and biases do not affect their research results, and that the biases of others do not affect them either. Nevertheless, as discussed in later chapters, complete "objectivity" is rarely possible in social work despite the many strategies that have been developed over the years to achieve it.

Suppose, for example, that a social worker is trying to help a mother interact more positively with her child. The worker, together with a colleague, may first observe the child and mother in a playroom setting, recording how many times the mother makes eye contact with the child, hugs the child, criticizes the child, makes encouraging comments, and so forth on a three-point scale (i.e., discouraging, neutral, encouraging). The social worker may perceive a remark that the mother has made to the child as "neutral," while the colleague thinks it was "encouraging."

As you will see throughout this book, in such a situation it is impossible to resolve the disagreement. If there were six objective observers, for example, five opting for "neutral" and only one for "encouraging," the one "encouraging observer" is more likely to be wrong than the other five, and it is very likely that the mother's remark was "neutral." As you know from Chapter 1, as more people agree on what they have observed, the less likely it becomes that the observation was distorted by bias, and the more likely it is that the agreement reached is "objectively true." As should be obvious by now, objectivity is largely a matter of agreement.

There are some things—usually physical phenomena—about which most people agree. Most people agree, for example, that objects fall when dropped, water turns to steam at a certain temperature, sea water contains salt, and so forth. However, there are other things—mostly to do with values, attitudes, and feelings—about which agreement is far more rare.

An argument about whether Beethoven is a better composer than Bach, for example, cannot be "objectively" resolved. Neither can a dispute about the rightness of capital punishment, euthanasia, or abortion. It is not surprising, therefore, that physical researchers, who work with physical phenomena, are able to be more "objective" than social work researchers, who work with human beings.

Striving Toward Reducing Uncertainty

Positivistic-orientated research studies try to totally rule out uncertainty. Since all observations and/or measurements in the social sciences are made by human beings, personal bias cannot be entirely eliminated, and there is always the possibility that an observation and/or measurement is in error, no matter how many people agree about what they saw or measured.

There is also the possibility that the conclusions drawn from even an accurate observation or measurement will be wrong. A huge number of people may agree, for example, that an object in the sky is a UFO when in fact it is a meteor. Even if they agree that it is a meteor, they may come to the conclusion—probably erroneously—that the meteor is a warning from an angry extraterrestrial person.

In the twentieth century, most people do not believe that natural phenomena have anything to do with extraterrestrial people. They prefer the explanations that modern researchers have proposed. Nevertheless, no researcher would say—or at least be quoted as saying—that meteors and extraterrestrial beings are not related for certain.

When utilizing the research method of knowledge development, nothing is certain. Even the best-tested theory is only tentative and accepted as true until newly discovered evidence shows it to be untrue or only partly true. All knowledge gained through the research method is thus provisional. Everything presently accepted as true is true only with varying degrees of probability.

Striving Toward Duplication

Positivistic researchers try to do research studies in such a way that the studies can be duplicated. Suppose, for a moment, you are running a 12-week intervention program to help fathers who have abused their children to manage their anger without resorting to physical violence. You have put a great deal of effort into designing your program, and believe that your intervention (the program) is more effective than other interventions currently used in other anger-management programs.

You develop a method of measuring the degree to which the fathers in your group have learned to dissipate their anger in nondamaging ways and you find that, indeed, the group of fathers shows marked improvement.

Improvement shown by one group of fathers, however, is not convincing evidence for the effectiveness of your program. Perhaps your measurements were in error and the improvement was not as great as you hoped for. Perhaps the improvement was a coincidence, and the fathers' behaviors changed because they had joined a health club and each had vented his fury on a punching bag. In order to be more certain, you duplicate your program and measuring procedures with a second group of fathers. In other words, you replicate your study.

After you have used the same procedures with a number of groups and obtained similar results each time, you might expect that other social workers will eagerly adopt your methods. As presented in the previous chapters, tradition dies hard. Other social workers have a vested interest in their interventions, and they may suggest that you found the results you did only because you wanted to find them.

In order to counter any suggestion of bias, you ask another, independent social worker to use your same anger-management program and measuring methods with other groups of fathers. If the results are the same as before, your colleagues in the field of anger management may choose to adopt your intervention method (the program). Whatever your colleagues decide, you are excited about your newfound program.

You wonder if your methods would work as well with women as they do with men, with adolescents as well as with adults, with Native Americans, Asians, or African Americans as well as with Caucasians, with mixed groups, larger groups, or groups in different settings. In fact, you have identified a lifetime project, since you will have to apply your program and measuring procedures repeatedly to all these different groups of people.

Striving Toward the Use of Standardized Procedures

Finally, a true-to-the-bone positivist researcher tries to use well-accepted standardized procedures. For a positivistic-oriented research study to be creditable, and before others can accept its results, they must be satisfied that your study was conducted according to accepted scientific standardized procedures. The allegation that your work lacks "objectivity" is only one of the criticisms they might bring.

In addition, they might suggest that the group of fathers you worked with was not typical of abusive fathers in general, and that your results are not therefore applicable to other groups of abusive fathers. It might be alleged that you did not make proper measurements, or you measured the wrong thing, or you did not take enough measurements, or you did not analyze your data correctly, and so on.

In order to negate these kinds of criticisms, over the years social work researchers have agreed on a set of standard procedures and techniques that are thought most likely to produce "true and unbiased" knowledge—which is what this book is all about.

Certain steps must be performed in a certain order. Foreseeable errors must be guarded against. Ethical behavior with research participants and colleagues must be maintained as outlined in Chapter 3. These procedures must be followed if your study is both to generate usable results and to be accepted as useful by other social workers.

STEPS WITHIN THE QUANTITATIVE RESEARCH APPROACH

The above discussion is only the philosophy behind the quantitative research approach to knowledge building. With this philosophy in mind, we now turn our attention to the eight general sequential steps (in a more-or-less straightforward manner) that all quantitative researchers follow as outlined in Figure 4.2.

These steps yield a very useful format for obtaining knowledge in our profession. The quantitative research approach as illustrated in Figure 4.2 is a "tried and tested" method of scientific inquiry. It has been used for centuries. As we now know, if data obtained within a research study are represented in the form of numbers, then this portion of the study is considered "quantitative." The numbers are then analyzed by descriptive and inferential statistics.

In a nutshell, most of the critical decisions to be made in a quantitative research study occur *before* the study is ever started. This means that the researcher is well aware of all the study's limitations before the study actually begins. It is possible, therefore, for a researcher to decide that a quantitative study has simply too many limitations and eventually comes to the conclusion not to carry it out.

Chapter 4: The Quantitative Research Approach

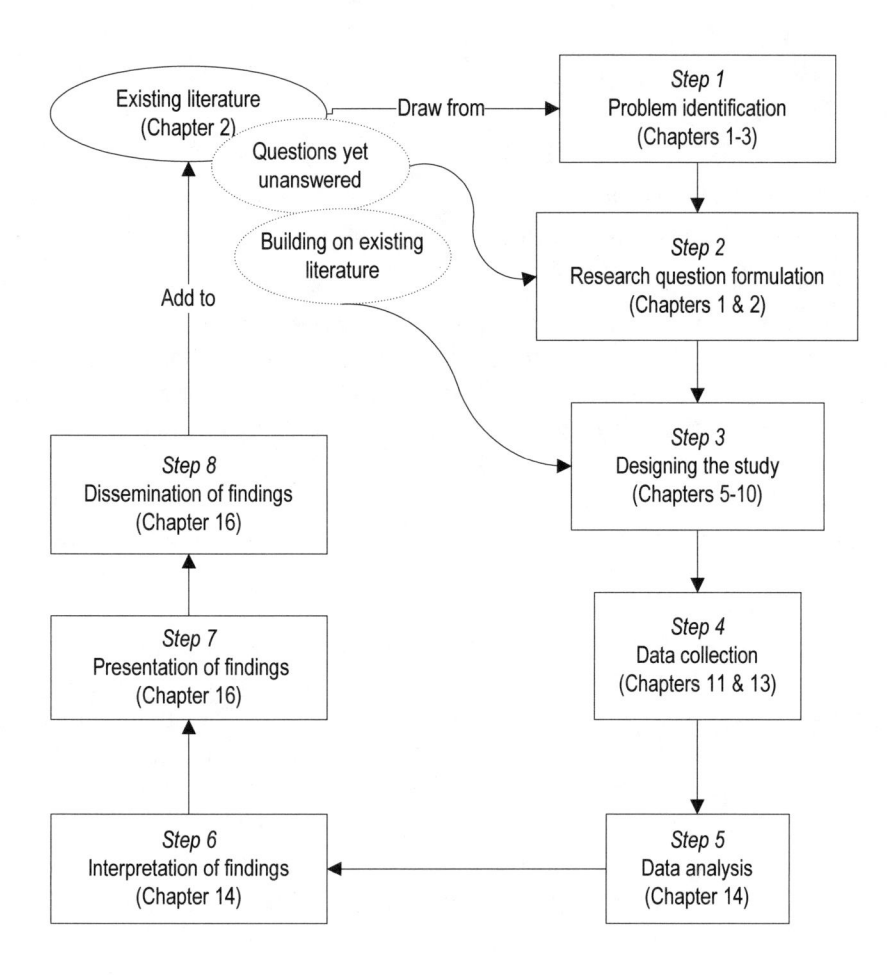

FIGURE 4.2
The Positivistic (Quantitative) Research Approach to Knowledge Building

Regardless of whether a proposed study is ever carried out or not, the process always begins with choosing a research topic and focusing the research question (topic of Chapter 2).

Steps 1 and 2: Developing the Research Question

As can be seen in Figure 2.2, the first two steps of the quantitative approach to knowledge development are to identify a general problem area to study (Step 1) and then to refine this general area into a research question that can be answered or a hypothesis that can be tested (Step 2).

These studies are usually deductive processes; that is, they usually begin with a broad and general query about a general social problem and then pare it down to a specific research question or hypothesis as illustrated in Figure 1.4. For instance, your general research problem may have started out with a curiosity about racial discrimination within public social service agencies. It could be written simply as:

General Problem Area:
Racial discrimination within public social service agencies

You may have noticed through your professional practice as a medical social worker in a local hospital, for example, that many of the patients within your hospital are from ethnic minority backgrounds who: (1) have high unemployment rates, (2) have a large proportion of their members living under the poverty level, and (3) have low levels of educational attainment. You believe that these three conditions alone should increase the likelihood of them utilizing the hospital's social service department where you work.

Conversely, and at the same time, you have also observed that there are more ethnic majorities than ethnic minorities who are seen in your hospital's social service department. Your personal observations may then lead you to question whether discrimination of ethic minorities exists when it comes to them having access to your hospital's social service department. You can easily test the possibility of such a relationship by using the quantitative research approach.

The next step in focusing your research question would be to visit the library and review the literature related to your two concepts:

- Racial discrimination within social services (Concept 1)
- Access to social service (Concept 2)

You would want to read the literature related to the two main concepts within the general research question—racial discrimination within social service agencies and access to them. You would want to learn about how various theories explain both of your main concepts in order to arrive at a meaningful research question. It may be, for example, that many ethnic minority cultures are unlikely to ask "strangers" for help with life's personal difficulties.

Furthermore, you may learn that most social service programs are organized using bureaucratic structures, which require new potential clients to talk to several strangers (e.g., telephone receptionist, waiting-room clerk, intake worker) before they are able to access social services. Given that you know, via the literature, that ethnic minorities do not like talking with strangers about their personal problems, and that social services are set up for people to deal with a series of strangers, you could develop a very simple quantitative research question: Do patients who come from ethnic minority backgrounds have difficulty accessing my hospital's social service department?

Task 1: Develop Concepts

Your simple straightforward general problem area has become much more specific via the construction of a research question. In your research question, for example, you have identified a person's ethnicity (Concept 1) and access to your hospital's social services (Concept 2) as your two concepts of interest.

What are concepts anyway? They are nothing more than ideas. When you speak of a client's ethnic background, for example, you have in mind the concept of *ethnicity*. When you use the word *ethnicity*, you are referring to the underlying idea that certain groups of people can be differentiated from other groups in terms of physical characteristics, customs, beliefs, language, and so on.

Take a female patient in your hospital, for example, who has just been referred to your social service department. She is a patient in the hospital, she is a woman, and she is now also your client. If she is married, she is a wife. If she has children, she is a mother. She may be a home owner, a committee member, an Asian, or a Catholic. She may be hostile, demanding, or compassionate. All of her characteristics are concepts. They are simply ideas that are all members of a society share—to a greater or lesser degree, of course.

Some concepts are perceived the same way by all of us. On the other hand, some concepts give rise to huge disagreements. The concept of being a mother, for example, involves the concept of children and, specifically, the concept of having given birth to a child. Today, most people would agree that giving birth to a child is only *one* way of defining a mother.

The idea of motherhood in Western society involves more than simply giving birth, however. Also involved in motherhood are the concepts of loving, of caring for the child's physical needs, of offering the child emotional support, of advocating for the child with others, of accepting legal and financial responsibility for the child, and of being there for the child in all circumstances and at all times.

Some of us could easily argue that a woman who does all of these things is a mother, whether she has given birth or not. Others would say that the biological mother is the *only* real mother even if she abandoned her child at birth. Like many other qualities of interest to social workers, ethnicity is a highly complex concept with many possible dimensions. Intelligence is another such concept, as are alienation, morale, conformity, and a host of others.

Task 2: Identify Variables within Concepts

You can now relate the concept of "the existence of different ethnic groups" to the patients who seek out your hospital's social service department. Some patients will belong to one ethnic group, some to another, some to a third, and so on. In other words, these folks *vary* with respect to which ethnic group they belong to. Any characteristic that can vary, logically enough, is called a *variable*. So, your variable name in this example is *ethnic group*.

Task 3: Put Value Labels on Variables

You now have a concept, ethnicity—and a related variable—ethnic group. Finally, you need to think about which particular ethnic groups will be useful for your study. Perhaps you know that Asians are patients within your hospital, and so are Caucasians, Hispanics, African Americans, and Native Americans. This gives you five categories, or *value labels,* of your ethnic group variable:

Value Labels for Ethnicity Variable:
- Asian
- Caucasian
- Hispanic
- African American
- Native American

During your quantitative study, or more accurately, "during the quantitative portion of your study," you will ask all of the hospital's patients which of the five ethnic groups they belong to; or perhaps these data will be recorded on the hospital's intake forms and you will not need to ask them. In any case, the resulting data will be in the form of numbers or percentages for each value label. You will have succeeded in measuring the variable *ethnic group* by describing it in terms of five value labels, or categories. Value labels do nothing more than describe a variable.

You will note that these five categories only provide one possible description. You could also have included Pacific Islanders, for example, if there were any receiving medical treatment in your hospital, and then you would have had six value labels of your ethnic group variable instead of only five. If you were afraid that not all clients receiving medical treatment would fit into one of these categories, then you could include a miscellaneous category, *other,* to be sure you had accounted for everyone.

By reviewing the literature and your knowledge of your social service unit, you have more or less devised a direction for your study in relation to your ethnicity concept. You have come up with a concept, a variable, and five value labels for your variable which are:

Chapter 4: The Quantitative Research Approach

Concept: Ethnicity
Variable: Ethnic group
Value Labels: Asian
Caucasian
Hispanic
African American
Native American

As you know, ethnicity is not the only concept of interest in your study. There is also *access to social work services* which is the idea that some people, or groups of people, are able to access social work services more readily than other people or groups.

You might think of access simply in terms of how many of the patients receiving medical treatment within your hospital actually saw a social worker. Clients will *vary* with respect to whether they saw a social worker or not and so you have the variable—*saw social worker*—and two value labels of that variable:

- *yes*, the patient saw a social worker
- *no*, the patient did not see a social worker

You could, for example, ask each patient upon leaving the hospital a very simple question such as:

Did you see a social worker while you were in the hospital?
- Yes
- No

If you wish to explore access in more depth, you might be interested in the factors affecting access. For example, perhaps your review of the literature has led you to believe that some ethnic groups tend to receive fewer referrals to social work services than other groups. If this is the case in your hospital, clients will vary with respect to whether they received a referral or not and you immediately have a second variable—*referral*—and two value labels of that variable:

- *yes*, the patient was referred
- *no*, the patient was not referred

Once again, this variable can take the form of a very simple question:

When you were a patient within the hospital, were you at anytime referred to the hospital's social services department?
- Yes
- No

However, there is more to accessing hospital social work services than just being referred. Perhaps, according to the literature, certain ethnic groups are more likely to follow up on a referral than other groups because of cultural beliefs around the appropriateness of asking non-family members for help.

In that case, you have a third variable, *follow up of referral*, with two value labels of its own:

- *yes*, the client followed up
- *no*, the client did not follow up

This also can be put into a question form such as:

If you were referred to social work services while you were a patient in the hospital, did you follow-up on the referral and actually see a social worker?
- Yes
- No

In addition, folks who do try to follow up on referrals may meet circumstances within the referral process that are more intimidating for some than for others. Perhaps they are obliged to fill out a large number of forms or tell their stories to many unfamiliar people before they actually succeed in achieving an appointment with a social worker. If this is the case, they *vary* with respect to how intimidating they find the process and you have a fourth variable, *feelings of intimidation around the referral process*. The value labels here are not so immediately apparent but you might decide on just three:

- not at all intimidated
- somewhat intimidated
- very intimidated

This also can be put into the form of a simple question that you would ask all patients who were referred to social services:

How intimidated were you when you were referred to the social services department?
- Not at all intimidated
- Somewhat intimidated
- Very intimidated

By reviewing the literature and your knowledge of your hospital's social service department, you have more or less devised a direction for your study in relation to your access concept. You have come up with a concept, four variables, and value labels for each one of the four variables which are as follows:

Concept: Access to social work services

First Variable: Saw social worker?
    Value Labels: Yes
            No
Second Variable: Referral?
    Value Labels: Yes
            No
Third Variable: Follow-up of referral?
    Value Labels: Yes
            No
Fourth Variable: Feelings of intimidation around the referral process
    Value Labels: Not at all intimidated
            Somewhat intimidated
            Very intimidated

Task 4: Define Independent and Dependent Variables

A simple quantitative research study may choose to focus on the relationship between only two variables, called a *bivariate relationship*. The study tries to answer, in general terms: Does Variable X affect Variable Y? Or, how does Variable X affect Variable Y? If one variable affects the other, the variable that does the affecting is called an *independent variable*, symbolized by X.

The variable that is affected is called the *dependent variable*, symbolized by Y. If enough is known about the topic, and you have a good idea of what the effect will be, the question may be phrased: If X occurs, will Y result? If Variable X affects Variable Y, whatever happens to Y will depend on X.

1. No Independent or Dependent Variables. Some quantitative research studies are not concerned with the effect that one variable might have on another. Perhaps it is not yet known whether two variables are even associated, and it is far too soon to postulate what the relationship between them might be. Your study might try to ascertain the answer to a simple question, such as, "How intimidated do *ethnic minority* patients feel when they are referred to my hospital's social service department?"

In this simple question, there is no independent variable; neither is there a dependent variable. There is only one variable—degree of intimidation felt by one group of people, the ethnic minorities. You could even include ethnic majorities as well and ask the question:

How intimidated do *all* patients feel when they are referred to my hospital's social service department?

Task 5: Construct a Hypothesis

There are many types of hypotheses where we will only briefly discuss two: (1) nondirectional, and (2) directional.

1. Nondirectional Hypotheses. A non-directional hypothesis (also called a two-tailed hypothesis), is simply a statement that says you expect to find a relationship between two or more variables. You are not willing, however, to "stick your neck out" as to the specific relationship between them. A nondirectional hypothesis for your each one of your access variables could be, for example:

- Nondirectional Research Hypothesis 1: *Saw a Social Worker*
  Ethnic minorities and ethnic majorities see hospital social workers differentially
- Nondirectional Research Hypothesis 2: *Referral*
  Ethnic minorities and ethnic majorities are referred to the hospital's social service department differentially
- Nondirectional Research Hypothesis 3: *Follow-up*
  Ethnic minorities and ethnic majorities vary to the degree they follow-up on referrals
- Nondirectional Research Hypothesis 4: *Intimidation*
  Ethnic minorities and ethnic majorities feel differently on how intimidated they were about the referral process

2. Directional Hypotheses. Unlike a nondirectional hypothesis, a directional hypothesis (also called a one-tailed hypothesis) specifically indicates the "predicted" direction of the relationship between two or more variables. The direction stated is based on an existing body of knowledge related to the research question.

You may have found out through the literature (in addition to your own observations), for example, that you have enough evidence to suggest the following directional research hypotheses:

- Directional Research Hypothesis 1: *Saw a Social Worker*

  Ethnic majorities see hospital social workers more than ethnic minorities

- Directional Research Hypothesis 2: *Referral*

  Ethnic majorities are referred to the hospital's social service department more than ethnic minorities

- Directional Research Hypothesis 3: *Follow-up*

  Ethnic majorities follow-up with social service referrals more than ethnic minorities

- Directional Research Hypothesis 4: *Intimidation*

  Ethnic minorities are more intimidated with the referral process than ethnic majorities

Task 6: Evaluate the Hypothesis

Our discussion on how to evaluate good hypotheses has be adapted and modified from Grinnell and Williams (1990), Williams, Tutty, and Grinnell (1995), Williams, Unrau, and Grinnell (1998), and Williams, Unrau, and Grinnell (2003). Hypotheses have to be relevant, complete, specific, and testable.

1. Relevance. It is hardly necessary to stress that a useful hypothesis is one that contributes to the profession's knowledge base. Nevertheless, some social work problem areas are so enormously complex that it is not uncommon for people to get so sidetracked in reading the professional literature that they develop very interesting hypotheses totally unrelated to the original problem area they wanted to investigate in the first place. The relevancy criterion is a reminder that, to repeat, the hypothesis must be directly related to the research question, which in turn must be directly related to the general problem area.

2. Completeness. A hypothesis should be a complete statement that expresses your intended meaning in its entirety. The reader should not be left with the impression that some word or phrase is missing. "Moral values are declining" is one example of an incomplete hypothesis.

Other examples include a whole range of comparative statements without a reference point. The statement, "Males are more aggressive," for example, may be assumed to mean "Men are more aggressive than women," but someone investigating the social life of animals may have meant, "Male humans are more aggressive than male gorillas."

3. Specificity. A hypothesis must be unambiguous. The reader should be able to understand what each variable contained in the hypothesis means and what relationship, if any, is hypothesized to exist between them. Consider, for example, the hypothesis, "Badly timed family therapy affects success." Badly timed family therapy may refer to therapy offered too soon or too late for the family to benefit; or to the social worker or family being late for therapy sessions; or to sessions that are too long or too short to be effective.

Similarly, "success" may mean resolution of the family's problems as determined by objective measurement, or it may mean the family's—or the social worker's—degree of satisfaction with therapy, or any combination of these.

With regard to the relationship between the two variables, the reader may assume that you are hypothesizing a negative correlation. That is, the more badly timed the therapy, the less success will be achieved. On the other hand, perhaps you are only hypothesizing an association. Bad timing will invariably coexist with lack of success.

Be that as it may, the reader should not be left to guess at what you mean by a hypothesis. If you are trying to be both complete and specific, you may hypothesize, for example:

> Family therapy that is undertaken after the male perpetrator has accepted responsibility for the sexual abuse of his child is more likely to succeed in reuniting the family than family therapy undertaken before the male perpetrator has accepted responsibility for the sexual abuse.

This hypothesis is complete and specific. It leaves the reader in no doubt as to what you mean, but it is also somewhat wordy and clumsy. One of the difficulties in writing a good hypothesis is that specific statements need more words than unspecific, or ambiguous statements.

4. Potential for Testing. The last criterion for judging whether a hypothesis is good and useful is the ease with which the truth of the hypothesis can be verified. Some statements cannot be verified at all with presently available measurement techniques. "Telepathic communication exists between identical twins," is one such statement.

A hypothesis of sufficient importance will often generate new data-gathering techniques, which will enable it to be eventually tested. Nevertheless, as a general rule, it is best to limit hypotheses to statements that can be tested immediately by current and available measurement methods.

Step 3: Designing the Research Study

As can be seen in Figure 2.2, Step 3 involves designing the research study a bit more. Having focused your general research question, and if appropriate, developed a hypothesis, you enter into the next phase of your study—designing the study. We begin with a word about sampling.

One objective of your research study may be to generate findings that can be generalized beyond your study's sample. As you will see in Chapter 8, the "ideal" sample when this is the goal is one that has been randomly selected from a carefully defined population. The topic of sampling will be discussed much more fully in Chapter 8.

As you shall see in this book, many research questions have at least one independent variable and one dependent variable. You could easily design your quantitative study where you have one independent variable, ethnicity (i.e., ethnic minority, ethnic majority) and one dependent variable, difficulty in accessing social services (i.e., yes, no). You could easily organize your variables in this way because you are expecting that a person's ethnicity is somehow related to his/her difficulty in accessing social services.

It would be absurd to say the opposite—that the degree of difficulty that folks have in accessing social services influences their ethnicity. You could write your directional hypothesis as follows:

Directional Hypothesis:
Ethnic minorities have more difficulty than ethnic majorities in accessing my hospital's social service department

Having set out your hypothesis in this way, you can plainly see that your research design will compare two groups (i.e., ethnic minorities, ethnic majorities) in terms of whether or not (i.e., yes, no) each group had difficulty accessing your hospital's social service department. Your research design is the "blueprint" for the study. It is a basic guide to deciding how, where, and when data will be collected.

How data are collected and where they are collected from is determined by the data collection method you choose (Chapters 11–13). *When* data are collected is dictated by the specific research design you select (Chapters 9 and 10). Clearly, there are many things for you to consider when developing your research design.

Step 4: Collecting the Data

Data collection is one step within any research design. Data collection is where you truly test out the operational definitions of your study's variables. There are three features of data collection that are key to all quantitative research studies:

1. *All of your variables must be measurable.* This means that you must precisely record the variable's frequency, and/or its duration, and/or its magnitude (intensity). Think about your ethnic minority variable for a minute. As noted earlier, you could simply operationalize this variable into two categories: ethnic minority and ethnic majorities:

   Are you an ethnic minority?
   - Yes
   - No

   Here you are simply measuring the presence (ethnic minority) or absence (ethnic majority) of a trait for each research participant within your study. You also needed to operationalize, or measure the difficulty in accessing the hospital's social services department variable. One again, you could have measured this variable in a number of ways. You chose, however to measure it where each person could produce a response to a simple question:

   Did you have difficulty in accessing our hospital's social services department?
   - Yes
   - No

2. *All of your data collection procedures must be objective.* That is, the data are meant to reflect a condition in the *real* world and should not be biased by the person collecting the data in any way. In your quantitative study, the research participants produced the data—not you, the researcher. That is,

Chapter 4: The Quantitative Research Approach

you only recorded the data that each participant individually provided for both variables:

- "Ethnic Minority" or "Ethnic Majority" for the ethnicity variable
- "Yes" or "No" for the access to social service variable

3. *All of your data collection procedures must be able to be duplicated.* In other words, your data collection procedures that you used to measure the variables must be clear and straightforward enough so that other researchers could use them in their research studies.

The three features of measurability, objectivity, and duplication within a quantitative research study are accomplished by using a series of standardized uniform steps that are applied consistently throughout a study's implementation. You must ensure that all of your research participants are measured in exactly the same way—in reference to their ethnicity and whether or not they had any difficulty in accessing hospital social services.

Steps 5 and 6: Analyzing and Interpreting the Data

There are two major types of quantitative data analyses: (1) descriptive statistics, and (2) inferential statistics.

Descriptive Statistics

As presented in Chapter 14, descriptive statistics describe your study's sample or population. Consider your ethnicity variable for a moment. You can easily describe your research participants in relation to their ethnicity by stating how many of them fell into each category label of the variable. Suppose, for example, that 50 percent of your sample were in the ethnic minority category and the remaining 50 percent were in the non-ethnic minority category as shown:

Value Label:
- Ethnic Minority.............50%
- Ethnic Majority.............50%

The preceding two percentages will give you a "picture" of what your sample looked like in relation to their ethnicity. A different picture could be produced where 10 percent of your sample were ethnic minorities and 90 percent were not as illustrated below:

Value Label:
- Ethnic Minority..............10%
- Ethnic Majority.............90%

The above describes only one variable—ethnicity. A more detailed picture is given when data for two variables are displayed at the same time (Table 4.1). Suppose, for example, that 70 percent of your research participants who were ethnic minorities reported that they had difficulty in accessing your hospital's social services, compared to 20 percent of those who were ethnic majorities.

Other descriptive information about your research participants could include variables such as average age, percentages of males and females, average income, and so on. Much more will be said about descriptive statistics in Chapter 14 when we discuss how to analyze quantitative data—data that are in the form of numbers.

Inferential Statistics

Inferential statistics determine the probability that a relationship between the two variables within your sample also existed within the population from which it was drawn. Suppose in your quantitative study, for example, you find a statistically significant relationship between your research participant's (your sample) ethnicity and whether they successfully accessed social services within your hospital setting.

The use of inferential statistics will permit you to say whether or not the relationship detected in your study's sample existed in the larger population from which it was drawn—and the exact probability that your finding is in error. Much more will also be said about inferential statistics in Chapter 14.

TABLE 4.1
Difficulty in Accessing Social Services?

Ethnicity	Yes	No
Ethnic Minority	70%	30%
Ethnic Majority	20%	80%

Chapter 4: The Quantitative Research Approach

Steps 7 and 8: Presentation and Dissemination of Findings

Quantitative findings are easily summarized in tables, figures, and graphs. When data are disseminated to lay people, we usually rely on straightforward graphs and charts to illustrate our findings. As discussed in-depth in Chapter 16, presentation of statistical findings is typically reserved for professional journals.

EXAMPLE OF THE QUANTITATIVE RESEARCH APPROACH

The following discussion presents the positivistic, or quantitative research process, in detail by delineating 13 steps—seven steps (Steps 1–7) for planning research studies and six steps (Steps 8–13) for carrying out research studies (Grinnell, Unrau, & Williams, 2008).

The most important thing to remember at this point is that all the steps are intertwined to various degrees and it is difficult to describe a single step in isolation of the remaining 12 steps. However, for the sake of instruction, we will describe each step in a sequential manner.

Step 1: What's the Problem?

Let's pretend for a moment that you are a MSW-level social worker in a large social service agency. The past 3 years of agency intake data reveal that many clients served experience similar social issues. More specifically, as illustrated below most clients seeking services from your agency regularly present with one or more of the following problems: (1) poverty, (2) domestic violence, (3) substance abuse, and (4) homelessness.

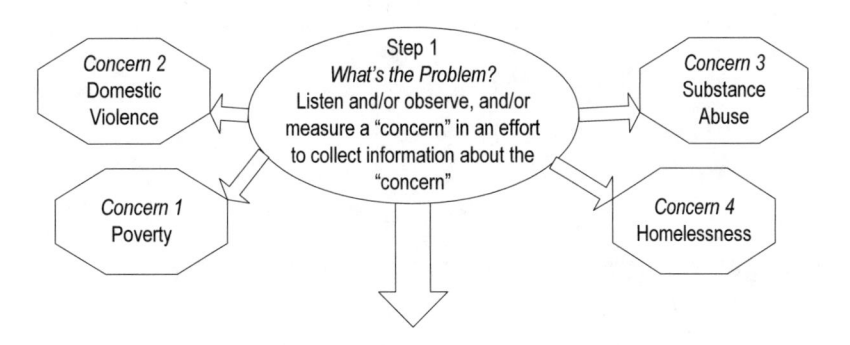

Your agency's Executive Director establishes a "Practice-Informed-by-Research" Committee that is "charged" with the following mandate: to design a study involving agency clientele so that workers at the agency will be better informed about the complex relationship between the issues of poverty, domestic violence, substance abuse, and homelessness. You are asked to chair the "Practice-Informed-by-Research" Committee. Gladly, you accept and begin leading your committee through the remaining 12 steps of the research process.

Step 2: Formulating Initial Impressions

As Chair of the "Practice-Informed-by-Research" Committee, your first step is to invite open discussion with your committee members about what they "know" about each of the four presenting problems, or issues, and the relationship between them. Drawing upon your task-group facilitation skills learned in one of your social work practice classes, you help your committee members tease out the source(s) of their collective wisdom (or knowledge) as a group.

Questions about possible relationships among the presenting problems are discussed and efforts are made to determine whether committee "answers" are based on authority, tradition, experience, intuition or research? During this process of discussion, the committee learns very important things, such as how much of their present knowledge set is based on empirical knowledge versus other ways of knowing; what biases, values or beliefs are active among committee members, and how the research method can be used to advance the current knowledge base among workers at the agency.

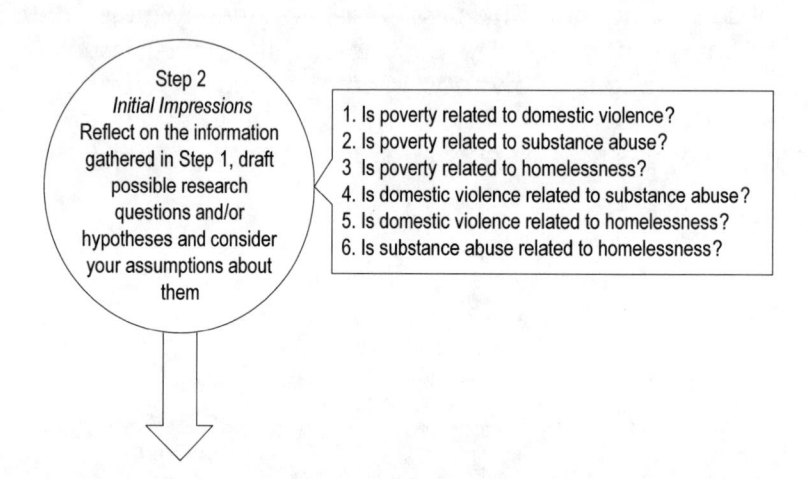

Step 2
Initial Impressions
Reflect on the information gathered in Step 1, draft possible research questions and/or hypotheses and consider your assumptions about them

1. Is poverty related to domestic violence?
2. Is poverty related to substance abuse?
3. Is poverty related to homelessness?
4. Is domestic violence related to substance abuse?
5. Is domestic violence related to homelessness?
6. Is substance abuse related to homelessness?

Step 3: Determining What Others Have Found?

Given your superb skills as a group facilitator, you have energized your committee members and stimulated their curiosity such that they are eager to read the latest research about poverty, domestic violence, substance abuse, and homelessness. Each committee member accepts a task to search the literature for articles and information that reports on the relationship between one or more of the four presenting problems of interest.

Committee members search a variety of sources of information such as professional journals, credible websites, and books (see Chapter 3). Using quality criteria to evaluate the evidence of information gathered, the committee synthesizes the knowledge gathered from empirical sources. If so desired, the committee might endeavor to systematically assess the literature gathered through meta-analyses.

With either a synthesis or meta-analysis of the literature gathered, the "Practice-Informed-by-Research" Committee is in an excellent position to make an informed statement about up-to-date knowledge about the relationship between poverty, domestic violence, substance abuse, and homelessness.

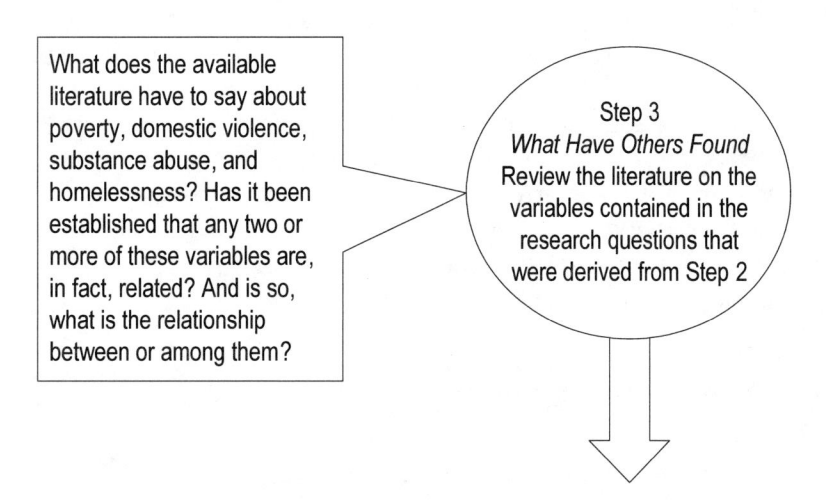

What does the available literature have to say about poverty, domestic violence, substance abuse, and homelessness? Has it been established that any two or more of these variables are, in fact, related? And is so, what is the relationship between or among them?

Step 3
What Have Others Found
Review the literature on the variables contained in the research questions that were derived from Step 2

Step 4: Refining the General Problem Area

By this point the excitement of your committee members has waned somewhat. As it turned out, the task of reviewing the literature (Step 3) was more onerous than anticipated. Indeed, the number of research studies investigating poverty, domestic violence, substance abuse, and homelessness was overwhelming. Committee members had to agree on the parameters of their search.

For example, a search for research on the topic of domestic violence required committee members to define "partner" in domestic violence and decide whether the research to be reviewed would include gay and lesbian partnerships, as well as heterosexual partners. Moreover, applying criteria to evaluate individual studies was not as straightforward as initially thought. Nevertheless, your committee trudged through and indeed was successful at finding relevant empirical articles.

Step 4
Refining the Problem Area
Select the key variables for focus in the investigation. Decide upon a question/hypothesis—exploratory, descriptive, explanatory—and corresponding research design

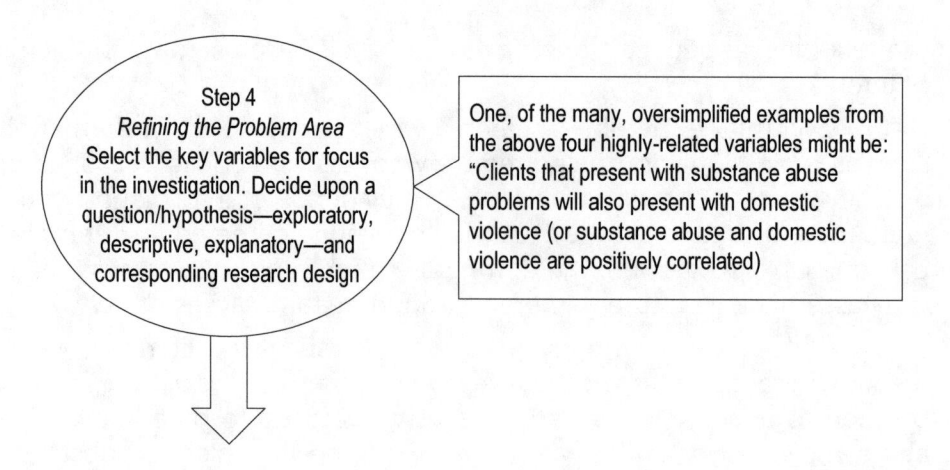

One, of the many, oversimplified examples from the above four highly-related variables might be: "Clients that present with substance abuse problems will also present with domestic violence (or substance abuse and domestic violence are positively correlated)"

To re-charge the committee and keep the momentum of the research process going, you work with committee members to set priorities for the remainder of your work together. In short, you decide what research question(s)—exploratory, descriptive, or explanatory—will be the focus of your investigation. In turn, you decide what research design is best suited to investigating the proposed research question. As the above suggests, the committee decides to focus on the descriptive relationship between only two presenting problem areas—domestic violence and substance abuse.

Step 5: Measuring the Variables

At Step 5 you review with committee members the work already accomplished in the previous four steps and note that planning thus far has largely been a conceptual exercise. Beginning with this Step 5—defining and deciding how to measure variables—you realize that the tasks for the committee become much more specific in nature, particularly as related to the research process.

With respect to measurement, committee members must develop operational definitions for the two presenting problems, or variables (i.e., substance abuse, & domestic violence) that will be studied and find measuring instruments for each variable.

What's more the committee wants to select measuring instruments that are not only valid and reliable but also relevant to their clientele and meaningful to their practice. As Chair of the "Practice-Informed-by-Research Committee," you suggest that committee members "refresh" themselves of the concepts of measurement by reading Chapters 6 and 7. As you can see, the end result of this step is the selection of the Domestic Violence Inventory (DVI) that will measure domestic violence and the Substance Abuse Questionnaire (SAQ) that will measure substance abuse.

Chapter 4: The Quantitative Research Approach

Step 5
Measuring the Variables
Decide on how to measure the two variables: substance abuse and domestic violence

State exactly how the variables will be measured:

Substance abuse: Substance Abuse Questionnaire: SAQ (http://www.bdsltd.com/index2.htm)

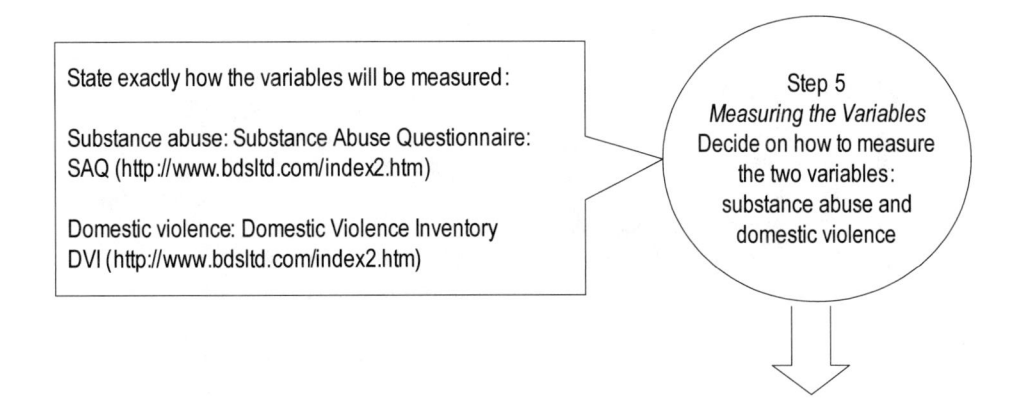

Domestic violence: Domestic Violence Inventory DVI (http://www.bdsltd.com/index2.htm)

Step 6: Deciding on a Sample

With a clear research hypothesis, a solid understanding of other research, and the selection of measuring instruments for key variables, the next step for the committee is to decide the sampling procedures for the study. At this point, research and practice knowledge come together to decide which parameters for the sample will be most meaningful. For example, it turns out that intake data show that most (but not all) clients presenting with either substance abuse or domestic violence problems at intake report that they have been living together for more than one year.

The intake data show that only a handful of coupled clients are new relationships of less than one year. Consequently, this characteristic of clients at the agency becomes a criterion to define the population, and consequently the sample, to be studied. Once all of the sampling criteria are established, the committee must decide the particular sampling method to use. The aim is to select a sample that is representative (e.g., similar in age, race, marital status, service history) of *all* clients that fit the eligibility criteria. Much more will be said about sampling in Chapter 8.

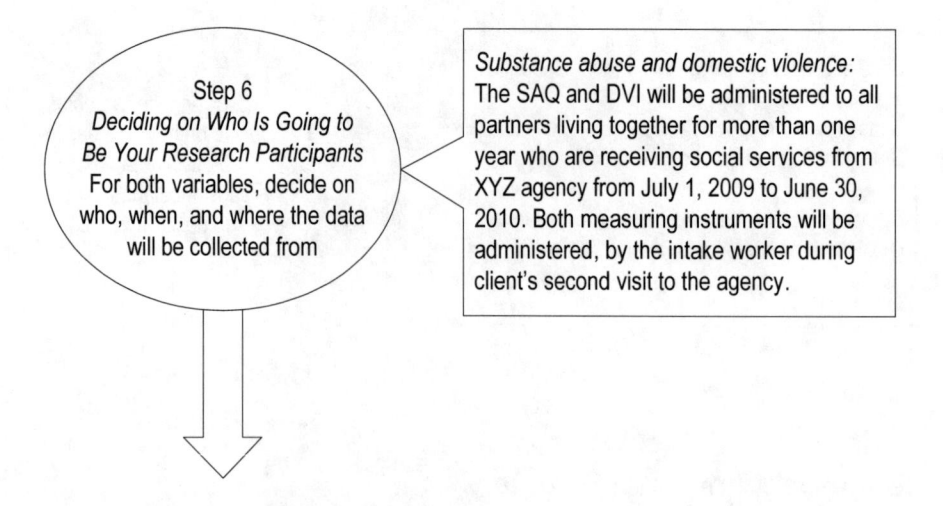

Part II: Approaches to Knowledge Development

Step 6
Deciding on Who Is Going to Be Your Research Participants
For both variables, decide on who, when, and where the data will be collected from

Substance abuse and domestic violence: The SAQ and DVI will be administered to all partners living together for more than one year who are receiving social services from XYZ agency from July 1, 2009 to June 30, 2010. Both measuring instruments will be administered, by the intake worker during client's second visit to the agency.

Step 7: Obeying Ethical Principles

As an ethical social work practitioner, who adheres to the NASW *Code of Ethics*, you are aware that not one speck of data for the study will be collected until proper procedures have developed and independently reviewed. After reading Chapter 3 of this text and Section 5 of the NASW *Code of Ethics*, the committee gets to work developing proper consent forms and research protocols, as well as getting all necessary ethics approvals.

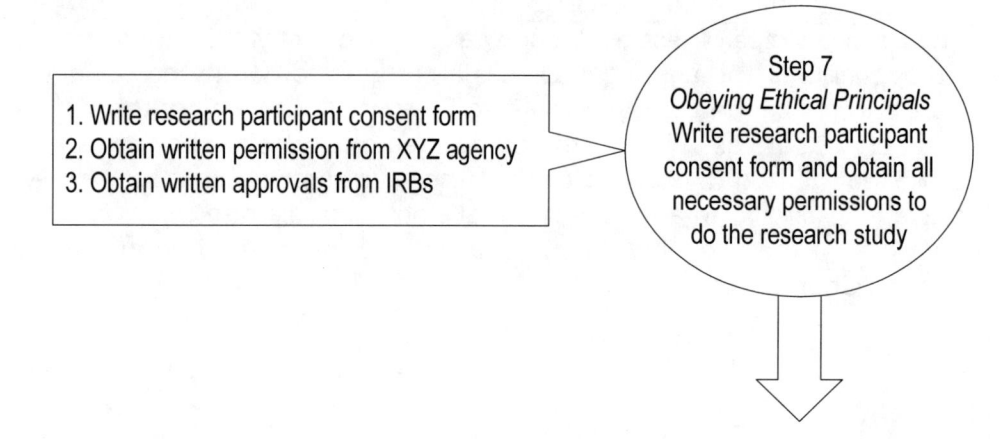

1. Write research participant consent form
2. Obtain written permission from XYZ agency
3. Obtain written approvals from IRBs

Step 7
Obeying Ethical Principals
Write research participant consent form and obtain all necessary permissions to do the research study

Step 8: Collecting the Data

As shown below, data collection is a step that indicates the study is underway. There are many different methods of data collection available and each has particular advantages and disadvantages. Continuing with our example, data collection involves administering SAQ and DVI instruments to sampled partners that have been living together for more than one year.

Once the study is underway, the protocols established for data collection should be carried out precisely so as to avoid errors. Moreover, data collection procedures should not be changed. If changes are necessary, then they must be approved by the ethical oversight bodies involved at Step 7 *before* those changes are implemented. Your committee's focus at this point is only to monitor that the study is being carried out as planned and to use research principles to trouble shoot any problems that arise. Chapters 11–13 present the various data-collection methods that could have been used to collect the data.

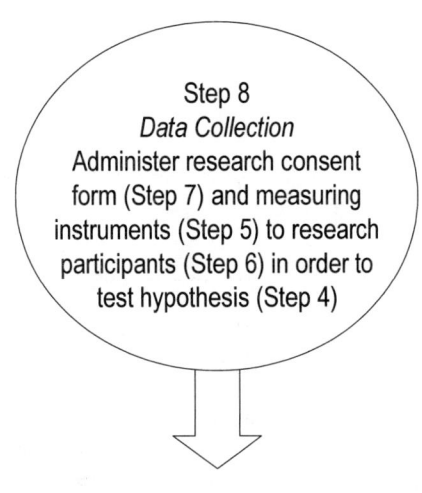

> **Step 8**
> *Data Collection*
> Administer research consent form (Step 7) and measuring instruments (Step 5) to research participants (Step 6) in order to test hypothesis (Step 4)

Step 9: Gathering and Analyzing the Data

Step 9 is an exciting one in the research process because it is at this step that you and your committee get the "answer" to your stated hypothesis—Clients who have substance abuse issues will also have domestic violence issues (and vice versa). Step 9 reveals two key facts or results from your study that indeed supports this hypothesis. Before it was possible to produce these results, your committee was hard at work ensuring that data were properly entered into an appropriate computer program and analyzed.

Since data analysis (quantitative and qualitative) requires advanced skills it may be that your committee hired a consultant to complete this step of the research process. Or perhaps you developed such skills (in this example, statistics would be required) in your graduate program and decide to perform this step yourself. Chapters 14 and 15 discuss data analysis in detail: Chapter 14 on quantitative analyses and Chapter 15 on qualitative analyses.

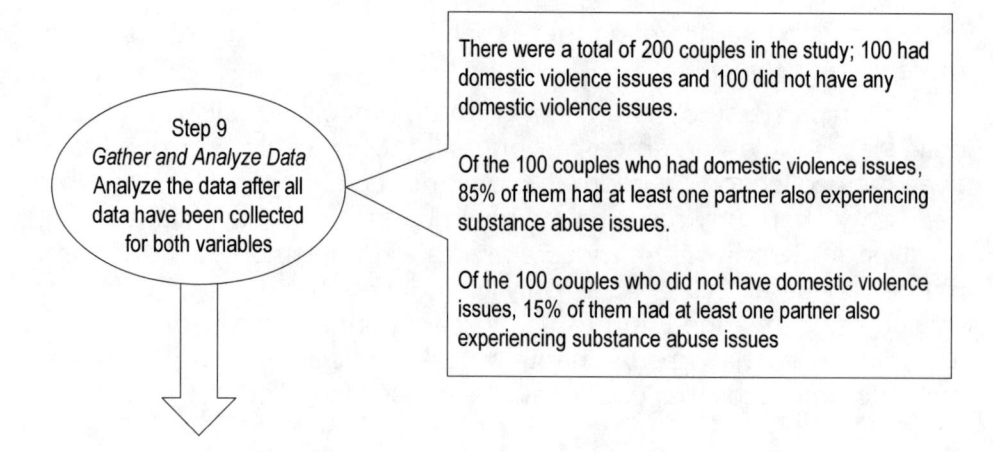

Step 9
Gather and Analyze Data
Analyze the data after all data have been collected for both variables

There were a total of 200 couples in the study; 100 had domestic violence issues and 100 did not have any domestic violence issues.

Of the 100 couples who had domestic violence issues, 85% of them had at least one partner also experiencing substance abuse issues.

Of the 100 couples who did not have domestic violence issues, 15% of them had at least one partner also experiencing substance abuse issues

Step 10: Interpreting the Data

Getting an answer to your research question and deciding the usefulness of the answer are two separate steps. In step 10, the committee will go beyond the reported "facts" of the study's findings to make a conclusive statement such as illustrated:

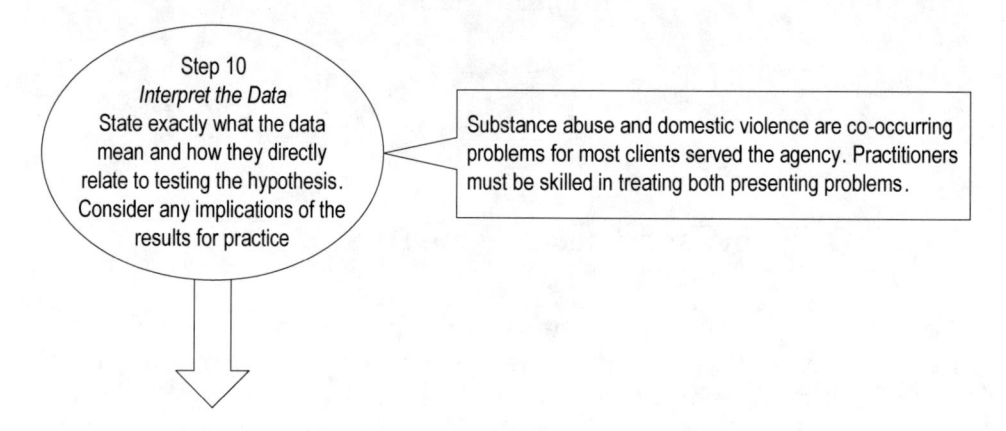

Step 10
Interpret the Data
State exactly what the data mean and how they directly relate to testing the hypothesis. Consider any implications of the results for practice

Substance abuse and domestic violence are co-occurring problems for most clients served the agency. Practitioners must be skilled in treating both presenting problems.

Step 11: Comparing Results with Others Findings

As Chair of the "Practice-Informed-by-Research Committee," you are aware that every research study is only a single piece in the puzzle of knowledge development. Consequently, you ask that the committee consider the findings of your study in context of other research that was reviewed in Step 3, as well as any new research published while your study was underway. By contrasting your findings with what others have found, your committee builds on the existing knowledge base.

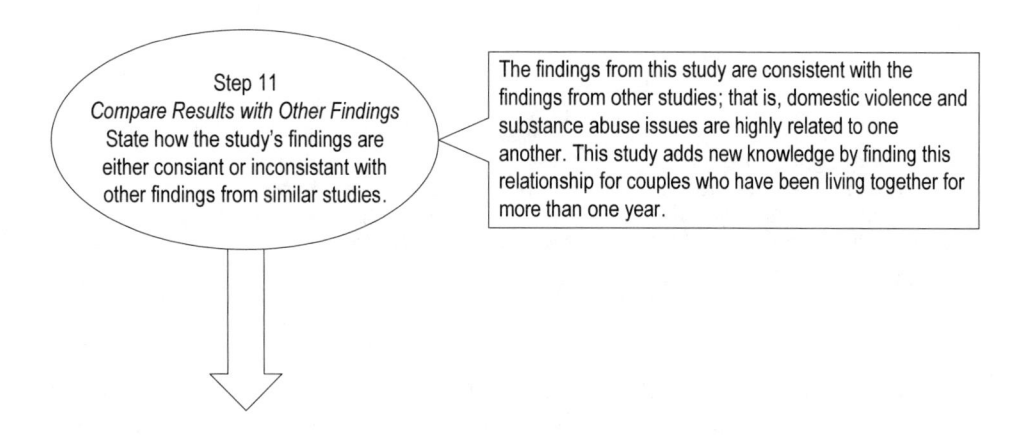

Chapter 4: The Quantitative Research Approach   **95**

Step 11
Compare Results with Other Findings
State how the study's findings are either consiant or inconsistent with other findings from similar studies.

> The findings from this study are consistent with the findings from other studies; that is, domestic violence and substance abuse issues are highly related to one another. This study adds new knowledge by finding this relationship for couples who have been living together for more than one year.

Step 12: Specifying the Study's Limitations

You also realize that no study is perfect and that includes yours. Many study limitations are predetermined by the particular research design that you used. For example, threats to internal and external validity are limitations commonly discussed in research reports and articles. Acknowledging your study's limitations can be a humbling experience. While you may be better informed at the conclusion of your study, you will not have found the "magic cure" that is sure to alleviate your clients' struggles and suffering in only one research study.

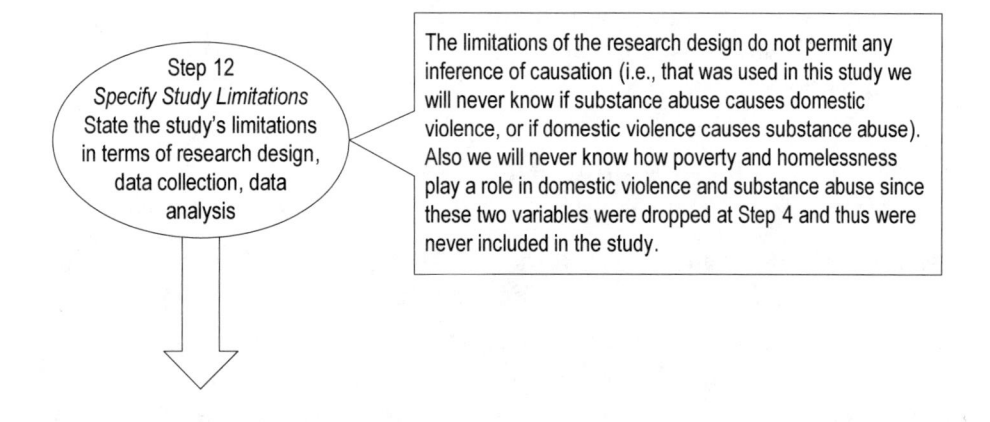

Step 12
Specify Study Limitations
State the study's limitations in terms of research design, data collection, data analysis

> The limitations of the research design do not permit any inference of causation (i.e., that was used in this study we will never know if substance abuse causes domestic violence, or if domestic violence causes substance abuse). Also we will never know how poverty and homelessness play a role in domestic violence and substance abuse since these two variables were dropped at Step 4 and thus were never included in the study.

Step 13: Writing and Disseminating the Study's Results

A final and important step in the research process is sharing the new knowledge learned in your study. In the academic world, dissemination most often refers to publication of the study in a peer-reviewed journal (like those reviewed in Steps 2 and 11). While some practitioners also publish in peer-reviewed journals, dissemination in the practice world more commonly refers to sharing findings at conferences or local meetings.

> **Step 13**
> *Dissemination of Results*
> Write and publish a research report that tells others what was found so the study's results can contribute to the existing literature for others to use in Step 3

Even with all the limitations of the study's design some data are better than none; that is, no matter how many threats to internal and external validity, as long as the study's procedures and limitations are clearly spelled-out in the final report the study's findings will be useful to others.

The idea of dissemination is to make your study results available to others who are working in the same area—whether as practitioners, researchers, policy makers or educators. As Chair of the "Practice-Informed-by-Research" Committee, you will want to share your study's results with others in your agency. The knowledge gained might be used to develop new procedures in the agency (e.g., intake, training, referrals). Whatever the route of dissemination, the aim is to have completed Steps 1 through 12 at the highest level of research integrity so that the dissemination of your results will be most useful to others.

SUMMARY

This chapter briefly discussed the process of the quantitative research approach to knowledge development. The following chapter presents how the qualitative research approach can be used within the research method using the same example and format of this chapter.

REFERENCES AND FURTHER READINGS

Alasuutari, P., Bickman, L., & Brannen, J. (2008). *The SAGE handbook of social research methods.* Thousand Oaks, CA: Sage.

Cresswell, J. (2008). *Research design: Qualitative, quantitative, and mixed methods* (3rd ed.). Thousand Oaks, CA: Sage.

de Vaus, D. (2006). *Research design* (4 vols.). Thousand Oaks, CA: Sage.

Grinnell, R.M., Jr., Unrau, Y.A., & Williams, M. (2008). Group-level designs. In R.M. Grinnell, Jr., & Y.A. Unrau (Eds.), *Social work research and evaluation: Foundations of evidence-based practice* (8th ed., pp. 179–204). New York: Oxford University Press.

Holosko, M. (2008). Evaluating quantitative research studies. In R.M. Grinnell, Jr., & Y.A. Unrau (Eds.), *Social work research and evaluation: Foundations of evidence-based practice* (8th ed., pp. 423–461). New York: Oxford University Press.

Krysik, J., & Grinnell, R.M., Jr. (1997). Quantitative approaches to the generation of knowledge. In R.M. Grinnell, Jr., & Y.A. Unrau (Eds.), *Social work research and evaluation: Quantitative and qualitative approaches* (5th ed., pp. 67–105). Itasca, IL: F.E. Peacock.

Teddlie, C., & Tashakkori, A. (2008). *Foundations of mixed methods research: Integrating quantitative and qualitative techniques in the social and behavioral sciences.* Thousand Oaks, CA: Sage.

Unrau, Y.A., Grinnell, R.M., Jr., & Williams, M. (2008). The quantitative research approach. In R.M. Grinnell, Jr., & Y.A. Unrau (Eds.), *Social work research and evaluation: Foundations of evidence-based practice* (8th ed., 61–81). New York: Oxford University Press.

Williams, M., Tutty, L.M., & Grinnell, R.M., Jr. (1995). *Research in social work: An introduction* (2nd ed.). Itasca, IL: F.E. Peacock.

Williams, M., Unrau, Y.A., & Grinnell, R.M., Jr. (1998). *Introduction to social work research.* Itasca, IL: F.E. Peacock.

Check Out Our Website

PairBondPublications.com

√ *Student Workbook Exercises*
√ *Chapter Power Point Slides*
√ *On-line Glossaries*
√ *General Links*
√ *Specific Chapter Links*

Chapter 5

The Qualitative Research Approach

WHAT IS THE INTERPRETIVE WAY OF THINKING?
    Multiple Realities
    Data vs. Information
    Subjects vs. Research Participants
    Values
PHASES WITHIN THE QUALITATIVE APPROACH
    Phases 1 and 2: Problem Identification and Question Formulation
    Phase 3: Designing the Research Study
    Phase 4: Collecting the Data
    Phases 5 and 6: Analyzing and Interpreting the Data
    Phases 7 and 8: Presentation and Dissemination of Findings
COMPARING APPROACHES
    Philosophical Differences
    Similar Features
USING BOTH APPROACHES IN A SINGLE STUDY
    What Do You Really Want To Know?
    Example of Using Both Approaches in a Single Study
SUMMARY
REFERENCES AND FURTHER READINGS

5

THE LAST CHAPTER PRESENTED a brief discussion of how the generation of social work knowledge is acquired through the use of research studies that gather quantitative data. This chapter is a logical extension of the last one in that we now focus our attention on how knowledge is developed through the qualitative research approach.

As we know from the last chapter, the quantitative approach to knowledge development is embedded within the "positivist way of thinking, or viewing the world." In direct contract to the quantitative approach, the qualitative approach to knowledge development is embedded within the "interpretive way of thinking, or viewing the world." We will now turn our attention to the "interpretive way of thinking" in detail below.

WHAT IS THE INTERPRETIVE WAY OF THINKING?

The interpretive approach to knowledge development is the second way of obtaining knowledge in our profession (see right-hand side of Figure 2.1). It basically discards the positivist notion that there is only one external reality waiting to be discovered. Instead, the qualitative research approach is based off of the interpretive perspective which states that reality is defined by the research participants' interpretations of their own realities. In sum, it is the *subjective* reality that is studied via the qualitative research approach rather than the *objective* one that would be studied by the quantitative approach.

As you will see in Chapters 11–13, the differences between the philosophy of the quantitative approach and the qualitative approach to knowledge development naturally leads to different data collection methods. Subjective reality, for example, cannot be explored through the data collection method of observation. Empiricism—the belief that science must be founded on observations and measurements, another tenet of positivism—is thus also discarded.

The qualitative approach says that the only real way to find out about the subjective reality of our research participants is to ask them, and the answer will come back in words, not in numbers. In a nutshell, qualitative research methods produce *qualitative* data in the form of text. Quantitative research methods produce *quantitative* data in the form of numbers.

Multiple Realities

As you know from a positivist standpoint as presented in the last chapter, if you did not accept the idea of a single reality, which was not changed by being observed and/or measured, and from which you—the researcher—were "detached from the study," then you were not a "real researcher" doing a "real research study." Thus, as a non-researcher doing a non-research study, your findings were not thought to be of much use.

Many of the supposed "non-researchers" whose views were not thought to be valid were women and/or came from diverse minority groups. Feminists, for example, have argued that there is *not* only one reality—there are many realities. They contend that men and women experience the world differently and so they both exist in different realities, constructed by them from their own perceptions. Similarly, people from various cultural groups view the world from the perspective of their own beliefs and traditions and also experience different realities.

As for the idea that reality is not changed by being observed and/or measured, feminists have argued that a relationship of some kind is always formed between the researcher and the research participant (subject), resulting in yet another mutual reality constructed between the two of them. In any study involving human research participants, there will thus be at least three realities:

- the researcher's reality
- the research participant's reality
- the mutual reality they (researcher and research participant) both created and share

Moreover, all three realities are constantly changing as the study proceeds and as further interactions occur. The positivist idea of a single unchanged and unchanging reality, some feminists have argued, was typically a male idea, probably due to the fact that men view human relationships being less important than women.

This is a low blow which will quite properly be resented by the many men who do in fact ascribe importance to relationships. But, that aside, perhaps the problem lies less with phenomenalism (a single unchanged reality as opposed to multiple changing realities) than it does with scientism (the idea that the physical and social sciences can be approached in the same way).

Chapter 5: The Qualitative Research Approach

Data vs. Information

Here, it is worth pausing for a moment to discuss what is meant by *data*. Data are plural; the singular is *datum* from the Latin *dare*, to give. A datum is thus something that is given, either from a quantitative observation and/or measurement or from a qualitative discussion with Ms. Smith about her experiences in giving birth at her home.

A number of observations and/or measurements with the quantitative approach, or a number of discussions with the qualitative approach, constitute *data*. Data are not the same thing as *information*, although the two words are often used interchangeably. The most important thing to remember at this point is that both approaches to the research method produce data. They simply produce different kinds of data.

Information is something you hope to get from the data once you have analyzed them—whether they are words or numbers. You might, for example, collect data about the home-birthing experiences of a number of women and your analysis might reveal commonalties between them: perhaps all the women felt that their partners had played a more meaningful role in the birthing process at home than would have been possible in a hospital setting.

The enhanced role of the partner is *information* that you, as a researcher, have derived from the interview *data*. In other words, data are pieces of evidence, in the form of words (qualitative data) or numbers (quantitative data), that you put together to give you information—which is what the research method is all about.

Subjects vs. Research Participants

Having dealt with what data are (don't ever write "data *is*"), let's go back to the implications of collecting data about people's subjective realities. Because it is the research participant's reality you want to explore, the research participant is a very important data source. The quantitative approach may seem like it tends to relegate the research participant to the status of an object or subject.

In a study of caesarian births at a hospital during a certain period, for example, Ms. Smith will not be viewed as an individual within the quantitative approach to knowledge development, but only as the 17th woman who experienced such a birth during that period.

Details of her medical history may be gathered without any reference to Ms. Smith as a separate person with her own hopes and fears, failings, and strengths. Conversely, a qualitative approach to caesarian births will focus on Ms. Smith's individual experiences. What *was* her experience? What did it mean to her? How did she interpret it in the context of her own reality?

Values

In order to discover the truth of Ms. Smith's reality, however, you must be clear about the nature of your own reality. We discussed in Chapter 1 about *value awareness* as one of the characteristics that distinguish the research method from the other ways of knowledge development. As you know, value awareness is the ability for you to put aside your own values when you are conducting research studies or when you are evaluating the results obtained by other researchers.

This is sometimes called *disinterestedness*. Researchers who are *disinterested* are ones who are able to accept evidence that runs against their own positions. From a hard-line quantitative perspective, this putting aside of values seems more akin to sweeping them under the carpet and pretending they don't exist. Researchers engaged in quantitative studies will deny that their own values are important.

They claim their values have nothing to do with the study. In a nutshell, their values cease to exist. On the other hand, qualitative researchers take a very different view. Their values are a part of their own realities and a part of the mutual reality that is constructed through their interaction with their research participants.

A qualitative researcher's values therefore must be acknowledged and thoroughly explored so that the mutual shaping of realities resulting from the interaction with their research participants may be more completely and honestly understood. The term *value awareness*, while important to both research approaches, is thus understood in different ways. To quantitative researchers, it means putting values aside so they don't affect the study. To qualitative researchers, it means an immersion in values so that their inevitable effect is understood and their research participants' realities emerge more clearly.

PHASES WITHIN THE QUALITATIVE APPROACH

Like quantitative researchers, qualitative researchers make a major commitment in terms of time, money, and resources when they undertake research studies. As can be seen in Figure 4.2, a quantitative study has eight basic sequential steps that must be followed to produce useful quantitative data. On the other hand, as can be seen in Figure 5.1, a qualitative study does not have these specific steps—the activities are more phases than steps as many of the phases highly interact with one another as the study progresses from one phase to the next.

As can be seen by comparing Figures 4.2 and 5.1, one of the major differences between the two research approaches is how they utilize the literature. In a quantitative study, for example, the literature is utilized mostly within the first three steps of the research method. In a qualitative study, the literature is heavily utilized in all of the phases.

Chapter 5: The Qualitative Research Approach

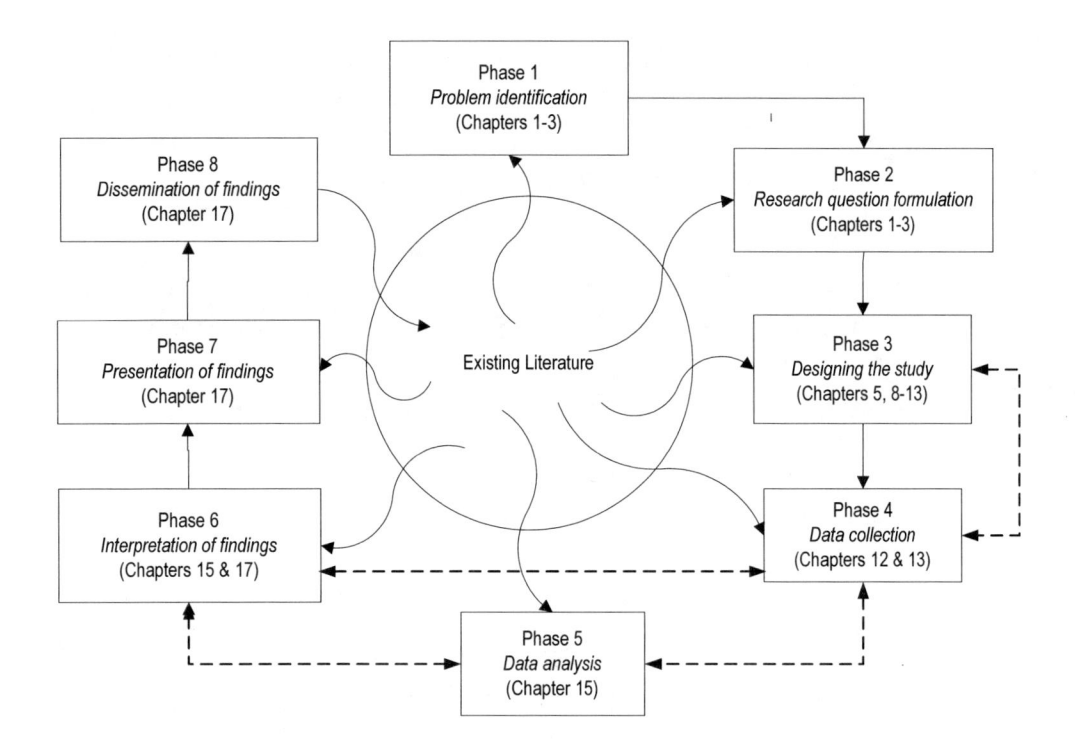

FIGURE 5.1

Phases of the Qualitative Research Process

The qualitative research approach is akin to exploring a "social problem maze" that has multiple entry points and paths. You have no way of knowing whether the maze will lead you to a place of importance or not but you enter into it out of your own curiosity and, perhaps, even conviction. You enter the maze without a map or a guide; you have only yourself to rely on and your notebook to record important events, observations, conversations, and impressions along the way.

You will begin your journey of an interpretive inquiry by stepping into one entrance and forging ahead. You move cautiously forward, using all of your senses in an effort to pinpoint your location and what surrounds you at any one time. You may enter into dead-end rooms within the maze and have to backtrack. You may also encounter paths that you did not think possible.

In some cases, you may even find a secret passageway that links you to a completely different maze. Tutty, Rothery, and Grinnell (1996) present a few characteristics that most qualitative research studies have in common:

- Research studies that are conducted primarily in the natural settings where the research participants carry out their daily business in a "non-research" atmosphere.
- Research studies where variables cannot be controlled and experimentally manipulated (though changes in variables and their effect on other variables can certainly be observed).
- Research studies in which the questions to be asked are not always completely conceptualized and operationally defined at the outset (though they can be).
- Research studies in which the data collected are heavily influenced by the experiences and priorities of the research participants, rather than being collected by predetermined and/or highly structured and/or standardized measurement instruments.
- Research studies in which meanings are drawn from the data (and presented to others) using processes that are more natural and familiar than those used in the quantitative method. The data need not be reduced to numbers and statistically analyzed (though counting and statistics can be employed if they are thought useful).

Phases 1 and 2: Problem Identification and Question Formulation

Qualitative studies are generally inductive and require you to reason in such a way that you move from a part to a whole or from a particular instance to a general conclusion. Let's return to the research problem introduced in the last chapter—racial discrimination within the social services. You begin the qualitative research process, once again, from your observations—ethnic minorities are among the highest groups for unemployment, poverty, and low education; however, Caucasians outnumber ethnic minorities when seeking assistance from social services.

You can focus your qualitative research question by identifying the key concepts in your question. These key concepts set the parameters of your research study—they are the "outside" boundaries of your maze. As in the quantitative research approach, you would want to visit the library and review the literature related to your key concepts. Your literature review, however, takes on a very different purpose. Rather than pinpointing "exact" variables to study, you review the literature to see how your key concepts are generally described and defined by previous researchers.

Going with the maze example for the moment, you might learn whether your maze will have rounded or perpendicular corners, or whether it will have multiple levels. The knowledge you glean from the literature assists you with ways of thinking that you hope will help you move through the maze in a way that you will arrive at a meaningful understanding of the problem it represents. Because you may never have been in the maze before, you must also be prepared to abandon what you "think you know" and accept new experiences as you go.

Let's revisit your research question—Do ethnic minorities have difficulty in accessing social services? In your literature review, you would want to focus on definitions and theories related to discrimination within the social services. In the quantitative research approach, you reviewed the literature to search for meaningful variables that could be measured. You do not want, however, to rely on the literature to define key variables in your qualitative study. Rather, you will rely upon the qualitative research process itself to identify key variables and how they relate to one another.

In rare occasions, hypotheses can be used in a qualitative research study. They can focus your research question even further. A hypothesis in a qualitative study is less likely to be outright "accepted" or "rejected," as is the case in a quantitative study. Rather, it is a "working hypothesis" and is refined over time as new data are collected.

Your hypothesis is changed throughout the qualitative research process based on the reasoning of the researcher—not on a statistical test. So all of this leads us to ask the question, "what do qualitative researchers actually do when they carry out a research study?" Neuman (2009) has outlined several activities that qualitative researchers engage in when carrying out their studies:

- Observes ordinary events and activities as they happen in natural settings, in addition to any unusual occurrences
- Is directly involved with the people being studied and personally experiences the process of daily social life in the field
- Acquires an insider's point of view while maintaining the analytic perspective or distance of an outsider
- Uses a variety of techniques and social skills in a flexible manner as the situation demands
- Produces data in the form of extensive written notes, as well as diagrams, maps, or pictures to provide detailed descriptions
- Sees events holistically (e.g., as a whole unit, not in pieces) and individually in their social context
- Understands and develops empathy for members in a field setting, and does not just record "cold" objective facts
- Notices both explicit and tacit aspects of culture
- Observes ongoing social processes without upsetting, disrupting, or imposing an outside point of view

- Is capable of coping with high levels of personal stress, uncertainty, ethical dilemmas, and ambiguity

Phase 3: Designing the Research Study

You can enter into a qualitative research study with general research questions or working hypotheses. However, you are far less concerned about honing-in on specific variables. Because qualitative research studies are inductive processes, you do not want to constrain yourself with preconceived ideas about how your concepts or variables will relate to one another. Thus, while you will have a list of key concepts, and perhaps loosely-defined variables, you want to remain open to the possibilities of how they are defined by your research participants and any relationships that your research participants may draw.

A qualitative study is aimed at an in-depth understanding of a few cases, rather than a general understanding of many cases, or people. In other words, the number of research participants in a qualitative study is much smaller than in a quantitative one. As we will see in Chapter 8, sampling is a process of selecting the "best-fitting" people to provide data for your study. Nonprobability sampling strategies are designed for this task because they purposely seek out potential research participants. More will be discussed about nonprobability sampling strategies in Chapter 8.

The qualitative research approach is about studying a social phenomenon within its natural context. As such, the "case study" is a major qualitative research design. A case can be a person, a group, a community, an organization, or an event. You can study many different types of social phenomena within any one of these cases. Any case study design can be guided by different qualitative research methods.

Grounded theory is a method that guides you in a "back and forth" process between the literature and the data you collect. Using grounded theory, you can look to the literature for new ideas and linkages between ideas that can bring meaning to your data. In turn, your data may nudge you to read in areas that you might not have previously considered.

Ethnography is a branch of interpretive research that emphasizes the study of a culture from the perspective of the people who live the culture. With your research example, you would be interested in studying the culture of social services, particularly with respect to how ethnic minorities experience it.

Phenomenology is another branch of interpretive research. It is used to emphasize a focus on people's subjective experiences and interpretations of the world. These subjective experiences include those of the researcher, as well as of the research participants. As researchers in your discrimination study, you would want to keep a careful account of your reactions and questions to the events you observe and the stories you hear. Your task is to search for meaningful patterns within the volumes of data (e.g., text, drawings, pictures, video recordings).

Phase 4: Collecting the Data

Qualitative researchers are the principal instruments of data collection (Franklin & Jordan, 2003). This means that data collected are somehow "processed" through the person collecting them. Interviewing, for example, is a common data collection method that produces text data. Data collection in the interview is interactive, where you can check out your understanding and interpretation of your participants' responses as you go along.

To collect meaningful text data, you want to be immersed into the context or setting of the study. You want to have some understanding, for example, of what it is like to be a client of social services before you launch into a dialogue with clients about their experiences of discrimination, if any, within the social services. If you do not have a grasp of the setting in which you are about to participate, then you run the risk of misinterpreting what is told to you.

Given that your general research question evolves in a qualitative study, the data collection process is particularly vulnerable to biases of the data collector. There are several principles to guide you in your data collection efforts:

- First, you want to make every effort to be aware of your own biases. In fact, your own notes on reactions and biases to what you are studying are used as sources of data later on, when you interpret the data (Chapter 15).
- Second, data collection is a two-way street. The research participants tell you their stories and, in turn, you tell them your understanding or interpretation of their stories. It is a process of check and balance.
- Third, qualitative data collection typically involves multiple data sources and multiple data collection methods. In your study, you may see clients, workers, and supervisors as potential data sources. You may collect data from each of these groups using interviews, observation, and existing documentation (data-collection methods).

Phases 5 and 6: Analyzing and Interpreting the Data

Collecting, analyzing, and interpreting qualitative data are intermingled. Let's say that, in your first round of data collection, you interview a number of ethnic minority clients about their perceptions of racial discrimination in the social services. Suppose they consistently tell you that to be a client of social services, they must give up many of their cultural values. You could then develop more specific research questions for a second round of interviews in an effort to gain more of an in-depth understanding of the relationship between cultural values and being a social service client.

Overall, the process of analyzing qualitative data is an iterative one. This means that you must read and reread the volumes of data that you collected. You simply look for patterns and themes that help to capture how your research participants are experiencing the problem you are studying.

The ultimate goal is to interpret data in such a way that the true expressions of research participants are revealed. You want to explain meaning according to the beliefs and experiences of those who provided the data. The aim is to "walk the walk" and "talk the talk" of research participants and not to impose "outside" meaning to the data they provided. Much more will be said about analyzing and interpreting qualitative data in Chapter 15.

Phases 7 and 8: Presentation and Dissemination of Findings

Qualitative research reports are generally lengthier than quantitative ones. This is because it is not possible to strip the context of a qualitative study and present only its findings. The knowledge gained from a qualitative endeavor is nested within the context from which it was derived. Furthermore, text data are more awkward and clumsy to summarize than numerical data.

You cannot rely on a simple figure to indicate a finding. Instead, you display text usually in the form of quotes or summary notes to support your conclusions. Much more will be said about the presentation and dissemination of qualitative studies in Chapter 17.

COMPARING APPROACHES

Philosophical Differences

By comparing the philosophical underpinnings of quantitative and qualitative research approaches, you can more fully appreciate their important differences. Each approach offers you a unique method to studying a social work-related problem; and the same research problem can be studied using either approach.

Suppose, you are interested in a broad social problem such as, racial discrimination. In particular, let's say you are interested in studying the social problem of racial discrimination within public social service programs. Let's now look at the major differences between the two approaches and see how your research problem, racial discrimination, could be studied under both approaches.

- *Perceptions of Reality*

    Quantitative. Ethnic minorities share similar experiences within the public social service system. These experiences can be described objectively; that is, a single reality exists outside any one person.

    Qualitative. Individual and ethnic group experiences within the public social service system are unique. Their experiences can only be described subjectively; that is, a single and unique reality exists within each person.

- *Ways of "Knowing"*

    Quantitative. The experience of ethnic minorities within public social services is made known by closely examining specific parts of their experiences. Scientific principles, rules, and tests of sound reasoning are used to guide the research process.

    Qualitative. The experience of ethnic minorities within public social services is made known by capturing the whole experiences of a few cases. Parts of their experiences are considered only in relation to the whole of them. Sources of knowledge are illustrated through stories, diagrams, and pictures that are shared by the people with their unique life experiences.

- *Value Bases*

    Quantitative. The researchers suspend all their values related to ethnic minorities and social services from the steps taken within the research study. The research participant "deposits" data, which are screened, organized, and analyzed by the researchers who do not attribute any personal meaning to the research participants or to the data they provide.

    Qualitative. The researcher *is* the research process, and any personal values, beliefs, and experiences of the researcher will influence the research process. The researcher learns from the research participants, and their interaction is mutual.

- *Applications*

    Quantitative. Research results are generalized to the population from which the sample was drawn (e.g., other minority groups, other social services programs). The research findings tell us, on the average, the experience that ethnic minorities have within the public social service system.

    Qualitative. Research results tell a story of a few individuals' or one group's experience within the public social service system. The research findings provide an in-depth understanding of a few people. The life context of each research participant is key to understanding the stories he or she tells.

Similar Features

So far we have been focusing on the differences between the two research approaches. They also have many similarities. First, they both use careful and diligent research processes in an effort to discover and interpret knowledge. They both are guided by systematic procedures and orderly plans.

Second, both approaches can be used to study any particular social problem. The quantitative approach is more effective than the qualitative approach in reaching a specific and precise understanding of one aspect (or part) of an already well-defined social problem. The quantitative approach seeks to answer research questions that ask about quantity, such as:

- Are women more depressed than men?
- Does low income predict one's level of self-concept?
- Do child sexual abuse investigation teams reduce the number of times an alleged victim is questioned by professionals?
- Is degree of aggression related to severity of crimes committed among inmates?

A qualitative research approach, on the other hand, aims to answer research questions that provide you with a more comprehensive understanding of a social problem from an intensive study of a few people. This approach is usually conducted within the context of the research participants' natural environments (Rubin & Babbie, 2008). Research questions that would be relevant to the qualitative research approach might include:

- How do women experience depression as compared to men?

- How do individuals with low income define their self-concept?
- How do professionals on child sexual abuse investigation teams work together to make decisions?
- How do federal inmates describe their own aggression in relation to the crimes they have committed?

As you will see throughout our book, not only can both approaches be used to study the same social problem, they both can be used to study the same research question. Which ever approach is used clearly has an impact on the type of findings produced to answer a research question (or to test a hypothesis).

USING BOTH APPROACHES IN A SINGLE STUDY

Given the seemingly contradictory philosophical beliefs associated with the two research approaches, it is difficult to imagine how they could exist together in a single research study. As is stands, most research studies incorporate only one approach. The reason may, in part, relate to philosophy, but practical considerations of cost, time, and resources are also factors.

It is not unusual, however, to see quantitative data used within a qualitative study or qualitative data in a quantitative study. Just think that, if you were to use a quantitative approach, there is no reason why you could not ask research participants a few open-ended questions to more fully explain their experiences. In this instance, your quantitative research report would contain some pieces of qualitative data to help bring meaning to the study's quantitative findings.

Let's say you want to proceed with a qualitative research study to examine your research question about discrimination within the public social service system. Surely, you would want to identify how many research participants were included, as well as important defining characteristics such as their average age, the number who had difficulty accessing social services, or the number who were satisfied with the services they received.

While it is possible to incorporate qualitative research activity into a quantitative study (and quantitative research activity into a qualitative study) the approach you finally select must be guided by your purpose for conducting the study in the first place. Ultimately, all research studies are about the pursuit of knowledge. Just what kind of knowledge you are after is up to you.

What Do You Really Want To Know?

As mentioned previously, both research approaches have their advantages and disadvantages, and both shine in different phases within the research method. Which approach you select for a particular study depends not on whether you are a positivist or an interpretivist but on what particular research question your study is trying to answer. Are you looking for descriptions or explanations? If the former, a qualitative study will be spot on, if the latter, a quantitative one will do the trick.

Human nature being what it is, we are always looking, in the end, for explanation. We want to know not only what reality is like but what its interconnections are and what we can do to change it to make our lives more comfortable and safer. However, first things first. Description comes before explanation. Before you can know whether poverty is related to child abuse, for example, you must be able to describe both poverty and child abuse as fully as possible. Similarly, if you want to know whether low self-esteem in women contributes to spousal abuse, you must know what self-esteem is and what constitutes abuse.

By now, because you are a social worker interested in people and not numbers, you may be ready to throw the whole quantitative research approach out of the window. But let's be sure that you don't throw the baby out with the bath water. Social work values dictate that you make room for different approaches to knowledge development, different opinions, differing views on what reality really is. Believe us, the two different approaches each have value in their own way, depending on what kind of data (quantitative and/or qualitative) you hope to gain from a particular research study.

Example of Using Both Approaches in a Single Study

Suppose, for example, you have an assumption that caesarian operations were being conducted too often and unnecessarily for the convenience of obstetricians rather than for the benefit of mothers and their babies. In order to confirm or refute this hunch (it has yet to be proven), you would need data on the number of caesarian births in a particular time frame, and how many of them were justified on the basis of medical need. Numbers would be required—quantitative data. The questions about how many and how often could not be answered solely by descriptions of Ms. Smith's individual experiences.

On the other hand, Ms. Smith's experiences would certainly lend richness to the part of your study that asked how far the hospital's services took the well-being of mothers into account. Many of the best research studies use quantitative and qualitative methods within the same study. It is important to remember that the former provides the necessary numerical data while the latter provides the human depth that allows for a richer understanding of the numbers in their particular context.

Sometimes, therefore, depending on the research question (assumption) to be answered, Ms. Smith will be seen as no more than a number. At other times, her individuality will be of paramount importance. If she is seen as a number, for example, her role will be passive. She will be one of a large number of persons.

On the other hand, if she is seen as an individual, her part in the research method will be far more active. It is *her* reality that you are now exploring. She will be front and center in a research method that is driven by her and not the researcher. Even the language will change. She is no longer a subject—or possibly an object—but a full and equal *participant*, along with the researcher.

SUMMARY

This chapter briefly discussed the qualitative research approach to knowledge building. We also highlighted a few differences and similarities between the two research approaches. These two complementary and respected research approaches are divergent in terms of their philosophical principles. Yet, they both share the following processes: choosing a general research topic, focusing the topic into a research question, designing the research study, collecting the data, analyzing and interpreting the data, and writing the report.

REFERENCES AND FURTHER READINGS

Barbour, R. (2007). *Introducing qualitative research: A student's guide to the craft of doing qualitative research.* Thousand Oaks, CA: Sage.

Corbin, J., & Strauss, A. (2008). *Basics of qualitative research: Techniques and procedures for developing grounded theory* (3rd ed.). Thousand Oaks, CA: Sage.

Cresswell, J.W. (2007). *Qualitative inquiry and research design* (2nd ed.). Thousand Oaks, CA: Sage.

Denzin, N.K., & Lincoln, Y.S. (Eds.). (2007). *Collecting and interpreting qualitative materials* (3rd ed.). Thousand Oaks, CA: Sage.

Denzin, N.K., & Lincoln, Y.S. (Eds.). (2007). *The landscape of qualitative research* (3rd ed.). Thousand Oaks, CA: Sage.

Denzin, N.K., & Lincoln, Y.S. (Eds.). (2007). *Strategies of qualitative inquiry* (3rd ed.). Thousand Oaks, CA: Sage.

Denzin, N.K. & Lincoln, Y.S., & Tuhiwai Smith, L. (Eds.). (2008). *Handbook of critical and indigenous methodologies*. Thousand Oaks, CA: Sage.

Epstein, I. (1988). Quantitative and qualitative methods. In R.M. Grinnell, Jr. (Ed.), *Social work research and evaluation* (3rd ed., pp. 185–198). Itasca, IL: F.E. Peacock.

Holliday, A. (2007). *Doing and writing qualitative research*. Thousand Oaks, CA: Sage.

Morris, T. (2008). *Social work research methods: Four alternative paradigms* (2nd ed.). Thousand Oaks, CA: Sage.

Neuman, W.L. (2009). *Understanding research*. Boston: Allyn & Bacon.

Padgett, D.K. (2008). *Qualitative methods in social work research* (2nd ed.) Thousand Oaks, CA: Sage.

Raines, J.C. (2008). Evaluating qualitative research studies. In R.M. Grinnell, Jr., & Y.A. Unrau (Eds.), *Social work research and evaluation: Foundations of evidence-based practice* (8th ed., pp. 445–461). New York: Oxford University Press.

Richards, L., & Morse, J.M. (2007). *READ ME FIRST for a user's guide to qualitative methods* (2nd ed.). Thousand Oaks, CA: Sage.

Rubin, A., & Babbie, E. (2008). *Research methods for social work* (6th ed.). Pacific Grove, CA: Wadsworth.

Silverman, D. (2007). *A very short, fairly interesting and reasonably cheap book about qualitative research*. Thousand Oaks, CA: Sage.

Silverman, D., & Marvasit, A. (2008). *Doing qualitative research: A comprehensive guide*. Thousand Oaks, CA: Sage.

Tutty, L.M., Rothery, M.L., & Grinnell, R.M., Jr. (Eds.). (1996). *Qualitative research for social workers: Phases, steps, and tasks*. Boston: Allyn & Bacon.

Williams, M., Unrau, Y.A., & Grinnell, R.M., Jr. (2005). The qualitative research approach. In R.M. Grinnell, Jr. & Y.A. Unrau (Eds.), *Social work research and evaluation: Quantitative and qualitative approaches* (7th ed., pp. 78–89). New York: Oxford University Press.

Part III

Measuring Variables

Chapter 6: Measuring Variables   116

Chapter 7: Measuring Instruments   134

Chapter 6

Measuring Variables

DESCRIBING VARIABLES
    Correspondence
    Standardization
    Quantification
    Duplication
CRITERIA FOR SELECTING A MEASURING INSTRUMENT
    Utility
    Sensitivity to Small Changes
    Non-Reactivity
    Reliability
        Test-Retest Method
        Alternate-Form Method
        Split-Half Method
        Observer Reliability (the reliability of the process)
    Validity
        Content Validity
        Criterion Validity
        Face Validity
RELIABILITY AND VALIDITY REVISITED
MEASUREMENT ERRORS
    Constant Errors
    Random Errors
SUMMARY
REFERENCES AND FURTHER READINGS

6

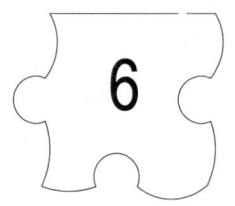

MEASURING VARIABLES IS THE CORNERSTONE of the positivist approach to the research method. Shining and formidable measuring instruments may come to mind, measuring things to several decimal places. The less scientifically inclined might merely picture rulers but, in any case, measurement for most of us means reducing something to numbers. As we know, these "somethings" are called variables.

However, the purpose of measuring a variable is to describe it as completely and accurately as possible. Often, the most complete and accurate possible description of a variable not only involves numbers but also involves words—numbers through the positivist approach and words through the interpretive approach.

DESCRIBING VARIABLES

When using the research method to do a positivist research study, we need to describe the variables we are studying as accurately and completely as possible for four reasons: (1) correspondence, (2) standardization, (3) quantification, and (4) duplication.

Correspondence

Correspondence means making a link between what we measure and/or observe and the theories we have developed to explain what we have measured and/or observed. For example, the concept of attachment theory can easily explain the different behaviors (variables) of small children when they are separated from—or reunited with—their mothers. Measuring and recording children's behaviors in this context provides a link between the abstract and the concrete—between attachment (an unspecific and nonmeasurable concept) and its indicators, or variables, such as a child's behaviors (a more specific and more measurable variable).

Standardization

Variables can be also complex and the more complex they are the more likely it is that people will interpret the exact same variable in different ways. Like concepts for sure, a single variable can at times mean different things to different people even when using the same words. "Self-esteem," for example, can mean different things to different people. However, the perceptions linked to self-esteem (that is, the empirical indicators of self-esteem) may be drawn together in the form of a measuring instrument, as they are in Hudson's *Index of Self-Esteem* (i.e., Figure 6.1).

You may or may not agree that all of the 25 items, or questions, contained in Hudson's *Index of Self-Esteem* together reflect what you mean by self-esteem—but at least you know what Hudson meant, and so does everyone else who is using his measuring instrument. By constructing this instrument, Hudson has *standardized* a complex variable so that everyone using his instrument will mean the same thing by the variable and measure it in the same way.

Moreover, if two or more different researchers use his instrument with the same research participants they ought to get approximately the same results. The use of the word "approximately" here means that we must allow for a bit of error—something discussed at the end of this chapter.

Quantification

Quantification means nothing more than defining the level of a variable in terms of a single number, or score. The use of Hudson's *Index of Self-Esteem*, for example, results in a single number, or score, obtained by following the scoring instructions. Reducing a complex variable like self-esteem to a single number has disadvantages in that the richness of the variable can never be completely captured in this way.

However, it also has advantages in that numbers can be used in statistics to search for meaningful relationships between one variable and another. For example, you might hypothesize that there is a relationship between two variables: self-esteem and marital satisfaction. Hudson has *quantified* self-esteem, allowing the self-esteem of any research participant to be represented by a single number. He has also done this for the variable of marital satisfaction. Since both variables have been broken-down to two numbers, you can use statistical methods (discussed in Chapter 14) to see if the relationship you hypothesized actually does exist.

Duplication

In the physical sciences, experiments are routinely replicated. For example, if you put a test-tube containing 25 ounces of solution into an oven to see what is left when the liquid evaporates, you may use five test-tubes containing 25 ounces each, not just one.

Chapter 6: Measuring Variables

Name: _____ Today's Date: _____

This questionnaire is designed to measure how you see yourself. It is not a test, so there are no right or wrong answers. Please answer each item as carefully and as accurately as you can by placing a number beside each one as follows:

1 = None of the time
2 = Very rarely
3 = A little of the time
4 = Some of the time
5 = A good part of the time
6 = Most of the time
7 = All of the time

1. ___ I feel that people would not like me if they really knew me well.
2. ___ I feel that others get along much better than I do.
3. ___ I feel that I am a beautiful person.
4. ___ When I am with others I feel they are glad I am with them.
5. ___ I feel that people really like to talk with me.
6. ___ I feel that I am a very competent person.
7. ___ I think I make a good impression on others.
8. ___ I feel that I need more self-confidence.
9. ___ When I am with strangers I am very nervous.
10. ___ I think that I am a dull person.
11. ___ I feel ugly.
12. ___ I feel that others have more fun than I do.
13. ___ I feel that I bore people.
14. ___ I think my friends find me interesting.
15. ___ I think I have a good sense of humor.
16. ___ I feel very self-conscious when I am with strangers.
17. ___ I feel that if I could be more like other people I would have it made.
18. ___ I feel that people have a good time when they are with me.
19. ___ I feel like a wallflower when I go out.
20. ___ I feel I get pushed around more than others.
21. ___ I think I am a rather nice person.
22. ___ I feel that people really like me very much.
23. ___ I feel that I am a likeable person.
24. ___ I am afraid I will appear foolish to others.
25. ___ My friends think very highly of me.

Copyright © 1993, Walter W. Hudson   Illegal to Photocopy or Otherwise Reproduce
3, 4, 5, 6, 7, 14, 15, 18, 21, 22, 23, 25.

FIGURE 6.1
Hudson's *Index of Self-Esteem*

AUTHOR: Walter W. Hudson

PURPOSE: To measure problems with self-esteem.

DESCRIPTION: The *ISE* is a 25-item scale designed to measure the degree, severity, or magnitude of a problems the client has with self-esteem. Self-esteem is considered as the evaluative component of self-concept. The *ISE* is written in very simple language, is easily administered, and easily scored. Because problems with self-esteem are often central to social to social and psychological difficulties, this instrument has a wide range of utility for a number of clinical problems.

The *ISE* has a cutting score of 30 (+ or –5), with scores above 30 indicating the respondent has a clinically significant problem and scores below 30 indicating the individual has no such problem. Another advantage of the *ISE* is that it is one of nine scales of the *Clinical Measurement Package* (Hudson, 1982), all of which are administered and scored the same way.

NORMS: This scale was derived from tests of 1,745 respondents, including single and married individuals, clinical and nonclinical populations, college students and nonstudents. Respondents included Caucasians, Japanese, and Chinese Americans, and a smaller number of members of other ethnic groups. Not recommended for use with children under the age of 12.

SCORING: For a detailed description on how to score the *ISE*, see: Bloom, Fischer, and Orme (2009) or go to: www.walmyr.com.

RELIABILITY: The *ISE* has a mean alpha of .93, indicating excellent internal consistency, and an excellent (low) S.E.M. of 3.70. The *ISE* also has excellent stability with a two-hour test-retest correlation of .92.

VALIDITY: The *ISE* has good know-groups validity, significantly distinguishing between clients judged by clinicians to have problems in the area of self-esteem and those known not to. Further, the *ISE* has very good construct validity, correlating well with a range of other measures with which it should correlate highly, e.g., depression, happiness, sense of identity, and scores on the *Generalized Contentment Scale* (depression).

PRIMARY REFERENCE: Hudson, W.W. (1982). *The Clinical Measurement Package: A Field Manual.* Chicago: Dorsey.

FIGURE 6.1 (Continued)
Basic Information about the *ISE*

Chapter 6: Measuring Variables

121

Then, you will have five identical samples of solution evaporated at the same time under the same conditions, and you will be much more certain of your results than if you had just evaporated one sample. The word *replication* means doing the same thing more than once at the same time.

In our profession, we can rarely replicate research studies but we can *duplicate* them. That is, a second researcher can attempt to confirm a first researcher's results by doing the same thing again later on, as much as is practically possible under the same conditions. Duplication increases certainty and it is only possible if the variables being studied have been *standardized* and *quantified*.

For example, you could duplicate another researcher's work on attachment only if you measured attachment in the same way. If you used different child behaviors to indicate attachment and you assigned different values to mean, say, weak attachment or strong attachment, you may have done a useful study but it would not be a duplicate of the first.

CRITERIA FOR SELECTING A MEASURING INSTRUMENT

Now that you know why you need to measure variables, let us go on to look at *how* you measure them in the first place. In order to measure a variable, you need a measuring instrument to measure it with—much more about this topic in the following chapter. Most of the measuring instruments used in social work are paper and pencil instruments like Figure 6.1.

Many other people besides Hudson have come up with ways of measuring self-esteem and if you want to measure self-esteem in your study, you will have to choose between the various measuring instruments that are available that measure self-esteem. The same embarrassment of riches applies to most of the other variables you might want to measure. Remember that a variable is something that varies between research participants. Participants will vary, for example, with respect to their levels of self-esteem.

What you need are some criteria to help you decide which instrument is best for measuring a particular variable in any given particular situation. There are five criteria that will help you to do this: (1) utility, (2) sensitivity to small changes, (3) non-reactivity, (4) reliability, and (5) validity.

Utility

In order to complete Hudson's *Index of Self-Esteem* (Figure 6.1), for example, a research participant must preferably be able to read. Even if you, as the researcher, read the items to the participants, they must be able to relate a number between 1 and 7 (where 1 = none of the time, and 7 = all of the time) to each of the 25 items, or questions.

Further, they must know what a "wallflower" is before they can answer Item 19. If the research participants in your study cannot do this for a variety of reasons, then no matter how wonderful Hudson's *Index of Self-Esteem* might be in other respects, it is not useful to you in your particular study.

Hudson's *Index of Self-Esteem* may take only a few minutes to complete but other instruments can take far longer. The Minnesota Multiphasic Personality Inventory (the MMPI), for example, can take three hours or more to complete, and some people may not have the attention span or the motivation to complete the task. In sum, a measuring instrument is not useful if your research participants are unable or unwilling to complete it—for whatever reasons.

If they do complete it, however, you then have to score it. While the simple measuring instrument contained in Figure 6.1 is relatively quick and simple to score, other instruments are far more complex and time-consuming. Usually, the simple instruments—quick to complete and easy to score—are less accurate than the more demanding instruments, and you will have to decide how far you are prepared to sacrifice accuracy for utility.

The main consideration here is what you are going to do with the measurements once you have obtained them. If you are doing an assessment that might affect a client's life in terms of treatment intervention, referral, placement, and so on, accuracy is paramount and you will need the most accurate instrument (probably the longest and most complex) that the client can tolerate. On the other hand, if you are doing an exploratory research study where the result will be a tentative suggestion that some variable may be related to another, a little inaccuracy in measurement is not the end of the world and utility might be more important.

Sensitivity to Small Changes

Suppose that one of your practice objectives with your 8-year old client, Johnny, is to help him stop wetting his bed during the night. One obvious indicator of the variable—bedwetting—is a wet bed. Thus, you hastily decide that you will measure Johnny's bedwetting behavior by having his mother tell you if Johnny has—or has not—wet his bed during the week; that is, Did he, or did he not wet his bed at least once during the week?

However, if Johnny has reduced the number of his bed-wetting incidents from five per week to only once per week, you will not know if your intervention was working well because just the one bedwetting incident per week was enough to officially count as "wetting the bed."

Chapter 6: Measuring Variables

In other words, the way you chose to measure Johnny's bedwetting behavior was sensitive to the large difference between wetting and not wetting in a given week, but insensitive to the smaller difference between wetting once and wetting more than once in a given week.

In order to be able to congratulate Johnny on small improvements, and of course to track his progress over time, you will have to devise a more sensitive measuring instrument; such as one that measures the number of times Johnny wets his bed per week. Often, an instrument that is more sensitive will also be less convenient to use and you will have to balance sensitivity against utility.

Non-Reactivity

A *reactive* measuring instrument is nothing more than an instrument that changes the behavior or feeling of a person that it was supposed to measure. For instance, you might have decided, in the example above, to use a device that rang a loud bell every time that Johnny had a bedwetting accident. His mother would then leap from sleep, make a check mark on the form you had provided, and fall back into a tormented doze. This would be a sensitive measure—though intrusive and thus less useful—but it might also cause Johnny to reduce his bedwetting behavior in accordance with behavior theory.

Clinically, this would be a good thing—unless he developed bell phobia—but it is important to make a clear distinction between an *instrument* that is designed to *measure* a behavior and an *intervention* that is designed to *change* it. If the bell wakes up Johnny so he can go to the bathroom and thus finally eliminate his bedwetting behavior, the bell is a wonderful *intervention*. It is not a good measuring instrument, however, as it has changed the very behavior it was supposed to measure. A change in behavior resulting from the use of a measuring instrument is known as a *reactive effect*.

The ideal, then, is a *non-reactive* measuring instrument that has no effect on the variable being measured. If you want to know, for example, whether a particular intervention is effective in raising self-esteem in girls who have been sexually abused, you will need to be sure that any measured increase in self-esteem is due to the intervention and *not* to the measuring instrument you happen to be using. If you fail to make a distinction between the measuring instrument and the intervention, you will end up with no clear idea at all about what is causing what.

Sometimes, you might be tempted to use a measuring instrument as a clinical tool. If your client responded to Hudson's *Index of Self-Esteem* Item 13 (Figure 6.1) that she felt she bored people all of the time, you might want to discuss with her the particular conversational gambits she feels are so boring in order to help her change them. This

> **BOX 6.1**
> **Major Types of Measurement Reliability and Questions Addressed by Each**
>
> | | |
> |---|---|
> | Test-Retest Method | Does an individual respond to a measuring instrument in the same general way when the instrument is administered twice? |
> | Alternate-Forms Method | When two forms of an instrument that are equivalent in their degree of validity are given to the same individual, is there a strong convergence in how that person responds? |
> | Split-Half Method | Are the scores on one half of the measuring instrument similar to those obtained on the other half? |

is perfectly legitimate so long as you realize that, by so doing, you have turned a measuring instrument into part of an intervention. Thus you must now find a different instrument to measure your client's self-esteem.

Reliability

A good measuring instrument is reliable in that it gives the same score over and over again provided that the measurement is made under the same conditions and nothing about the research participant has changed. A reliable measuring instrument is obviously necessary since, if you are trying to track the increase in a client's self-esteem, for example, you need to be sure that the changes you see over time are due to changes in the client, not to inaccuracies in the measuring instrument.

Researchers are responsible for ensuring that the measuring instruments they use are reliable. Hence, it is worth looking briefly at the four main methods used to establish the reliability of a measuring instrument: (1) test-retest, (2) alternate form, (3) split half, and (4) observer reliability.

Test-Retest Method

The test-retest method of establishing reliability involves administering the same measuring instrument to the same group of people on two separate occasions. The two sets of results are then compared to see how similar they are; that is, how well they *correlate*. We will discuss *correlation* more fully in Chapter 14. For now, it is enough to say that correlation in this context can range from 0 to 1, where 0 means no correlation at all between the two sets of scores and 1 means a perfect correlation.

Chapter 6: Measuring Variables

Generally, a correlation of 0.8 means that the instrument is reasonably reliable and 0.9 is very good. Note that there is a heading RELIABILITY in Figure 6.1 which means that Hudson's *Index of Self-Esteem* has "excellent stability with a two-hour test-retest correlation of 0.92." The "two-hour' bit means, of course, that the two administrations of the instrument took place two hours apart. The problem with completing the same instrument twice is that the answers given on the first occasion may affect the answers given on the second. This is known as a *testing effect*. For example, Ms. Smith might remember what she wrote the first time and write something different just to enliven the proceedings. She may be less anxious, or more bored or irritated the second time, just because there was a first time, and these states might affect her answers.

Obviously, the greater the testing effects, the less reliable the instrument. Moreover, the closer together the tests, the more likely testing effects become because Ms. Smith is more likely to remember the first occasion. Hence, if the instrument is reliable over an interval of two hours and you want to administer it to your study participants on occasions a day or a month apart, it should be even more reliable with respect to testing effects.

However, people may change their answers on a second occasion for reasons other than testing effects: they are having a good day or a bad day; or they have a cold; or there is a loud pneumatic drill just outside the window. Such external factors are more likely to change over a day or a month than they are over two hours, so you might expect an instrument tested over a two-hour gap to be somewhat less reliable when used over a longer gap. Generally, researchers using instruments that have already been tested for reliability tend to accept the reliability information provided, and usually it is safe to do so.

However, a word of caution is in order. Sometimes, clients complete the same measuring instrument every few weeks for a year or more as a way of monitoring their progress over time. The more often an instrument is completed, the more likely it is to generate testing effects. Hence, social service programs that use instruments in this way need to be sure that the instruments they use are still reliable under the conditions in which they are to be used.

Alternate-Form Method

The second method of establishing the reliability of a measuring instrument is the alternate-form method. As the same suggests, an alternate form of an instrument is a second instrument that is as similar as possible to the original except that the wording of the items contained in the second instrument has changed. Administering the original form and then the alternate form reduces testing effects since the respondent is less likely to base the second set of answers on the first.

However, it is time-consuming to develop different but equivalent instruments, and they must still be tested for reliability using the test-retest method, both together as a pair, and separately as two distinct instruments.

Split-Half Method

The split-half method involves splitting one instrument in half so that it becomes two shorter instruments. Usually, all the even-numbered items, or questions, are used to make one instrument while the odd-numbered items makeup the other. The point of doing this is to ensure that the original instrument is internally consistent; that is, it is homogeneous, or the same all the way through, with no longer or more difficult items appearing at the beginning or the end.

When the two halves are compared using the test-retest method, they should ideally yield the same score. If they did give the same score when one half was administered to a respondent on one occasion and the second half to the same respondent on a different occasion, they would have a perfect correlation of 1. Again, a correlation of 0.8 is thought to be good and a correlation of 0.9 very good. Figure 6.1, under the RELIABILITY section, shows that Hudson's *Index of Self-Esteem* has an internal consistency of 0.93.

Observer Reliability (the reliability of the process)

Sometimes, behaviors are measured by observing how often they occur, or how long they last, or how severe they are. The results are then recorded on a straightforward simple form. Nevertheless, this is not as easy as it sounds since the behavior, or variable, being measured must first be very carefully defined and people observing the same behavior may have different opinions as to how severe the behavior was or how long it lasted, or whether it occurred at all.

The level of agreement or correlation between trained observers therefore provides a way of establishing the reliability of the process used to measure the behavior. Once we have established the reliability of the process, we can use the same method to assess the reliability of other observers as part of their training. The level of agreement between observers is known as *inter-rater reliability*.

**BOX 6.2
Major Types of Measurement Validity
and Questions Addressed by Each**

Content Validity	Does the measuring instrument adequately measure the major dimensions of the variable under consideration?
Criterion Validity	Does the individual's measuring instrument score predict the probable behavior on a second variable (criterion-related measure)?
Face Validity	Does the measuring instrument appear to measure the subject matter under consideration?

Validity

A measuring instrument is valid if it measures what it is supposed to measure—and measures it accurately. If you want to measure the variable assertiveness, for example, you don't want to mistakenly measure aggression instead of assertiveness. There are several kinds of validity—in fact, we should really refer to the *validities* of an instrument—and we will look at the following three: (1) content validity, (2) criterion validity, and (3) face validity.

Content Validity

Think for a moment about the variable self-esteem. In order to measure it accurately, you must first know what it is; that is, you must identify all the indicators (questions contained in the measuring instrument) that make up self-esteem, such as feeling that people like you, feeling that you are competent, and so on—and on and on and on... It is probably impossible to identify *all* the indicators that contribute to self-esteem.

It is even less likely that everyone (or even most people) will agree with all the indicators identified by someone else. Arguments may arise over whether "feeling worthless," for example, is really an indicator of low self-esteem or whether it has more to do with depression, which is a separate variable altogether. Furthermore, even if agreement could be reached, a measuring instrument like Hudson's *Index of Self-Esteem* would have to include at least one item, or question, for *every* agreed-upon indicator. If just one was missed, for example, "sense of humor"—then the instrument would not be accurately measuring self-esteem.

Because it did not include all the possible content, or indicators, related to self-esteem, it would not be *content valid*. Hudson's *Index of*

Self-Esteem, then, is not perfectly content valid because it is not possible to cover every indicator related to self-esteem in just 25 items. Longer instruments have a better chance of being content valid (perhaps one could do it in 25 *pages* of items) but, in general, perfect content validity cannot be achieved in any measuring instrument of a practical length. Content validity is a matter of "more or less" rather than "yes or no" and it is, moreover, strictly a matter of opinion.

For example, experts differ about the degree to which various instruments are content valid. It is therefore necessary to find some way of *validating* an instrument to determine how well it is, in fact, accurately measuring what it is supposed to measure. One such way is through a determination of the instrument's *criterion validity*.

Criterion Validity

An instrument has criterion validity if it gives the same result as a second instrument that is designed to measure the same variable. A client might complete Hudson's *Index of Self-Esteem*, for example, and achieve a score indicating high self-esteem. If the same client then completes a second instrument also designed to measure self-esteem and again achieves a good score, it is very likely that both instruments are, in fact, measuring self-esteem.

Not only do they have good criterion validity in that they compare well with each other, but probably each instrument also has good content validity. If the same client does not achieve similar scores on the two instruments, however, then neither of them are criterion valid, probably one is not content valid, and both will have to be compared with a third instrument in order to resolve the difficulty. There are two categories of criterion validity: (1) concurrent validity and (2) predictive validity.

1. Concurrent Validity. Concurrent validity deals with the present. For example, suppose you have an instrument (say a reading test) designed to distinguish between children who need remedial reading services and children who do not. In order to validate the measuring instrument, you ask the class teacher which children she thinks need remedial reading services. If the teacher and your instrument both come up with the same list of children, your instrument has criterion validity. If not, you will need to find another comparison: a different reading test or the opinion of another teacher.

2. Predictive Validity. Predictive validity deals with the future. Perhaps you have an instrument (say a set of criteria) designed to predict which students will achieve high grades in their social work programs. If the students your instrument identified had indeed achieved high grades by the end of their BSW programs, and the others had not, your instrument would have predictive validity.

In sum, criterion validity, whether concurrent or predictive, is determined by comparing the instrument with another designed to measure the same variable.

Face Validity

Face validity, in fact, has nothing to do with what an instrument actually measures but only with what it *appears* to measure to the one who is completing it. Strictly speaking, it is not a form of validity. For example, suppose that you are taking a course on social work administration. You have a lazy instructor who has taken your final exam from a course he taught for business students last semester. The exam in fact quite adequately tests your knowledge of administration theory but it does not seem relevant to you because the language it uses relates to the business world not to social work situations.

You might not do very well on this exam because, although it has content validity (it adequately tests your knowledge), it does not have face validity (an appearance of relevance to the respondent). The moral here is that a measuring instrument should not only *be* content valid, to the greatest extent possible; it should *appear* content valid to the person who completes it.

RELIABILITY AND VALIDITY REVISITED

Before we leave reliability and validity, we should say something about the relationship between them. If an instrument is not reliable, it cannot be valid. That is, if the same person completes it a number of times under the same conditions and it gives different results each time, it cannot be measuring anything accurately.

However, if an instrument *is* reliable, that does not necessarily mean it is valid. It could be reliably and consistently measuring something other than what it is supposed to measure, in the same way that people can be reliably late, or watches can be reliably slow. The relationship between validity and reliability can be illustrated with a simple analogy. Suppose that you are firing five rounds from a rifle at three different targets, as illustrated in Figure 6.2:

- In Figure 6.2a, the bullet holes are scattered, representing a measuring instrument that is neither reliable nor valid.

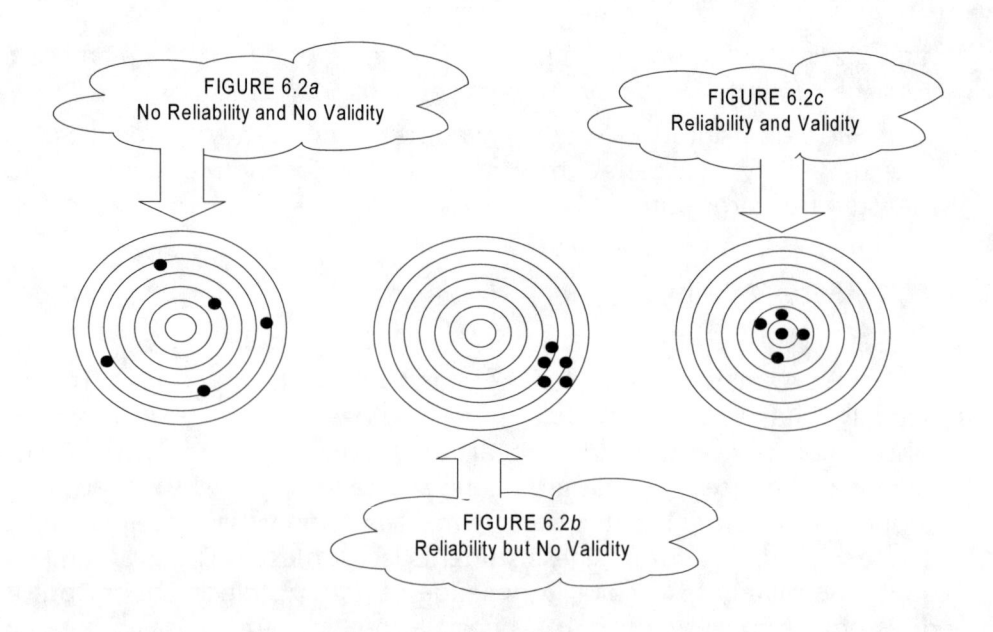

FIGURE 6.2
Target Illustrating the Validity Reliability Relationship

- In Figure 6.2b, you have adjusted your sights, and now all the bullet holes are in the same place but not in the center as you intended. This represents a measuring instrument that is reliable but not valid.
- In Figure 6.2c, all the shots have hit the bull's eye: the instrument is both reliable and valid.

MEASUREMENT ERRORS

No matter how good the reliability and validity of a measuring instrument, no measurement is entirely without error. You can make two errors when you measure variables. Your measurements can contain: (1) constant errors, and/or (2) random errors.

Constant Errors

Constant errors, as the name suggests, are those errors that stay constant throughout the research study. They stay constant because they come from an unvarying source. That source may be the measuring instruments used, the research participants, or the researchers themselves. Since we have already spent some time discussing the limitations of measuring instruments, we will focus this discussion on errors caused by the researchers and their research participants.

Research participants, with all the best will in the world, may still have personal styles that lead to error in research results. If they are being interviewed, for example, they may exhibit *acquiescence* (a tendency to agree with everything the researcher says, no matter what it is) or *social desirability* (a tendency to say anything that they think makes them look good) or *deviation* (a tendency to seek unusual responses).

If they are filling out a self-administered instrument, like Hudson's *Index of Self-Esteem*, they may show *error of central tendency*, always choosing the number in the middle and avoiding commitment to either of the ends. Moreover, they have personal characteristics with respect to gender, age, ethnic background and knowledge of the English language which remain constant throughout the study and may affect their answers.

Researchers also have personal styles and characteristics. Interviewers can affect the answers they receive by the way they ask the questions, by the way they dress, by their accent, mannerisms, gender, age, ethnic background, even by their hairstyles. Observers who are watching and rating research participants' behavior can commit various sins in a constant fashion, for example:

- *Contrast error*—to rate others as opposite to oneself with respect to a particular characteristic
- *Halo effect*—to think that a participant is altogether wonderful or terrible because of one good or bad trait. Or to think that the trait being observed must be good or bad because the participant is altogether wonderful or terrible
- *Error of leniency*—to always give a good report
- *Error of severity*—to always give a bad report
- *Error of central tendency*—observers, like participants, can choose always to stay comfortably in the middle of a rating scale and avoid both ends

Since these errors are constant throughout the study, they are sometimes recognized and steps can be taken to deal with them. A different interviewer or observer might be found, for example, or allowances might be made for a particular participant's characteristics or style.

Random Errors

Random errors that are not constant are difficult to find and make allowances for. Random errors spring out of the dark, wreak temporary havoc, and go back into hiding. It has been suggested that eventually they cancel each other out and, indeed, they might. They might not, as well, but there is little researchers can do about them except to be aware that they exist. There are three types of random error:

- *Transient qualities of the research participant*—things such as fatigue or boredom, or any temporary personal state that will affect the participant's responses
- *Situational factors*—the weather, the news, the pneumatic drill outside the window, or anything else in the environment that will affect the participant's responses
- *Administrative factors*—anything relating to the way the instrument is administered, or the interview conducted or the observation made. These include transient qualities of the researcher (or whoever collects the data) as well as sporadic stupidity like reading out the wrong set of instructions

SUMMARY

Measurement serves as a bridge between theory and reality. Our variables within a positivist study must be operationalized in such a way that they can be measured. Which measuring instrument we select will depend on why we need to make the measurement and under what circumstances it will be done. Provided that a number of instruments can be found that seem to meet our needs, the next step is to evaluate them.

Measurement error refers to variations in instrument scores that cannot be attributed to the variable being measured. Basically, all measurement errors can be categorized as constant errors or random errors. The next chapter is a logical extension of this one as it presents the many different types of measuring instruments that are available for our use.

REFERENCES AND FURTHER READINGS

Bloom, M., Fischer, J., & Orme, J. (2009). *Evaluating practice: Guidelines for the accountable professional* (6th ed.). Englewood Cliffs, NJ: Prentice-Hall.

Bostwick, G.J., Jr., & Kyte, N.S. (2005). Measurement. In R.M. Grinnell, Jr., & Y.A. Unrau (Eds.), *Social work research and evaluation: Quantitative and qualitative approaches* (7th ed., pp. 104–117). New York: Oxford University Press.

Engel, R., & Schutt, R. (2008). Conceptualization and measurement. In R.M. Grinnell, Jr., & Y.A. Unrau (Eds.), *Social work research and evaluation: Foundations of evidence-based practice* (8th ed., pp. 105–133). New York: Oxford University Press.

Hudson, W.W. (1982). *The Clinical Measurement Package: A Field Manual.* Chicago: Dorsey.

Salkind, N.J. (2006). *Tests and measurement for people who (think they) hate tests and measurements.* Thousand Oaks, CA: Sage.

Check out our Website for useful links and chapters (PDF) on:
- introductory measurement concepts
  Go to: www.pairbondpublications.com
  Click on: Student Resources
      Chapter-by-Chapter Resources
      Chapter 6

Check Out Our Website

PairBondPublications.com

√ *Student Workbook Exercises*
√ *Chapter Power Point Slides*
√ *On-line Glossaries*
√ *General Links*
√ *Specific Chapter Links*

Chapter 7

Measuring Instruments

ALL VARIABLES ARE NOT MEASURED THE SAME
    Nominal Variables
    Ordinal Variables
    Interval Variables
    Ratio Variables
QUESTIONS TO ASK BEFORE MEASURING A VARIABLE
    Why Do We Want to Make the Measurement?
    What Do We Want to Measure?
    Who Will Make the Measurement?
    What Format Do We Require?
    Where Will the Measurement Be Made?
    When Will the Measurement Be Made?
TYPES OF MEASURING INSTRUMENTS
    Journals or Diaries
    Logs
    Inventories
    Checklists
    Summative Instruments
STANDARDIZED MEASURING INSTRUMENTS
    Locating Standardized Measuring Instruments
        Publishers
        Professional Books and Journals
SUMMARY
REFERENCES AND FURTHER READINGS

7

AS WE KNOW FROM THE PRECEDING CHAPTERS in this book, on a general level, variables describe concepts and value labels describe variables. For example, the concept "ethnicity" can be described in terms of the existence of ethnic groups. "Ethnic group" is a variable since research participants will *vary* with respect to which ethnic group they belong to. The ethnic groups they *do* belong to—Asian, Caucasian, Hispanic, and so on—constitute the value labels of the variable "ethnic group."

ALL VARIABLES ARE NOT MEASURED THE SAME

The particular value labels a variable can take determine how the variable is measured. There are four types of variables: (1) nominal variables, (2) ordinal variables, (3) interval variables, and (4) ratio variables.

Nominal Variables

Nominal variables are the least precise type of variable when it comes to measurement. It is one whose value labels are different in kind only. The value labels for nominal variables constitute only different and mutually exclusive categories. "Gender," for example, is a variable with two value labels—male and female—which are different from one another and has two mutually exclusive categories. A researcher determining (measuring) the gender of research participants, for example, might include the nominal-level variable on a self-report questionnaire:

   What is your gender? (Circle one number below)
   1. Male
   2. Female

The number 1 immediately left of "male" indicates the first choice available to the respondents but it can also be used by the researchers as a form of shorthand when they enter the data into a computer. The numbers 1 and 2 have no quantitative meaning: females are not twice as good as males, for example. The numbers 1 and 2 are merely labels for the two categories. The same question could be written as follows:

What is your gender? (Circle one number below)
1. Female
2. Male

"Ethnic group" is another nominal-level variable because its value labels (e.g., Asian, Caucasian) are different in kind only. "Place of birth" is another. Being born in Toronto is different from being born in Chicago. "Marital status" is yet another nominal-level variable. Married is different from widowed, single, or divorced. Value labels for nominal-level variables do not have to be numbers as they can be letters, such as:

What is your gender? (Circle one letter below)
A. Female
B. Male

Ordinal Variables

As we know, nominal variables have value labels that are not better or worse than each other, or higher or lower: they are just different. Ordinal variables, on the other hand, have value labels that can be expressed as numerical values and thus that can be ranked, from low to high, or from high to low. For example, a social worker might be asked to rate a child's progress on the ordinal variable:

How much improvement has your client accomplished over the last month? (circle one number below)
1. Not improved
2. Slightly improved
3. Somewhat improved
4. Greatly improved

In reference to the four values this variable can take (i.e., 1, 2, 3, 4), "somewhat" is more than "slightly" and "greatly" is more than "somewhat" but we cannot tell how much more. Neither can we say that the amount of improvement between "slightly" and "somewhat" is the same as the amount of improvement between "somewhat" and "greatly."

In other words, the distances between the values (or numbers) are not known; and, since they are not known, they can never be assumed to be equal to each other.

Interval Variables

Unlike nominal- and ordinal-level variables described above, the distance between the values of interval variables *are* known. Not only are they known, they are equal to each other. A centigrade thermometer, for example, is an example of a measuring instrument that measures the variable "temperature." Here, the values of the thermometer are the degrees on the thermometer scale. If the temperature is 30 degrees centigrade on Wednesday, that is 10 degrees more than the 20 degrees it was on Tuesday, and 5 degrees less than the 35 degrees it was on Monday morning.

Because the distances between values are known and equal, it is possible to add and subtract the values. Such computations are important if you want to use statistics to determine the relationship between variables, and they are not possible with nominal and ordinal variables. For example, you cannot subtract "Asian" from "Caucasian" (with a nominal variable) or add "somewhat" to "greatly" (with an ordinal variable).

Ratio Variables

Ratio variables have values where 0 (zero) means the absence of the variable being measured. The centigrade thermometer mentioned above does not measure temperature at a ratio level because the 0 on the scale does not mean a complete absence of temperature; it just indicates the temperature at which water freezes. Similarly, a measuring instrument that measures intelligence would have to be at the interval level because even the dullest person does not have a complete absence of intelligence.

However, it is quite possible to have zero income or zero children or to have been married zero times. Moreover, someone who has 4 children has twice as many children as someone who has 2 children. It is the 0 on the scale that allows the "twice as many" to be calculated.

Using an intelligence measuring instrument, where there is no 0, it is not correct to say that a person with an IQ of 120 is twice as intelligent as a person with an IQ of 60; nor can we say that 60 is half of 120. The presence of 0 on a measuring instrument allows us to perform multiplication and division with ratio variables whereas all we are allowed with interval variables is addition and subtraction.

QUESTIONS TO ASK BEFORE MEASURING A VARIABLE

What we want is some method of making a selection from the huge array of measuring instruments that exist. There are six questions we can ask ourselves to help us make our choice. When we have answered these six questions, we will be able to distinguish the kind of instrument we need from the kind we do not need; hopefully, this will eliminate a large number of all those instruments lying in wait in the library. Let's take a simple example to illustrate how the use of measuring instruments can be used in the research process. Let's say we want to know if there is a relationship between the two variables: depression and sleep patterns.

Why Do We Want to Make the Measurement?

The first question is: *Why do we want to make the measurement?* At first glance, this does not seem too difficult. We just want to measure, or operationalize, depression and sleep patterns in order to study the relationship between the two variables, if any. But things are not quite as simple as they appear. There are three general reasons for using measuring instruments, and we need to select the one that applies to our study.

The first of these reasons is *assessment and diagnosis*. We are not really assessing or diagnosing anyone, as we might be in clinical practice, so this reason can be eliminated. The second reason is *evaluation of practice effectiveness,* and we are not doing that either. The last reason is *applied research*. Determining relationships between two variables certainly counts as research and we hope to eventually apply it to our practice, so we will select that one.

The point of making this selection is to discover how accurate we need the instrument to be. If our measurement is going to affect someone's life, for example, it has to be as accurate as we can possibly make it. We might be doing an assessment that will be used in making decisions about treatment interventions, referrals, placements, and so forth. On the other hand, if our measurement will not affect anyone's life directly, we can afford to be a little less rigid in our requirements. How much less rigid depends on what we are doing.

If this is to be a beginning research study (e.g., exploratory) in a relatively unexplored field, the result will only be a tentative suggestion that some variable is possibly related to some other variable. For instance, sleep patterns are related to depression. A little inaccuracy in measurement in this case is not the end of the world.

When a little more is known in our subject area (e.g., descriptive, explanatory), we might be able to formulate and test a more specific hypothesis; for example, depressed people spend less time in delta-- wave NREM sleep than nondepressed people. In this case, we obviously should be able to measure sleep patterns accurately enough to distinguish between delta-wave NREM sleep and other kinds of sleep.

All in all, then, how accurate our measurement needs to be depends on our purpose. Because we are only doing a beginning study, we can afford to be relatively inaccurate. We have now answered the first question: *Why do we want to make the measurement?*

What Do We Want to Measure?

The second question is, *What do we want to measure?* We know the answer to that one also; we want to measure depression and sleep patterns. Or, to put it another way, we want to operationalize depression and sleep patterns. But here again, it is not so simple as it seems. Not only are measuring instruments more accurate or less accurate, they can be *wideband* or *narrowband*. *Wideband* instruments measure a broad trait or characteristic. A trait is pretty much the same thing as a characteristic and it means some aspect of character such as bravery, gaiety, or depression.

Logically enough, *narrowband* instruments measure just a particular aspect of a particular trait. A narrowband instrument, for example, might tell us how Uncle Fred feels about his daughter moving to Moose Jaw but it will not give us an overall picture of Uncle Fred's depression.

A wideband instrument, on the other hand, will give an overall picture of Uncle Fred's depression but it will not tell us how he feels about his daughter moving to Moose Jaw. In our particular study, we are not interested too much in how Uncle Fred feels about his daughter moving to Moose Jaw. If he is one of our research participants, we just want to know about his overall depression so that we can relate it to his sleep patterns. We need a wideband instrument then, which does not have to be absolutely smack-on accurate.

Who Will Make the Measurement?

The third question asks: *Who will make the measurement?* We will—that seems obvious enough. However, it is not always social workers who make measurements, sometimes clients make them. In our study, it will be the people who participate in the study who will fill out the instrument to measure depression. Sometimes family members make the measurements, or then again, teachers, specially trained outside observers, or the staff members in an institution.

The point is that different kinds of people require different kinds of measuring instruments. An instrument that could be completed easily and accurately by a trained social worker might prove too difficult for Uncle Fred, who has arthritis and cataracts and a reading level of around the fourth grade. In our simple research study, we will have to take care that the instrument we choose is easy for our research participants to understand. So far, then, we need a wideband instrument, easy to complete, and not necessarily smack-on accurate.

What Format Do We Require?

The fourth question is, *What format do we require?* A format is the way our questions will look on the page. They may appear as an inventory, such as:

> List below the things that make you feel depressed.
> _____
> _____
> _____

Or a checklist such as:

> Check below all the things that you sometimes feel:
> _____ My mother gets on my nerves.
> _____ My father does not understand me.
> _____ I do not get along very well with my sister.
> _____ I think I hate my family sometimes.

Or a scale such as:

> How satisfied are you with your life? (Circle one number below.)
> 1. Very unsatisfied
> 2. Somewhat satisfied
> 3. Satisfied
> 4. More than satisfied
> 5. Very satisfied

More often than not, measuring instruments contain a number of items, or questions, that when totaled, yield more accurate results than just asking one question. These measuring instruments are called summative scales and will be discussed more thoroughly later in this chapter.

After careful thought, we decide that a wideband, easy-to-complete, not-smack-on-accurate summative scale would do the job. However, we have not finished yet. Instruments may be unidimensional or multidimensional. A unidimensional instrument only measures one variable, for example, self esteem (e.g., Figure 6.1). On the other hand, a multidimensional instrument measures a number of variables at the same time. A multidimensional instrument is nothing more than a number of unidimensional instruments stuck together. For example, Figure 7.1 is a multidimensional instrument that contains three unidimensional instruments:

1. Relevance of received social services (Items 1–11)
2. The extent to which the services reduced the problem (Items 12–21)
3. The extent to which services enhanced the client's self-esteem and contributed to a sense of power and integrity (Items 22–34)

Where Will the Measurement Be Made?

Question five asks: *Where will the measurement be made?* Well, probably in our good friend's sleep laboratory. At first glance, it may seem that it does not matter where the measurement is made but, in fact, it matters a great deal. For example, we might have a child who throws temper tantrums mostly in school. In this case, measurements dealing with temper tantrums should obviously be made at school.

We can see that an instrument that is to be completed in a railway station might differ from an instrument that is to be used in the comparative serenity of our office. It should be shorter, say, and simpler, possibly printed on paper that glows in the dark so that if it gets torn out of our hands, we can chase after it more easily. We decide that our research participants will probably be equally depressed everywhere, more or less, and the measurements will take place in the laboratory.

When Will the Measurement Be Made?

Our last question, question six, is: *When will the measurement be made?* Probably some time in August if it all goes well. But no, the month of the year is not what is meant by *when*. *When* refers to the time or times during the study when a measurement is made.

As will be seen in Chapter 9, there are certain research designs in which we measure a client's problem, do something to change the problem, and then measure the problem again to see if we have changed it. This involves two measurements, the first and second measurements of the problem. In research jargon we represent these measurements as *O*s (*O* stands for *O*bservation). Whatever we do to change the problem—usually a social work treatment—is represented by *X*.

SOCIAL SERVICE SATISFACTION SCALE

Using the scale from one to five described below, please indicate at the left of each item the number that comes closest to how you feel.

1 Strongly agree
2 Agree
3 Undecided
4 Disagree
5 Strongly disagree

1. ___ The social worker took my problems very seriously.
2. ___ If I had been the worker, I would have dealt with my problems in just the same way.
3. ___ The worker I had could never understand anyone like me.
4. ___ Overall the agency has been very helpful to me.
5. ___ If friends of mine had similar problems I would tell them to go to the agency.
6. ___ The social worker asks a lot of embarrassing questions.
7. ___ I can always count on the worker to help if I'm in trouble.
8. ___ The agency will help me as much as it can.
9. ___ I don't think the agency has the power to really help me.
10. ___ The social worker tries hard but usually isn't too helpful.
11. ___ The problem the agency tried to help me with is one of the most important in my life.
12. ___ Things have gotten better since I've been going to the agency.
13. ___ Since I've been using the agency my life is more messed up than ever.
14. ___ The agency is always available when I need it.
15. ___ I got from the agency exactly what I wanted.
16. ___ The social worker loves to talk but won't really do anything for me.
17. ___ Sometimes I just tell the social worker what I think she wants to hear.
18. ___ The social worker is usually in a hurry when I see her.
19. ___ No one should have any trouble getting some help from this agency.
20. ___ The worker sometimes says things I don't understand.
21. ___ The social worker is always explaining things carefully.
22. ___ I never looked forward to my visits to the agency.
23. ___ I hope I'll never have to go back to the agency for help.
24. ___ Every time I talk to my worker I feel relieved.
25. ___ I can tell the social worker the truth without worrying.
26. ___ I usually feel nervous when I talk to my worker.
27. ___ The social worker is always looking for lies in what I tell her.
28. ___ It takes a lot of courage to go to the agency.
29. ___ When I enter the agency I feel very small and insignificant.
30. ___ The agency is very demanding.
31. ___ The social worker will sometimes lie to me.
32. ___ Generally the social worker is an honest person.
33. ___ I have the feeling that the worker talks to other people about me.
34. ___ I always feel well treated when I leave the agency.

FIGURE 7.1
Reid-Gundlach Social Service Satisfaction Scale

Chapter 7: Measuring Instruments

In short, in research designs we represent the dependent variables by Os, and the independent variables by Xs. If our research design is such that we make an initial measurement of the dependent variable (O_1); introduce an independent variable (X); and then measure the same dependent variable again (O_2) our design would look like: $O_1\ X\ O_2$.

TYPES OF MEASURING INSTRUMENTS

The type of measuring instrument you choose to measure your variables within your research study depends on your research situation—the question you are asking, the kind of data you need, the research participants you have selected, and the time and money you have available. Every measuring instrument you consider must be evaluated in terms of the five criteria discussed in the previous chapter: is it valid, reliable, useful, sensitive, and non-reactive?

In general, there are many different types of measuring instruments. We will only discuss five: (1) journals or diaries, (2) logs, (3) inventories, (4) checklists, and (5) summative instruments.

Journals or Diaries

Journals or diaries are a useful means of data collection when you are undertaking an interpretive study. Perhaps you are asking a composition question, "What are women's experiences of home birth?" and you want your research participants to keep a record of their experiences from early pregnancy to post-delivery. With respect to the five criteria mentioned in the previous chapter, a journal is *valid* in this context to the extent that it completely and accurately describes the relevant experiences and omits the irrelevant experiences.

This can only be achieved if the women keeping them have reasonable language skills, can stick pretty much to the point (will they include a 3-page description their cats or their geraniums?) and are they willing to complete their journals on a regular basis.

A word is in order here about *retrospective data*: that is, data based on someone's memory of what occurred in the past. There is some truth to the idea that we invent our memories. At least, we might embellish or distort them, and a description is much more liable to be accurate if it is written immediately after the event it describes rather than days or weeks later.

The journal is *reliable* insofar as the same experience evokes the same written response. Over time, women may tire of describing again an experience almost identical to the one they had last week and they may either omit it (affecting validity) or change it a little to make it more interesting (again affecting validity) or try to write it in a different way (affecting reliability). Utility very much depends on whether the woman likes to write and is prepared to continue with what may become an onerous task. Another aspect of utility relates to your own role as researcher. Will you have the time required to go through each journal and perform the kind of qualitative analysis described in Chapter 15?

Sensitivity has to do with the amount of detail included in the journal. To some degree this reflects completeness and is a validity issue, but small changes in women's experiences as the pregnancy progresses cannot be tracked unless the experiences are each described in some detail.

Journals are usually very reactive. Indeed, they are often used as therapeutic tools simply because the act of writing encourages the writer to reflect on what has been written, thus achieving deeper insights which may lead to behavior changes. Reactivity is not desirable in a measuring instrument. On the other hand, a qualitative study seeks to uncover not just the experiences themselves but the meaning attached to them by the research participant, and meaning may emerge more clearly if the participant is encouraged to reflect.

Researchers themselves keep journals while conducting interpretive studies. Journal keeping by the researcher is discussed under qualitative data analysis in Chapter 15.

Logs

A log is nothing more than a structured kind of journal, where the research participant is asked to record events related to particular experiences or behaviors in note form. Each note usually includes headings: the event itself, when and where it happened, and who was there. A log may be more valid than a journal in that the headings prompt the participant to include only relevant information with no discursive wanderings into cats or geraniums.

The log may be more reliable because it is more likely that a similar experience will be recorded in a similar way. It may be more useful because it takes less time for the participant to complete and less time for the researcher to analyze. It is usually less sensitive to small changes because it includes less detail, and it may be somewhat less reactive depending on the extent to which it leads to reflection and change.

Inventories

An inventory is a list made by the research participants. For example, the following is an inventory designed to measure depression:

List below the things that make you feel depressed.

_____

_____

This is valid to the degree that the list is complete, and sensitive in that the addition or omission of items over time are indicative of change. It is useful if the participant is prepared to complete it carefully and truthfully, it is probably fairly reactive in that it provokes thought; and it is reliable in that the same experience should always result in the same entries on the list.

Checklists

A checklist is a list prepared by the researcher. For example, a checklist designed to measure depression would include more items than shown but would follow the format:

Check below all the things that you have felt during the past week.
____ A wish to be alone
____ Sadness
____ Powerlessness
____ Anxiety

With respect to the five evaluative criteria presented in the previous chapter, the same considerations apply to a checklist as to an inventory except that validity may be compromised if the researcher does not include all the possibilities that are relevant to the participant in the context of the study.

Summative Instruments

On a general level, inventories and checklists ask for yes or no answers. In other words, they are *dichotomous* in that research participants can only respond in one of two ways: *yes*, this occurred or *no*, this did not occur. Summative measuring instruments provide a greater range of responses, usually asking how frequently or to what degree a particular item, or question, applies. For example, the depression checklist shown above may be presented in the form of a summated instrument, as follows:

Indicate how often you have experienced the following feelings by circling the appropriate number.

	Never	Rarely	Sometimes	Often
A wish to be alone	1	2	3	4
Sadness	1	2	3	4
Powerlessness	1	2	3	4
Anxiety	1	2	3	4

The words "never" and "often" are known as *anchors* and serve to describe the meanings attached to their respective values. Participants circle a number for each item and the scores are summed, or totaled. With only the four items shown (an actual depression scale would contain many more), the lowest possible score is 4 and the highest possible score is 16.

This instrument is an example of a *summative instrument*. A summative measuring instrument is any instrument that allows the researcher to derive a sum or total score from a number of items, or questions. Most, but not all, summated measuring instruments are designed so that low scores indicate a low level of the variable being measured (depression, in this case) and high scores indicate a high level.

STANDARDIZED MEASURING INSTRUMENTS

Standardized measuring instruments are used widely in social work because they have usually been extensively tested and they come complete with information on the results of that testing. Figure 6.1 in Chapter 6 is an excellent example of a summative standardized measuring instrument in that it provides information about itself in six areas: purpose, description, norms, scoring, reliability, and validity.

Purpose is a simple statement of what the instrument is designed to measure. *Description* provides particular features of the instrument, including its length and often its *clinical cutting score*. The clinical cutting score is different for every instrument and is the score that differentiates respondents with a clinically significant problem from respondents with no such problem. In Hudson's Index of Self-Esteem, for example, people who score above 30 (plus or minus 5 for error) have a clinically significant problem with self-esteem and people who score less than 30 do not.

The section on *norms* tells you who the instrument was validated on. The *Index of Self-Esteem*, for example (see NORMS in Figure 6.1), was tested on 1,745 respondents including single and married individuals, clinical and non-clinical populations, college students and nonstudents, Caucasians, Japanese and Chinese Americans, and a smaller number of other ethnic groups. It is important to know this because people with different characteristics tend to respond differently to the sort of items contained in Hudson's *Index of Self-Esteem*.

For instance, a woman from a culture that values modesty might be unwilling to answer that she feels she is a beautiful person all of the time (Item 3). She might not know what a wallflower is (Item 19) and she might be very eager to assert that she feels self-conscious with strangers (Item 16) because she thinks that women ought to feel that way. It is therefore very important to use any measuring instrument *only* with people who have the same characteristics as the people who participated in testing the instrument. As another example, instruments used with children must have been developed using children.

Scoring gives instructions about how to score the instrument. *Reliability* and *validity* we have discussed already. Summated standardized instruments are usually reliable, valid, sensitive, and non-reactive. It is therefore very tempting to believe that they must be useful, whatever the research situation.

More often than not, they *are* useful—provided that what the instrument measures and what the researcher *wants* to measure are the same thing. If you want to measure family coping, for example, and come across a wonderful standardized instrument designed to measure family cohesion, you must resist the temptation to convince yourself that family cohesion is what you really wanted to measure in the first place.

Locating Standardized Measuring Instruments

Once you know what variable you want to measure, the next consideration is locating appropriate standardized measuring instruments from which to choose. The two general sources for locating such instruments are commercial or professional publishers and the professional literature.

Publishers

Numerous commercial and professional publishing companies specialize in the production and sale of standardized measuring instruments for use in the social services. They can be easily found on the Web. The cost of instruments purchased from a publisher varies considerably, depending on the instrument, the number of copies needed, and the publisher.

The instruments generally are well developed and their psychometric properties are supported by the results of several research studies. Often they are accompanied by manuals that include the normative data for the instrument. As well, publishers are expected to comply with professional standards such as those established by the American Psychological Association. These standards apply to claims made about the instrument's rationale, development, psychometric properties, administration, and interpretation of results.

Standards for the use of some instruments have been developed to protect the interests of clients. Consequently, purchasers of instruments may be required to have certain qualifications, such as possession of an advanced degree in a relevant field. A few publishers require membership in particular professional organizations. Most publishers will, however, accept an order from a social work student if it is co-signed by a qualified person, such as an instructor, who will supervise the use of the instrument.

Professional Books and Journals

Standardized measuring instruments are most commonly described in human service journals. The instruments usually are supported by evidence of their validity and reliability, although they often require cross-validation and normative data from more representative samples and sub-samples. More often than not, however, the complete instrument cannot be seen in the articles that describe them. They usually contain a few items that can be found in the actual instrument.

Locating instruments in journals or books is not easy. Of the two most common methods, computer searches of data banks and manual searches of the literature, the former is faster, unbelievably more thorough, and easier to use. Unfortunately, financial support for the development of comprehensive data banks has been limited and intermittent.

Another disadvantage is that many articles on instruments are not referenced with the appropriate indicators for computer retrieval. These limitations are being overcome by the changing technology of computers and information retrieval systems. Several services now allow for a complex breakdown of measurement need; data banks that include references from over 1,300 journals, updated monthly, are now available from a division of *Psychological Abstracts Information Services* and from *Bibliographic Retrieval Services*.

Nevertheless, most social workers will probably rely on manual searches of references such as *Psychological Abstracts*. Although the reference indices will be the same as those in the data banks accessible by computer, the literature search can be supplemented with appropriate seminal (original) reference volumes.

SUMMARY

In this chapter we have briefly looked at measurement and measuring instruments. The next chapter discusses how to select research participants to include in your study. You will then administer your measuring instruments to your selected research participants.

REFERENCES AND FURTHER READINGS

Bloom, M., Fischer, J., & Orme, J. (2009). *Evaluating practice: Guidelines for the accountable professional* (6th ed.). Englewood Cliffs, NJ: Prentice-Hall.

Fischer, J., & Corcoran, K. (2007). *Measures for clinical practice* (3rd ed.). *Volume 1: Couples, families, and children.* Volume 2: *Adults.* New York: Oxford University Press.

Hudson, W.W. (1982). *The clinical measurement package: A Field Manual.* Belmont, CA: Wadsworth.

Nurius, P.S., & Hudson, W.W. (1993). *Human services: Practice, evaluation, and computers.* Pacific Grove, CA: Brooks/Cole.

Jordan, C., & Franklin, C. (2003). *Clinical assessment for social workers: Quantitative and qualitative methods* (2nd ed.). Chicago: Lyceum Books.

Jordan, C., Franklin, C., & Corcoran, K. (1993). Standardized measuring instruments. In R.M. Grinnell, Jr. (Ed.), *Social work research and evaluation* (4th ed., pp. 198–220). Itasca, IL: F.E. Peacock.

Jordan, C., Franklin, C., & Corcoran, K. (2001). Measuring instruments. In R.M. Grinnell, Jr. (Ed.), *Social work research and evaluation: Quantitative and qualitative approaches* (6th ed., pp. 151–180). Itasca, IL: F.E. Peacock.

Salkind, N.J. (2006). *Tests and measurement for people who (think they) hate tests and measurements.* Thousand Oaks, CA: Sage.

Check out our Website for useful links and chapters (PDF) on:
- Finding measuring instruments on the Web
  Go to: www.pairbondpublications.com
  Click on: Student Resources
             Chapter-by-Chapter Resources
             Chapter 7

Check Out Our Website

PairBondPublications.com

√ *Student Workbook Exercises*
√ *Chapter Power Point Slides*
√ *On-line Glossaries*
√ *General Links*
√ *Specific Chapter Links*

Part IV

Sampling and the Logic of Research Design

Chapter 8: Selecting Research Participants   152

Chapter 9: Case-Level Research Designs   166

Chapter 10: Group-Level Research Designs   186

Chapter 8

Selecting Research Participants

DEFINING A POPULATION
SAMPLING PROCEDURES
    Probability Sampling
        Simple Random Sampling
        Systematic Random Sampling
        Stratified Random Sampling
        Cluster Random Sampling
    Nonprobability Sampling
        Availability Sampling
        Purposive Sampling
        Quota Sampling
        Snowball Sampling
SAMPLE SIZE
SUMMARY
REFERENCES AND FURTHER READING

8

IN THE LAST CHAPTER measuring instruments were presented—how to select them, where to find them, and how to evaluate them. This chapter discusses who will complete them. In short, we discuss how to select the actual people, or research participants, who will take part in our research study.

Let's continue with the previous example that was used in the last chapter where we are trying to discover whether there is a relationship between two variables—depression and sleep patterns. At this point, it is obviously necessary to find a few people so we can measure their depression levels (Variable 1) and sleep patterns (Variable 2).

One way of doing this is to phone our trusty social worker friend, Ken, who works with people who are depressed and ask if he will please find us some clients, or research participants, who do not mind sleeping in a laboratory once a week with wires attached to their heads. "Certainly," says Ken crisply. "How many do you want?" "Er . . ." we say, . . . "how many have you got?" "Oh, lots," says Ken with cheer. "Now, do you want them young or old, male or female, rich or poor, acute or chronic, severely or mildly depressed . . . ?" "Oh and by the way," he adds, "all of them are refugees from Outer Ganglinshan."

We decide that we need to think about this some more. Suppose that the sleep patterns of Outer Ganglinshanians who are depressed, for example, are different from the sleep patterns of Bostonians who are depressed.

Suppose even further, that males who are depressed have different sleep patterns from females who are depressed, adolescents who are depressed from seniors who are depressed, or Anabaptists who are depressed from Theosophists who are depressed. The list of possibilities is endless. We had hoped to find out only if there was a relationship between sleep patterns and depressed *people.*

One solution, of course, would be to assemble all the people who are depressed in the Western Hemisphere in the hope that Anabaptists and Theosophists and so forth were properly represented. This leaves out the Eastern Hemisphere, but we have to eliminate something since our sleep laboratory has only ten beds.

There is another difficulty too. Not only are we restricted to ten sleeping people but, on our limited budget, we will be hard pressed to pay them bus fare from their homes to our lab. This being the case, we have to forget about the Western Hemisphere and concentrate on the area served by the city transit, within which there are probably very few Anabaptists who are depressed and no Theosophists at all to speak of. How are we going to select ten people from this small area and still be fair to the Theosophist population? The answer is simple: We do this through the use of sampling frames.

DEFINING A POPULATION

A *population* is the totality of persons or objects with which our research study is concerned. If, for example, we are interested in all the people who are depressed in the Western Hemisphere, then our population is all the people who are depressed in the Western Hemisphere. If we decide to restrict our study to all the people who are depressed in the area served by city transit, then our population is all the people who are depressed in the area served by city transit. However we decide to define our population, the results of our study will apply only to that population from which our sample was drawn.

We cannot do our study within the confines of city transit and then apply the results to the whole of the Western Hemisphere. When our population has been defined, the next step is to make a list of all the people (or other units such as case files) included in that population.

Such a list is called a *sampling frame.* For example, if our population is to be all the people who are depressed in the area served by city transit, then we must create a list, or sampling frame, of all the people who are depressed in the area served by city transit. Obtaining a sampling frame is often one of the hardest parts of a research study.

We probably will not be able to generate a list of all the people who are depressed in the area served by city transit since there is not a single person or social service program who knows about all of them. It might be better to restrict our study to all the depressed people treated by Ken—or some other source of people who are depressed who will provide us with a list.

Suppose, then, that we have decided to restrict our study to all the people who are depressed treated by Ken. We may think that "all the people who are depressed treated by Ken" is a reasonable definition of our population. But a population has to be defined exactly. "All the people who are depressed treated by Ken" is possibly not very exact, especially when we consider that Ken has been treating them for over twenty years.

The ones treated eighteen and nineteen years ago have doubtless disappeared by now over the far horizon; indeed, the fates of those treated a mere year ago might be equally veiled in mystery. Perhaps we should redefine our population as "all the people who are depressed treated by Ken over the last three months" or; better yet, as "all the people who are depressed treated by Ken from January 1, 2010 to March 31, 2010."

SAMPLING PROCEDURES

If we can make a 3-month list of all the people who are depressed treated by Ken from January 1 to March 31, 2010, we have our sampling frame. The next thing is to use it to select our sample. This process is called *sampling* and the people (or other units such as case files) picked out make up a *sample*. There are two main ways of selecting samples: (1) *probability* sampling, which requires a sampling frame, and (2) *non-probability* sampling, which does not require a sampling frame.

Probability Sampling

The first main category of sampling procedures is probability sampling. This method of sampling is used more in positivist studies than interpretive ones. A *probability sample* is one in which all the people (or units in the sampling frame) have the same known probability of being selected for the sample. By probability, we mean chance, such as the probability of winning a lottery. The selection is based on some form of random procedure of which there are four main types: (1) simple random sampling, (2) systematic random sampling, (3) stratified random sampling, and (4) cluster random sampling.

Simple Random Sampling

The first type of probability sampling is simple random sampling. Suppose now that there are 100 names in our sampling frame; that is, Ken has seen 100 different clients who are depressed from January 1, 2010 to March 31, 2010. We assign each one of these clients a number; the first one on the list will be 001, the second 002, and so on until 100 is reached. Then, we take a book of random numbers, open the book at random, and pick a digit on the page also at random. The first half page of such a table is shown in Table 8.1.

Suppose the digit we happen to pick on is 1, the second digit in the number in the sixth row from the bottom in the second column from the left. (That whole number is 81864 and is highlighted in **bold**.) The two digits immediately to the right of 1 are 8 and 6; thus, we have 186. We take three digits in total because 100, the highest number on our sampling frame, has three digits. The number 186 is more than 100 so we ignore it.

TABLE 8.1
Partial Page of a Random Numbers Table

02584	75844	50162	44269	76402	33228	96152	76777
66791	44653	90947	61934	79627	81621	74744	98758
44306	88222	30967	57776	90533	01276	30525	66914
01471	15131	38577	03362	54825	27705	60680	97083
65995	81864	19184	61585	19111	08641	47653	27267
45567	79547	89025	70767	25307	33151	00375	17564
27340	30215	28376	47390	11039	39458	67489	48547
02584	75844	56012	44269	76402	33228	96152	76777
66791	44653	90497	61934	79627	81621	74744	98758
44306	80722	30317	57776	90533	01276	30525	66914
65995	**81864**	19184	61585	19131	08641	47653	27267
45567	79547	89025	70767	25307	33151	00375	17564
27340	30215	23456	47390	11039	39458	67489	48547
02471	10721	30577	03362	54825	27705	60680	97083
60791	40453	90227	61934	79627	81621	74744	98758
43316	87212	36967	57576	90533	01276	30525	66914

Going down the column, 954 is also more than 100 so we ignore it also. The next one, 021, is less than 100, so we can say that we have selected number 021 on our sampling frame to take part in our study. After 072 and 045 there are no more numbers less than 100 in the second column (middle three digits) so we go to the third column (middle three digits). Here we discover seven more people (i.e., 016, 094, 096, 049, 031, 057, 022).

We go down the columns, picking out numbers, until it occurs to us that we do not really know how many numbers should be selected. Our sleep laboratory has accommodation for ten but then, if we do a different ten each night of the week, that is seventy. Sundays off, to keep us sane, is sixty. Union hours, and we are looking at closer to thirty-five. There has to be a better way than union hours to figure out sample size; but, before looking at sample size, we should examine three more probability sampling procedures.

Systematic Random Sampling

The second type of probability sampling is systematic random sampling. Here, the size of our population is divided by the desired sample size to give us our sampling interval. To state it more simply, if we only want half the population to be in the sample, we select every other person. If only a third of the population is needed to be in the sample, we pick every third person; a quarter, every fourth person; a fifth, every fifth person; and so on.

The problem with this is that we need to know our sample size in order to do it. We will not really know what our sample size ought to be until later in this chapter. At this point, we can only make a guess. Let us assume that we will work at our sleep lab every weeknight. Ten sleepers every night for five nights gives us a sample size of five nights times ten sleepers which equals fifty. The idea of a sampling interval can be expressed mathematically in the following way: Suppose that our population size is 100 and we have set our sample size—the number of people taking part in our study—at fifty. Then, dividing the former by the later (100/50 = 2) provides the size of the sampling interval, which in this case is two.

If our sample was only going to be one-fourth of the population instead of one-half, our sampling interval would be four instead of two. We might start at the fourth person on our sampling frame and pick out, as well, the eighth, twelfth, sixteenth persons, and so on.

The problem with this method is that everyone does not have the same chance of being selected. If we are selecting every other person, starting with the second person, we select the fourth, sixth, and so on. But this means that the third and the fifth never get a chance to be chosen. This procedure introduces a potential bias that calls for caution.

Suppose, for example, we have applied for a credit card and the credit card company examines our bank account every thirtieth day. Suppose, further, that the thirtieth day falls regularly on the day after we pay this month's rent and on the day before we receive last month's paycheck. In other words, our bank account, while miserable at all times, is particularly low every thirtieth day. This is hardly fair because we never get a chance to show them how rich we are on days other than the thirtieth.

If we do not have to worry about this sort of bias, then a systematic sample is largely the same as a simple random sample. The selection is just a bit easier because we do not have to bother with random numbers as presented in Table 8.1.

Stratified Random Sampling

The third type of probability sampling is stratified random sampling. If, for example, our study of depression and sleeping patterns is also concerned with religious affiliation, we can look at our population and count how many of them are Christians, Jews, Muslims, Buddhists, Hindus, Theosophists, and so forth. Suppose that, in our population of 100, we found 40 Jews, 19 Christians, 10 Muslims, 10 Buddhists, 10 Hindus, 10 Sikhs, and 1 lone Theosophist.

Now, we can sample our religious categories, or strata, either proportionally or disproportionally. If we sample our population proportionally, we will choose, say, one-tenth of each category to make up our sample; that is, we will randomly select 4 Jews, 1.9 Christians, 1

Muslim, 1 Buddhist, 1 Hindu, 1 Sikh, and .1 of a Theosophist. This comprises a total sample of 10, as illustrated in Table 8.2.

However; the 1.9 Christians and the .1 of a Theosophist present a difficulty. In this case, it is necessary to sample disproportionately. For example, it may be preferable to choose one member of each religious affiliation for a total sample of seven. In this case, the sampling fraction is not the same for each category; it is 1/40 for Jews, 1/19 for Christians, 1/10 for Muslims, Buddhists, Hindus, and Sikhs, and 1 for Theosophists. A total sample of only seven is not a good idea for reasons that will be discussed shortly.

Our population is divided into religious categories only if we believe that religious affiliation will affect either depression or sleep patterns.; that is, if we really believe that Buddhists who are depressed have different sleep patterns than Muslims who are depressed, everything else being equal. However, we must admit that we do not believe this.

There is nothing in the literature or our past experience to indicate anything of the sort. And anyway, our friend's clients who are depressed, from whom the sample will be drawn, will not include anything so interesting as Buddhists and Muslims. Probably the best we can hope for is a few odd sects and the town atheist.

This method, though, could be used to look at the sleep patterns of different age groups. Our population could be divided quite sensibly into eight categories: those aged 10 to 20, 21 to 30, 31 to 40, and so on until we reach 81 to 90. We might then sample proportionately by randomly selecting one-tenth of the people in each category. Or we might sample disproportionately by selecting, say, six people from each category, regardless of the number of people in the category.

It might be preferable to sample disproportionately, for example, because there are a small number of people in the 71 to 80 and 81 to 90 categories, but it is our belief that advanced age significantly affects sleep patterns. Therefore, we want to include in our sample more than the one or two elderly people who would be included if we took one-tenth of each category.

Categorizing people in terms of age is fairly straightforward. They have only one age and they usually know what it is. Other types of categories, though, are more complex. Psychological labels, for instance, can be uncertain so that people fall into more than one category and the categories themselves are not homogeneous; that is, the categories are not made up of people who are all alike.

There is no point in using stratified random sampling unless the categories are both homogeneous and different from each other. In theory, the more homogeneous the categories are, the fewer people will be needed from each category to make up our sample. Suppose, for example, we had invented robots that were designed to perform

TABLE 8.2
Stratified Random Sample Example

Category	Number	1/10 Proportionate Sample	Number and (Disproportionate Sampling Fractions) for a Sample of 1 per Category
Jews	40	4.0	1 (1/40)
Christians	19	1.9	1 (1/19)
Muslims	10	1.0	1 (1/10)
Buddhists	10	1.0	1 (1/10)
Hindus	10	1.0	1 (1/10)
Sikhs	10	1.0	1 (1/10)
Theosophists	1	.1	1 (1)
Totals....	100	10.0	7

various tasks. Those robots designed to be electricians were all identical and so were all the plumbers, doctors, lawyers, and so forth. Of course, each kind of robot was different from every other kind.

If we then wanted to make some comparison between all lawyer and all doctor robots, we would only need one of each since same-kind robots were all the same; or, in other words, the robot categories were completely homogeneous. People categories are never completely homogeneous, but the more homogeneous they are, the fewer we need from each category to make comparisons. The fewer we need, the less our study will cost.

However, we must take care not to spend the money we save in this manner on the process of categorization. If people already are in categories, as they might be in a hospital, very good. If they are easy to categorize, say by age, very good too. If they are not already categorized and are difficult to categorize, it might be more appropriate to use another sampling method.

There is one more point to be considered. The more variables we are looking at, the harder it is to create homogeneous strata. It is easy enough, or example, to categorize people as Buddhists or Hindus, or as aged between 21 and 30 or 31 and 40. It is not nearly so easy if they have to be, say, between 21 and 30 *and* Buddhist.

Cluster Random Sampling

The fourth type of probability sampling is cluster random sampling. This is useful if there is difficulty creating our sampling frame. Suppose, for a moment, we want to survey all the people in the area served by city transit to see if they are satisfied with the transit system.

We do not have a list of all these people to provide a sampling frame. There is a list, however, of all the communities in the city served by the transit system and we can use this list as an alternative sampling frame.

First, we randomly select a community, or cluster, and survey every person living there. We will be certain what this community thinks about the transit system as we talked to every member of it. But there is still the possibility that this community, in which there are a large number of families with children, has a different opinion from that of a second community, which consists largely of senior citizens.

Perhaps we ought to survey the second community as well. Then there is a third community inhabited largely by penniless writers, artists, and social work students who might have yet a different opinion. Each community is reasonably homogeneous; that is, the people in the community are very much like each other. But the communities themselves are totally unlike; that is, they are heterogeneous with respect to each other.

One of the problems here is that we may not be able to afford to survey everyone in the three clusters. But we could compromise. Perhaps we could survey not every street in each cluster but only some streets taken at random—and not every house on our chosen streets, but only some houses, also taken at random. This way, we survey more clusters but fewer people in each cluster.

The only difficulty here is that, since we are not surveying everyone in the community, we might happen to select people who do not give us a true picture of that community. In a community of people with small children, for example, we might randomly select a couple who does not have children and does not intend to, and whose home was demolished, moreover, to make way for the transit line. Such untypical people introduce an error into our results that is due to our sampling procedure and is therefore called a sampling error.

We will now turn our attention to the next category of sampling procedures—nonprobability sampling.

Nonprobability Sampling

In nonprobability sampling, not all the people in the population have the same probability of being included in the sample and, for each one of them, the probability of inclusion is unknown. This form of sampling is often used in exploratory studies where the purpose of the study is just to collect as much data as possible. There are four types of nonprobability sampling procedures: (1) availability, (2) quota, (3) purposive, and (4) snowball.

TABLE 8.3
Quota Sampling Matrix of Body Weight by Age

	Body Weight	
Age	Obese	Non-obese
Young	Category A Obese/Young	Category B Non-obese/Young
Old	Category C Obese/Old	Category D Non-obese/Old

Availability Sampling

The first type of nonprobability sampling is availability sampling. It is the simplest of the four nonprobability sampling procedures and is sometimes called *accidental sampling*. As its name suggests, it involves selecting for our sample the first people or units who make themselves available to us. We might survey people, for example, who pass us in a shopping mall. Or we might base our study on the caseload of a particular social worker. Or we might just seize upon the first fifty of Ken's clients who are depressed who agree to sleep in our lab with wires attached to their heads.

Purposive Sampling

The second type of nonprobability sampling is purposive sampling. This type is used when we want to purposely choose a particular sample. For example, if we are testing a questionnaire that must be comprehensible to less well-educated people while not offending the intelligence of better-educated people, we might present it both to doctoral candidates and to people who left school cheerfully as soon as they could.

In other words, we purposely choose the doctoral candidates and the happy school leavers to be in our two subsamples. There is nothing random about it. In the same way, if we know from previous studies that there is more family violence in the city than in the country, we might restrict our sample to those families who live in cities. Purposive sampling is used a lot in qualitative research studies.

Quota Sampling

The third type of nonprobability sampling is quota sampling. In this type of sampling, we decide, on the basis of theory, that we should include in our sample so many of a certain type of person. Suppose we wanted to relate the sleep patterns of people who are depressed to body weight and age. We might decide to look at extremes; for example, obese and young, non-obese and young, obese and old, and non-obese and old. This gives us four categories (i.e., A, B, C, D) as illustrated in Table 8.3.

We might want, for example, fifteen people in each category or, for some reason, we might need more elderly people than young. In this case, we may decide on ten people in each of Categories A and B and twenty people in each of Categories C and D. Whatever quotas we decide on, we only have to find enough people to fill them who satisfy the two conditions of age and weight. We might discover all the obese people at a weight loss clinic, for example, or all the young, non-obese ones at a clinic for anorexics. It does not matter where or how we find them so long as we do.

Snowball Sampling

The fourth type of nonprobability sampling is snowball sampling. If a follow-up study is to be conducted on a self-help group that broke up two years ago, for example, we might find one member of the group and ask that member to help us locate other members. The other members will then find other members and so on until the whole group has been located. The process is a bit like telling one person a secret, with strict instructions not to tell it to anyone else. Like purposive sampling, snowball sampling is commonly used in qualitative research studies.

SAMPLE SIZE

Before our sample can be selected, we obviously have to decide on how many people are needed to take part in our study; in other words, we have to decide on our sample size. The correct sample size depends on both our population and research question. If we are dealing with a limited population, for example, such as the victims of some rare disease, we might include the whole population in our study. Then, we would not take a sample. Usually, however, the population is large enough that we do need to take a sample, the general rule being, the larger the sample, the better.

As far as a minimal sample size is concerned, experts differ. Some say that a sample of 30 will allow us to perform basic statistical procedures, while others would advise a minimum sample size of 100. In fact, sample size depends on how homogeneous our population is with respect to the variables we are studying.

Recalling all those categories of robot doctors, lawyers, and so forth considered awhile back, if our population of robot doctors is all the same, that is, homogeneous, we only need one robot doctor in our sample. On the other hand, if the factory messed up and our robot doctors have emerged with a wide range of medical skills, we will need a large sample to tell us anything about the medical skills of the entire robot doctor population. In this case, of course, medical skill is the variable we are studying.

Sample size must also be considered in relation to the number of categories required. If the sample size is too small, there may be only one or two people in a particular category; for example 1.9 Christians and .1 of a Theosophist. This should be anticipated and the sample size adjusted. The situation can also be handled using the disproportionate stratified random sampling procedure. This procedure is the one where we selected one person from each religious category, thus neatly avoiding our 1.9 Christians and .1 of a Theosophist.

There are many formulas available for calculating sample size but they are complicated and difficult to use. Usually, a sample size of one-tenth of the population is considered sufficient to provide reasonable control over sampling error. The same one-tenth convention also applies to categories of the population; we can include one-tenth of each category in our sample.

SUMMARY

In the course of a research study, the entire population that we are interested in is usually too big to work with and, in any case, we rarely have enough money to include everyone. For these reasons, it is necessary to draw a sample. The idea is that the sample represents the population from which it was drawn; that is, the sample is identical to the population with respect to every variable that we are interested in.

When the sample is truly representative of the population, we can generalize the results received from the sample back to the population from which it was drawn. No sample is ever totally representative of the population from which it was drawn. Now that we know how to draw a sample from a population, we will turn our attention to the use of research designs that use samples—the topic of the next chapter.

REFERENCES AND FURTHER READING

Dattalo, P. (2008). *Sample size: Balancing power, precision, and practicality.* New York: Oxford University Press.

Hektner, J.M., Schmidt, J.A., & Csikszentmihalyi, M. (2007). *Experience sampling method: Measuring the quality of everyday life.* Thousand Oaks, CA: Sage.

Schutt, R.K. (2008). Sampling. In R.M. Grinnell, Jr., & Y.A. Unrau (Eds.), *Social work research and evaluation: Foundations of evidence-based practice* (8th ed., pp. 135–156). New York: Oxford University Press.

Weinbach, R.W., & Grinnell, R.M., Jr. (2010). *Statistics for social workers* (8th ed.). Boston: Allyn & Bacon.

Check out our Website for useful links and chapters (PDF) on:
- introductory sampling concepts
- Web-based sampling guides
  Go to: www.pairbondpublications.com
  Click on: Student Resources
  　　　　　Chapter-by-Chapter Resources
  　　　　　Chapter 8

Check Out Our Website

PairBondPublications.com

√ *Student Workbook Exercises*
√ *Chapter Power Point Slides*
√ *On-line Glossaries*
√ *General Links*
√ *Specific Chapter Links*

Chapter 9

Case-Level Research Designs

CASE-LEVEL RESEARCH DESIGNS
EXPLORATORY CASE-LEVEL DESIGNS
    A Design
    B Design
        BB₁ Design
        BC Design
DESCRIPTIVE CASE-LEVEL DESIGNS
    AB Design
    ABC and *ABCD* Designs
EXPLANATORY CASE-LEVEL DESIGNS
    Reversal Designs
        ABA and *ABAB* Designs
        BAB Design
        BCBC Design
    Multiple-Baseline Designs
        More than One Case
        More than One Setting
        More than One Problem
SUMMARY
REFERENCES AND FURTHER READING

9

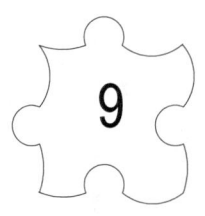

IN THE LAST CHAPTER WE DISCUSSED how to select research participants for research studies. This chapter is a logical extension of the pervious one in that we now look at the different ways of designing or setting up our studies in which research participants are included.

On a basic level, a research design is essentially a plan for conducting the entire research study from beginning to end. All research plans—interpretive or positivistic—are formulated in order to answer the following basic questions:

- When, or over what period, should the research study be conducted?
- What variables need to be measured?
- How should the variables be measured?
- What other variables need to be accounted for or controlled?
- From whom should the data be collected?
- How should the data be collected?
- How should the data be analyzed?
- How should the results of the study be disseminated?

As you should know by now, the above questions are highly interrelated and are directly related to the question we are trying to answer or the hypothesis we are testing.

If you are exploring the concept of bereavement, for example, you will collect data from bereaved people and perhaps involved social workers, and you will need to measure variables related to bereavement, such as grief, anger, depression, and levels of coping. You might need to measure these variables over a period of months or years, and the way you measure them will suggest appropriate methods of how you will analyze your data. Decisions about how best to accomplish these steps depends on how much we already know about the bereavement process: that is, where your bereavement research questions fall on the knowledge level continuum as presented in Figure 2.3.

Over the years, researchers have developed a kind of shorthand, representing research designs in terms of letters and numbers, dividing them into categories, and giving them names. Let us now look at how this research shorthand is applied to research questions that can be loosely classified as case-level research designs.

CASE-LEVEL RESEARCH DESIGNS

Some research studies are conducted in order to study one individual, or case, some to study groups of people (including families, organizations, and communities) and some to study social artifacts (such things as birth practices or divorces). The individual, group, or artifact being studied is called the *unit of analysis*.

If you are exploring the advantages and disadvantages of home birth, for example, you might be asking questions from women who have experienced homebirth, but the thing you are studying—the *unit of analysis*—is the social artifact, home birth. Conversely, if you are a social work practitioner, studying the impact of home birth on a particular client, the unit of analysis is the client or individual; and if you are studying the impact on a group of women, the unit of analysis is the group of women.

On a very general level, case-level designs are more "practice orientated" than group-level designs. That is, they are used more by social work "practitioners" than by social work "researchers." Research studies conducted to evaluate treatment interventions with social work clients are called case-level designs. They are also called *single-subject designs, single-case experimentations*, or *idiographic research*.

They are used to fulfill the major purpose of social work practice: to improve the situation of a client system—*an* individual client, *a* couple, *a* family, *a* group, *an* organization, or *a* community. Any of these client configurations can be studied with a case-level design. In short, they are used to study *one* individual or *one* group intensively, as opposed to studies that use two or more groups of research participants.

Case-level designs can provide information about how well a treatment intervention is working, so that alternative or complementary interventive strategies can be adopted if necessary. They can also indicate when a client's problem has been resolved. Single-case studies can be used to monitor client progress up to, and sometimes beyond, the point of termination.

They can also be used to evaluate the effectiveness of a social work program as a whole by aggregating or compiling the results obtained by numerous social workers serving their individual clients within the program. A family therapy program might be evaluated, for example, by combining family outcomes on a number of families that have been seen by different social workers.

A *case-level* design is represented in terms of the letters *A, B, C,* and *D*. Let us see how this works, first in relation to exploratory research questions (see Figure 2.3).

EXPLORATORY CASE-LEVEL DESIGNS

Suppose you have a client—Cecilia—whose underlying problem, you believe, is her high anxiety level. She will be the "case" in your case-level design. Before you go ahead with an intervention designed to decrease her social anxiety—which Cecilia doesn't need your intervention if your belief is wrong—you will have to answer the simple question, "Does Cecilia really have a clinically significant problem with anxiety?" In other words, does the "anxiety problem" you think she has in fact really exists in the first place?

In order to answer this question, you select a measuring instrument to measure the variable, anxiety, that is valid, reliable, sensitive, non-reactive, and useful in this particular situation. Say you choose the Interaction and Audience Anxiousness Scale (*IASS*). On this particular measuring instrument higher scores indicate higher social anxiety levels and the clinical cutting score is 40. You administer it to Cecilia and she scores 62 as shown in Figure 9.1.

Scores above the clinical cutting score indicate a clinically significant problem. You might think "Ah-Hah! She has a problem since her score was 22 points higher than the "minimum clinical cutting score" and rush forward with your intervention.

On the other hand, any social work intervention has the potential to harm as well as help (in the same way that any medication does) and you first want to be sure that this is a persisting problem and not just a reflection of Cecilia's high anxiety today. You might also want to be sure that her high anxiety problem will not go away by itself. Doctors usually do not treat conditions that resolve themselves, given time, and the same is true for social workers.

In order to see if Cecilia meets these two criteria for treatment—first, the problem is persisting and, second, it is either stable at an unacceptable level or getting worse—you will need to administer the same measuring instrument two or three times more at intervals of, say, a week. You might then graph your results as shown in Figure 9.1. This figure constitutes a baseline measure of Cecilia's social anxiety level over a seven-week period and is a very simple example of an *A design*.

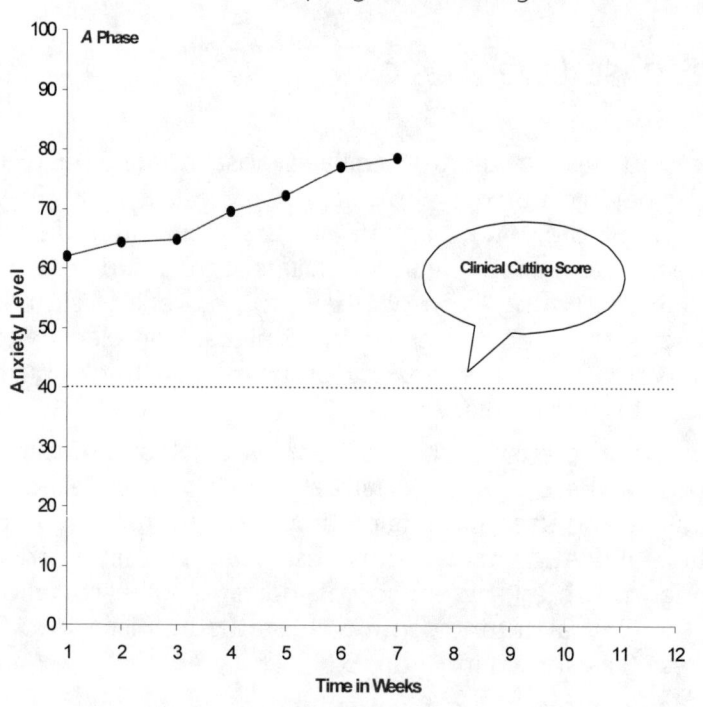

FIGURE 9.1
A Design: Cecilia's Anxiety Scores for the First Seven Weeks
(Without Intervention)

A Design

With the risk of sounding a bit ridiculous, the letter *A* simply designates "a research study" where the intention is to establish, via measurement, a baseline for an individual client's problem. Perhaps "research study" is a grandiose term to describe a routine assessment but the word *research* does mean *to look again* and you are indeed looking again at Cecilia's potential problem in order to see if her problem exists in the first place.

Three data points are the minimum number needed to show any kind of trend and some experts maintain that you need no less than seven. However, in a clinical situation the client's need for intervention is the primary factor and you will have to use your judgment to decide how long you ought to continue to gather baseline data before you intervene.

Figure 9.1 above indicates a worsening problem, needing intervention, since the scores are generally getting higher as time goes on. Remember, higher scores mean higher levels of the problem. Had the scores been generally getting lower, the problem would have been improving by itself and no intervention would be indicated. If the scores fell more or less on a horizontal line, intervention would be indicated if the line was above the clinical cutting score and not if the line was below it.

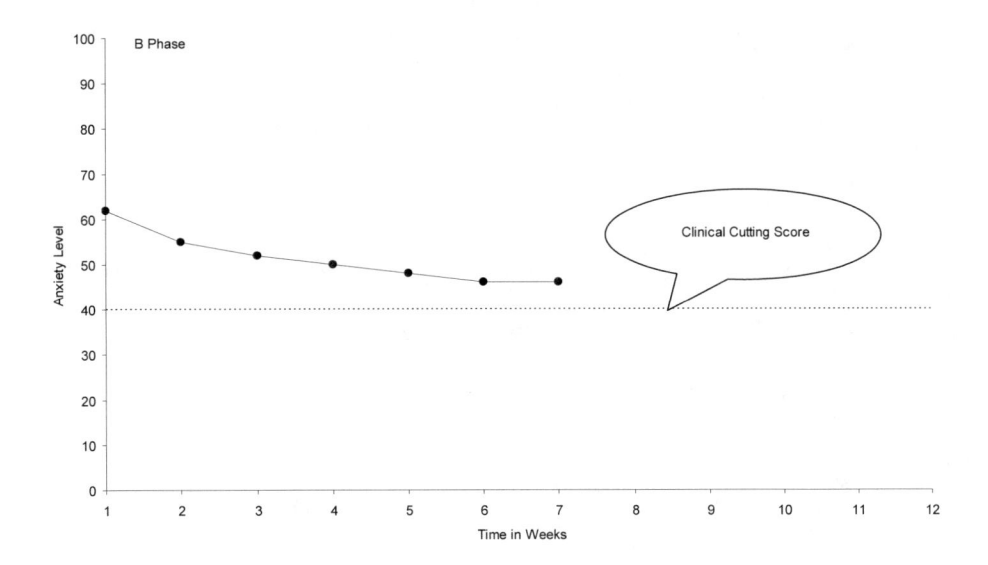

FIGURE 9.2
B Design: Bob's Anxiety Scores for the First Seven Weeks
(With Intervention)

B Design

The second type of exploratory case-level research design is the *B* design. As we have seen, an *A* design answers the question "Does the problem exist?" The *A* design also answers another type of exploratory question, "Does the problem exist at different levels over time?" In other words, "Is the problem changing *by itself*?"

A *B* design also addresses the question, "Is the problem changing?" but here we want to know whether the problem is changing *while an intervention is being applied*. Bob is Cecilia's friend and has also come to you complaining that he experiences a great deal of anxiety in social situations. He is nervous when he speaks to his boss or when he meets people for the first time and the prospect of giving public presentations at work appalls him.

You decide that you will measure Bob's anxiety level using the same standardized measuring instrument as you did with Cecilia (Interaction and Audience Anxiousness Scale, *IASS*). As you know about this particular standardized measuring instrument, higher scores indicate higher anxiety levels and the clinical cutting score for the *IASS* is 40. Bob, like Cecilia scores 62. This one score is more of a base point rather than a baseline but you decide that it would be inappropriate to collect baseline data over time in Bob's case as he is experiencing a great deal of discomfort at work, is highly nervous in your presence (you are a stranger, after all), and probably will not be able to bring himself to seek help in the future if he does not receive some kind of intervention now.

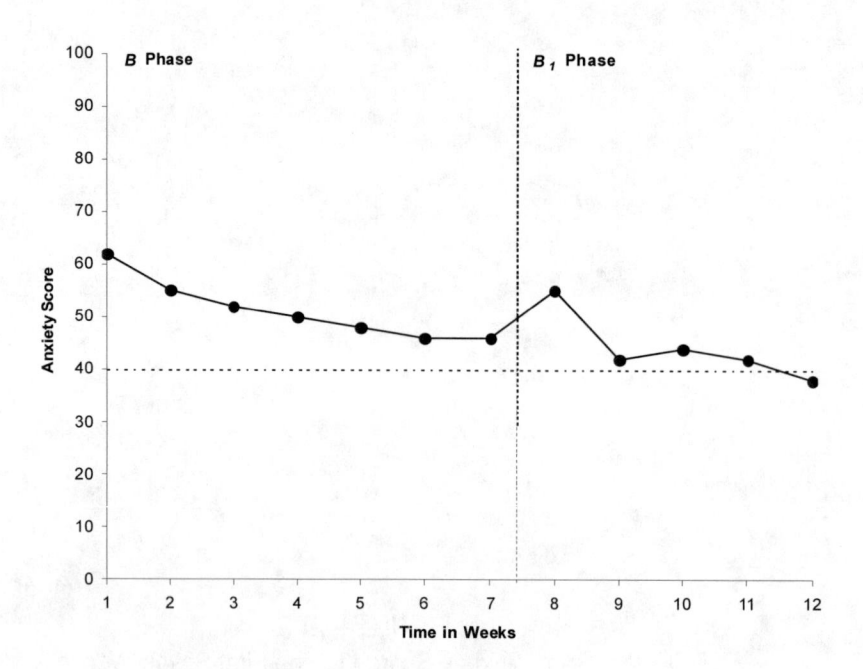

FIGURE 9.3
BB₁ Design: Bob's Anxiety Scores for *B* Phase and *B₁* Phase

You therefore begin your intervention, engaging Bob to the extent that he returns the following week, when you administer the *IASS* again. Now he scores 55. In the third week, he scores 52, as shown in Figure 9.2. On the other hand, you might make the clinical decision that it is worth trying a variation on your intervention: you might apply it more *frequently* by having Bob come twice a week instead of once, or you might apply it more *intensively* by increasing the amount of time Bob is expected to spend each evening on relaxation exercises. You can graph the changes that occur while you are applying the variation, as shown in Figure 9.3.

Figure 9.2 above is a simple example of a *B* design, in which you track change in the problem level at the same time as you are intervening. You do not know, from this graph, whether your intervention *caused* the change you see. Anything else could have caused it. Perhaps Bob is having a weekly massage to reduce muscle tension, or his boss has been fired, or the public presentation that he was suppose to do has been postponed.

Therefore, you cannot use the *B* design to answer explanatory research questions which come quite high on the knowledge continuum (refer to Figure 2.3).

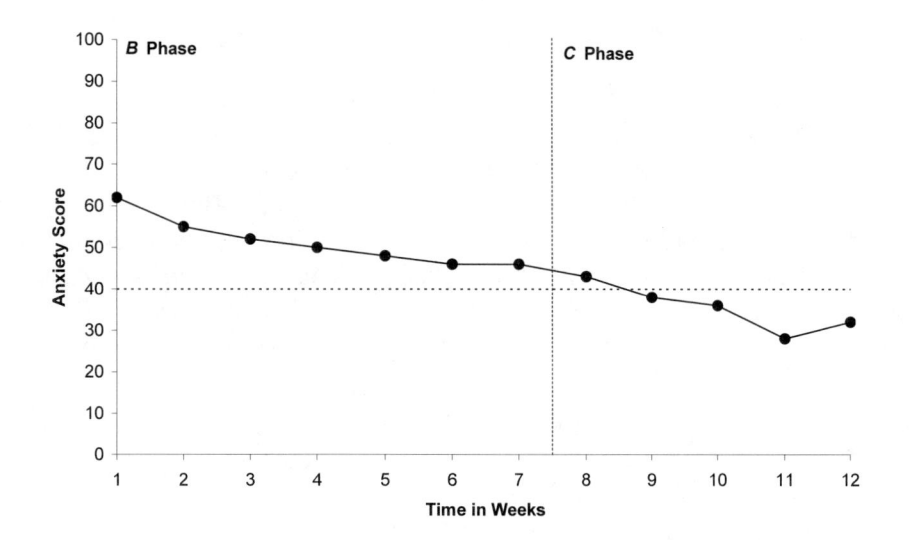

FIGURE 9.4
BC Design: Bob's Anxiety Scores for B Phase and C Phase

BB₁ Design

Figure 9.2 shows that Bob's anxiety level has improved but it has not fallen below the clinical cutting score. Moreover, it does not look as though it will because it has been relatively stable around 46 for the last three weeks. It may be that Bob is a naturally anxious person and no intervention, however inspired, will reduce his problem to below clinically significant levels.

Figure 9.3 shows two *phases*: the B phase and the B₁ phase, separated by the vertical dotted line that runs between Weeks 7 and 8. Week 7 marks the end of your original intervention, designated by the B intervention, and all the scores obtained by Bob while you were applying the B intervention constitute the B phase. (If it seems odd to call the first intervention B instead of A, remember that A has been used already to designate baseline scores.) Week 8 marks the beginning of the variation on your original intervention, designated B₁, and all the scores obtained by Bob while you were applying the variation constitute the B₁ phase. The B and B₁ phases together constitute the *BB₁ design*.

When you look at the score of 38 that Bob achieved by Week 12, you might be tempted to think "Hallelujah! My specific intervention *did* work. All Bob needed was a bit more of it." However, the same considerations apply to the BB₁ design as apply to the B design.

You cannot be sure that there is any relationship between your intervention and Bob's decreased anxiety, far less that one was the cause of the other.

BC Design

The last exploratory case-level research design is the *BC* design. Let us go back in time, to the point where you decided that it was worth trying a variation on your *B* intervention with Bob. Suppose you had decided instead to try an entirely different intervention, designated as *C* because it is a *different* intervention, following immediately after *B*. Now you implement the *C* intervention, administering the *IASS* every week, graphing your results, and creating a *C* phase after the *B* phase as shown in Figure 9.4.

Again, the *B* phase in Figure 9.4 is copied from Figure 9.2, and after the *C* intervention you see that Bob has succeeded in reducing his anxiety level to below the clinical cutting score of 40. Repressing your hallelujahs, you realize that there is still no sure relationship between your intervention and Bob's success. Indeed, the waters are becoming more murky because even if your intervention was in fact related to Bob's success you would still not know whether it was the *C* intervention that did the trick, or a delayed reaction to *B*, or some combination of *B* and *C*.

DESCRIPTIVE CASE-LEVEL DESIGNS

There are two kinds of case-level research designs that center around answering descriptive research questions: (1) *AB* designs, and (2) *ABC* and *ABCD* designs.

AB Design

An *AB* design is simply an *A*—or baseline phase—followed by a *B* or intervention phase. Returning to Cecilia, let's say she has a problem with low self-esteem and that you have already completed a four-week baseline phase with her, as shown in Figure 9.5, and that phase alone answered the two simple exploratory questions, "Does the problem exist?" and "Does the problem exist at different levels over time?" Now you implement a *B* intervention and find, to your pleasure, that Cecilia's self-esteem level approaches the clinical cutting score of 30 and falls below it at Weeks 9 and 10.

What you really want to know, of course, is whether there is any relationship between your *B* intervention and Cecilia's success. You are now in a better position to hypothesize that there is since you know that Cecilia was getting worse during the four weeks of the baseline phase and began to improve the week after you started your intervention. *Something* happened in Week 5 to set Cecilia on the road to recovery, and it would be very coincidental if that something were not

Chapter 9: Case-Level Research Designs

your intervention. However, coincidences do happen, and you cannot be certain that your intervention *caused* the change you see unless you can eliminate all the other coincidental happenings that might have caused it.

Hence, the *AB* design cannot answer explanatory research questions, but the change between the baseline data (getting worse) and the intervention data (getting better) is enough to indicate that there may be some relationship between your intervention and Cecilia's improvement.

The moral to the story is *always collect baseline data if you can* since social work ethics requires you to be reasonably sure an intervention is effective before you try it again with another client.

ABC and *ABCD* Designs

As discussed above, you can always follow a *B* phase with a *C* phase if the *B* intervention did not achieve the desired result. An *A* phase followed by a *B* phase followed by a *C* phase constitutes an *ABC* design, and if there is a *D* intervention as well, you have an *ABCD* design. So long as there is a baseline, you can conclude fairly safely that there is a relationship between the results you see and the intervention you implemented.

However, if you have more than one intervention, you will not know which intervention—or combination of interventions—did the trick, and the more interventions you try the more murky the waters become. Since a single intervention often comprises a package of practice techniques (e.g., active listening plus role play plus relaxation exercises), it is important to write down exactly what you did so that later on you will remember what the *B* or *C* or *D* interventions were.

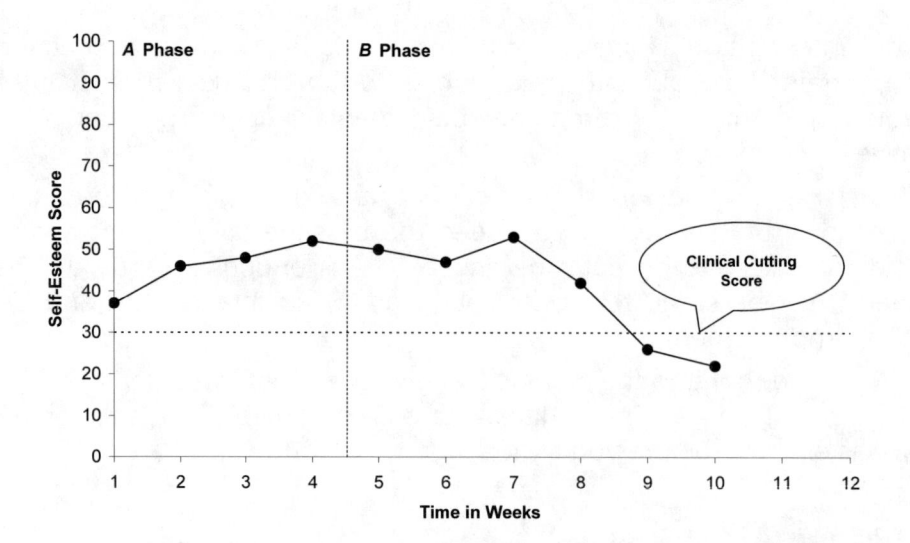

FIGURE 9.5
AB Design: Cecilia's Self-Esteem Scores for A Phase and B Phase

EXPLANATORY CASE-LEVEL DESIGNS

As we have seen, if you want to show that a particular intervention caused an observed result, you must eliminate everything else that may have caused it: in other words, you must control for intervening variables. There are two types of case-level designs that can answer causality, or explanatory research questions: (1) *reversal* designs, and (2) *multiple-baseline* designs.

Reversal Designs

The first type of explanatory case-level designs are the reversal designs. There are three kinds of case-level reversal designs: (1) *ABA* and *ABAB* designs, (2) *BAB* designs, and (3) *BCBC* designs.

ABA and *ABAB* Designs

Look back to Figure 9.5 which illustrates Cecilia's success in getting her self-esteem score below the clinical cutting score in Weeks 9 and 10. In Week 11, you decide that you will withdraw your intervention related to Cecilia's self-esteem since she seems to be doing well, but you will continue to monitor her self-esteem levels to ensure that treatment gains are maintained. Ongoing monitoring of problems that appear to be solved is something of a luxury in our profession.

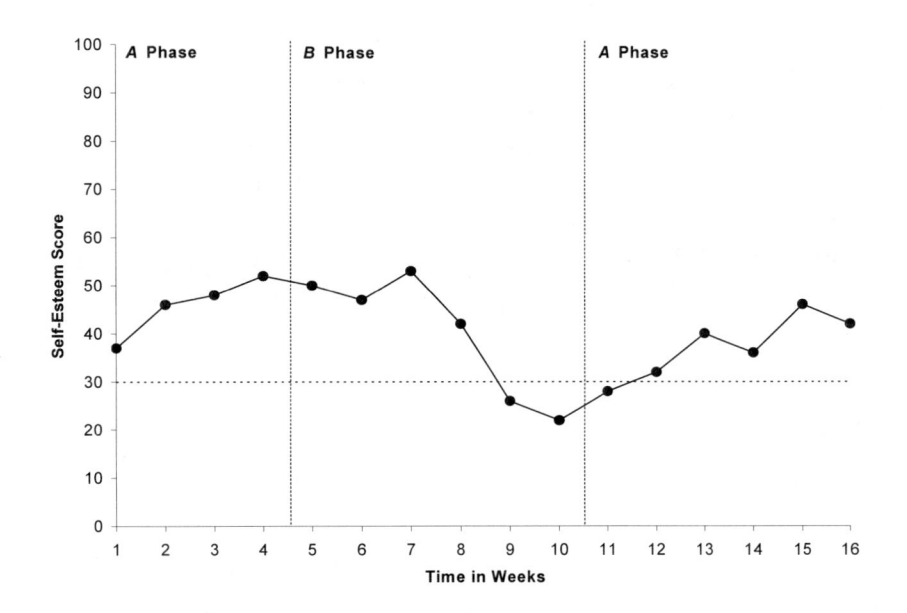

Chapter 9: Case-Level Research Designs

FIGURE 9.6
ABA Design: Cecilia's Self-Esteem Scores
for *A* Phase, *B* Phase, and *A* Phase

Too often our approach is crisis-oriented; follow-up tends to be ignored in the light of other, more pressing problems, and the result may well be a recurrence of the original problem because it had not been solved to the extent that the social worker thought.

However, with Cecilia you follow up. In Week 11, as shown in Figure 9.6, the score hovers at the clinical cutting score. In Week 12, it goes up a little, in Week 13 it jumps, and in Weeks 14 and 15, it is no better. Figure 9.6 illustrates an *ABA* design where the client's scores are displayed first without an intervention (the first *A* phase), then with an intervention (the *B* phase), then without an intervention again (the second *A* phase).

The scores are not as high in the second *A* phase as they were in the first *A* phase and this is to be expected since some of the strategies Cecilia learned in the *B* phase should remain with her even though the intervention has stopped.

However, from a research point of view, the very fact that her scores increased again when you stopped the intervention makes it more certain that it was your intervention that caused the improvement you saw in the *B* phase. Cecilia's improvement when the intervention started might have been a coincidence, but it is unlikely that her regression when the intervention stopped was also a coincidence.

Your certainty with respect to causality will be increased even further if you reintroduce the *B* intervention in Week 16 and Cecilia's score begins to drop again as it did in the first *B* phase.

Now you have implemented two *AB* designs one after the other with the same client to produce an *ABAB* design. This design is illustrated in Figure 9.7. It is sometimes called a *reversal* design or a *withdrawal* design. Causality is established with an *ABAB* design because the same intervention has been shown to work twice with the same client and you have baseline data to show the extent of the problem when there was no intervention.

BAB Design

Let's now return to Bob, with whom you implemented a *B* intervention to reduce his social anxiety as shown in Figure 9.2. When Bob's social anxiety level has fallen beneath the clinical cutting score, you might do the same thing with Bob as you did with Cecilia: withdraw the intervention and continue to monitor the problem, creating an *A* phase after the *B* phase. If the problem level worsens during the *A* phase, you intervene again in the same way as you did before, creating a second *B* phase and an overall design of *BAB*.

We have said that causality is established with an *ABAB* design because the same intervention has worked twice with the same client. We cannot say the same for a *BAB* design, however, as we do not really know that our intervention "worked" the first time. Since there was no initial baseline data (no first *A* phase), we cannot know whether the resolution of the problem on the first occasion had anything to do with the intervention.

The problem may have resolved itself or some external event (intervening variable) might have resolved it. Nor can we know the degree to which the problem changed during the first *B* phase (intervention), since there was no baseline data with which to compare the final result. An indication of the amount of change can be obtained by comparing the first and last scores in the *B* phase but the first score may have been an unreliable measure of Bob's problem.

Bob may have felt less or more anxious that day than usual, and a baseline is necessary to compensate for such day-to-day fluctuations. Since the effectiveness of the intervention on the first occasion is unknown, there can be no way of knowing whether the intervention was just as effective the second time it was implemented, or less or more effective. All we know is that the problem improved twice, following the same intervention, and this is probably enough to warrant using the intervention again with another client.

BCBC Design

A *BCBC* design, as the name suggests, is a *B* intervention followed by a *C* intervention implemented twice in succession.

Chapter 9: Case-Level Research Designs

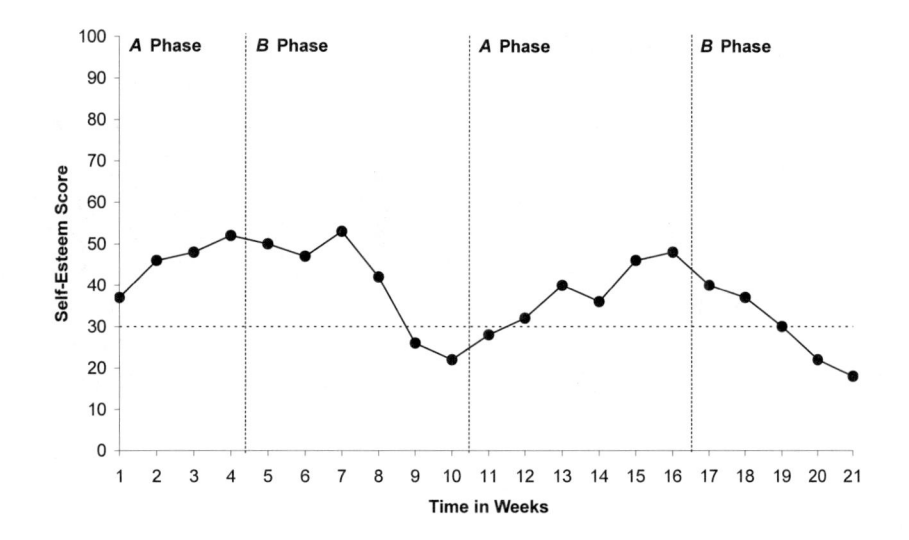

FIGURE 9.7
ABAB Design: Cecilia's Self-Esteem Scores
for Two *A* Phases and Two *B* Phases

The point of doing this is to compare the effectiveness of two interventions—*B* and *C*. It is unlikely that a social worker would implement this design with a client since, if the problem improved sufficiently using *B*, you would not need *C* and, if you did need *C*, you would hardly return to *B* whether or not *C* appeared to do the trick.

However, if the problem has nothing to do with a client's welfare but is concerned instead with a social work program's organizational efficiency, say, as affected by organizational structure, you might try one structure *B* followed by a different structure *C* and then do the same thing again in order to show that one structure really has proved more effective in increasing efficiency when implemented twice *under the same conditions.*

Multiple-Baseline Designs

The second type of explanatory case-level designs are the multiple-baseline designs. Multiple-baseline designs are like *ABAB* designs in that the *AB* design is implemented more than once. However, whereas *ABAB* designs apply to one case with one problem in one setting, multiple-baseline designs can be used with more than one case, with more than one setting, or within more than one problem.

More than One Case

Suppose that, instead of Bob with his social anxiety problem, you have three additional clients with anxiety problems, Breanne, Warren, and Alison. All three are residents in the same nursing home. You use the same measuring instrument to measure anxiety (the *IASS*) in all three cases and you give all three clients the same intervention. However, you vary the number of weeks over which you collect baseline data, allowing the baseline phase to last for six weeks in Breanne's case, eight weeks in Warren's case, and nine weeks for Alison. You plot your results as shown in Figure 9.8.

Breanne starts to show improvement in Week 7, the week you began your intervention. Had that improvement been due to some intervening variable—for example, some anxiety-reducing change in the nursing home's routine—you would expect Warren and Alison to also show improvement.

The fact that their anxiety levels continue to be high indicates that it was your intervention, not some other factor that caused the improvement in Breanne. Causality is demonstrated again in Week 9 when you begin to intervene with Warren and Warren improves but Alison does not. Your triumph is complete when Alison, given the same intervention, begins to improve in Week 10.

More than One Setting

Another way to conduct a multiple-baseline study is with one client in a number of settings. Suppose that your objective is to reduce the number of a child's temper tantrums at home, in school, and at the daycare center where the child goes after school. The same intervention is offered by the parents at home, the teacher in school, and the worker at the daycare center.

They are also responsible for measuring the number of tantrums that occur each day. The baseline phase continues for different lengths of time in each setting as shown in Figure 9.9. If the child improves at home after the intervention begins in Week 7 but continues to throw numerous tantrums in school and at daycare, the indication is that it was the intervention that caused the improvement at home.

More than One Problem

A third way to conduct a multiple-baseline study is to use the same intervention to tackle different target problems. Suppose that Joan is having trouble with her daughter, Anita. In addition, Joan is having trouble with her in-laws and with her boss at work.

Chapter 9: Case-Level Research Designs

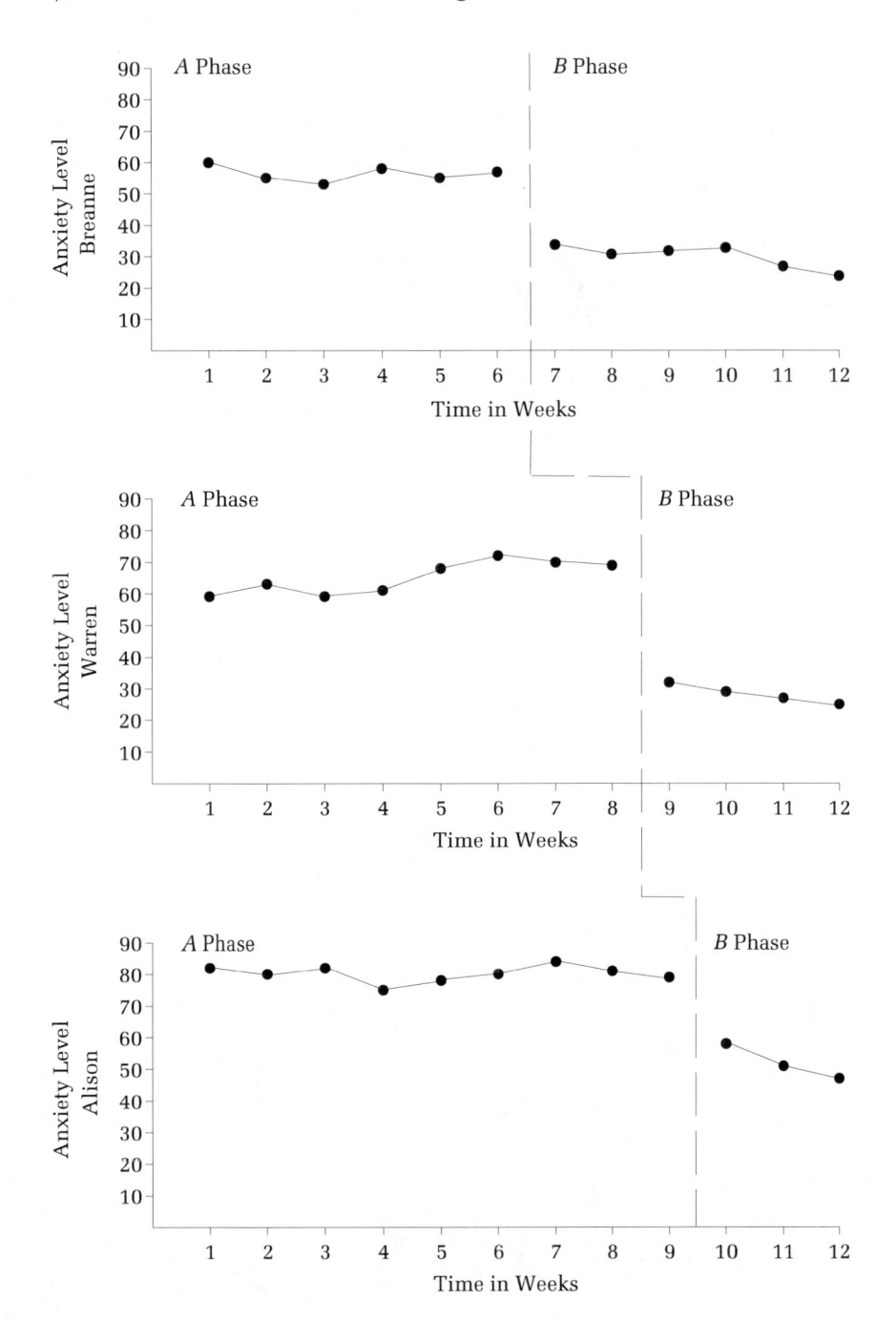

FIGURE 9.8
Multiple-Baseline Design Across Clients:
Magnitude of Anxiety Levels for Three Clients

FIGURE 9.9
Multiple-Baseline Design Across Settings:
Number of Temper Tantrums for One Client in Three Settings

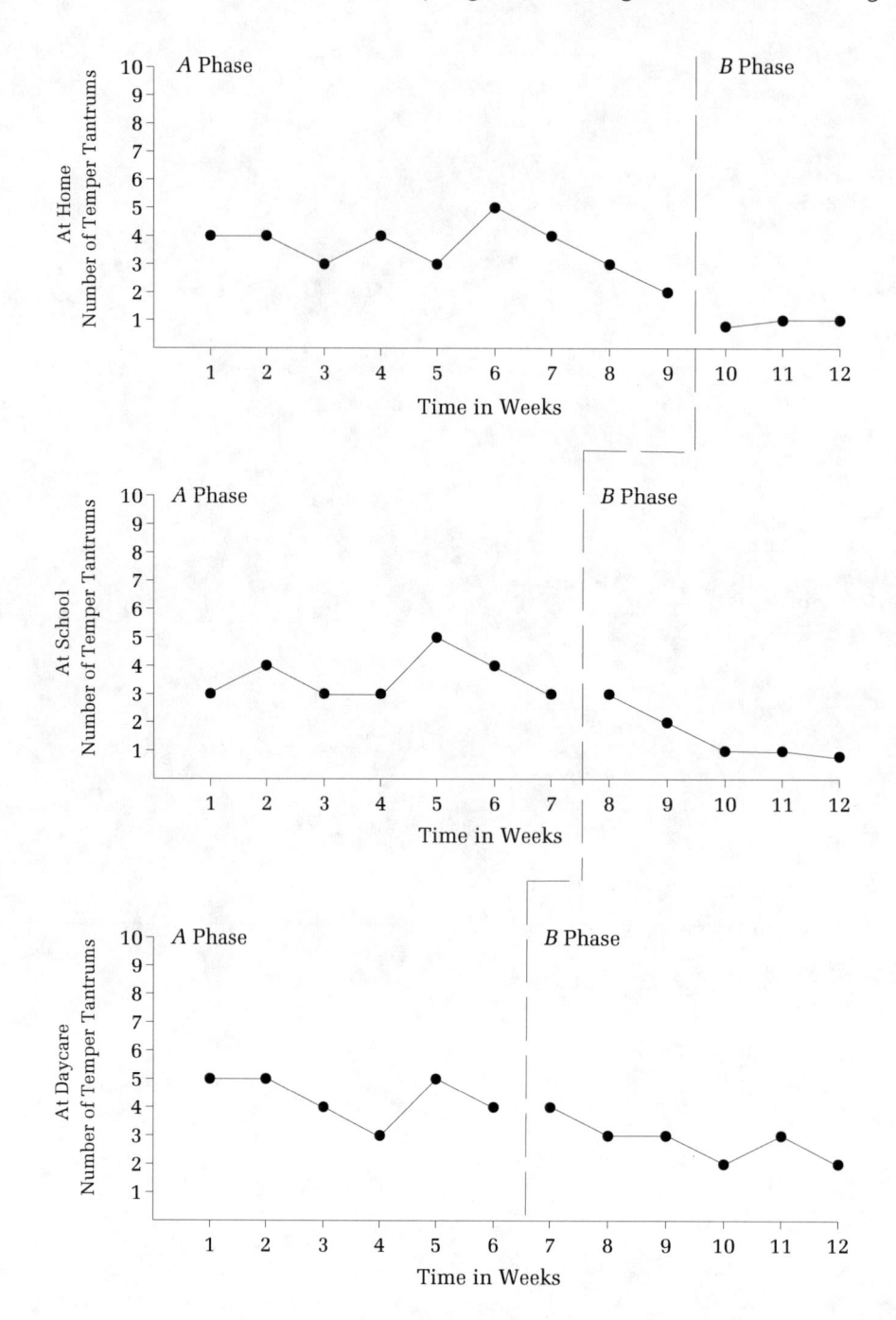

Chapter 9: Case-Level Research Designs **183**

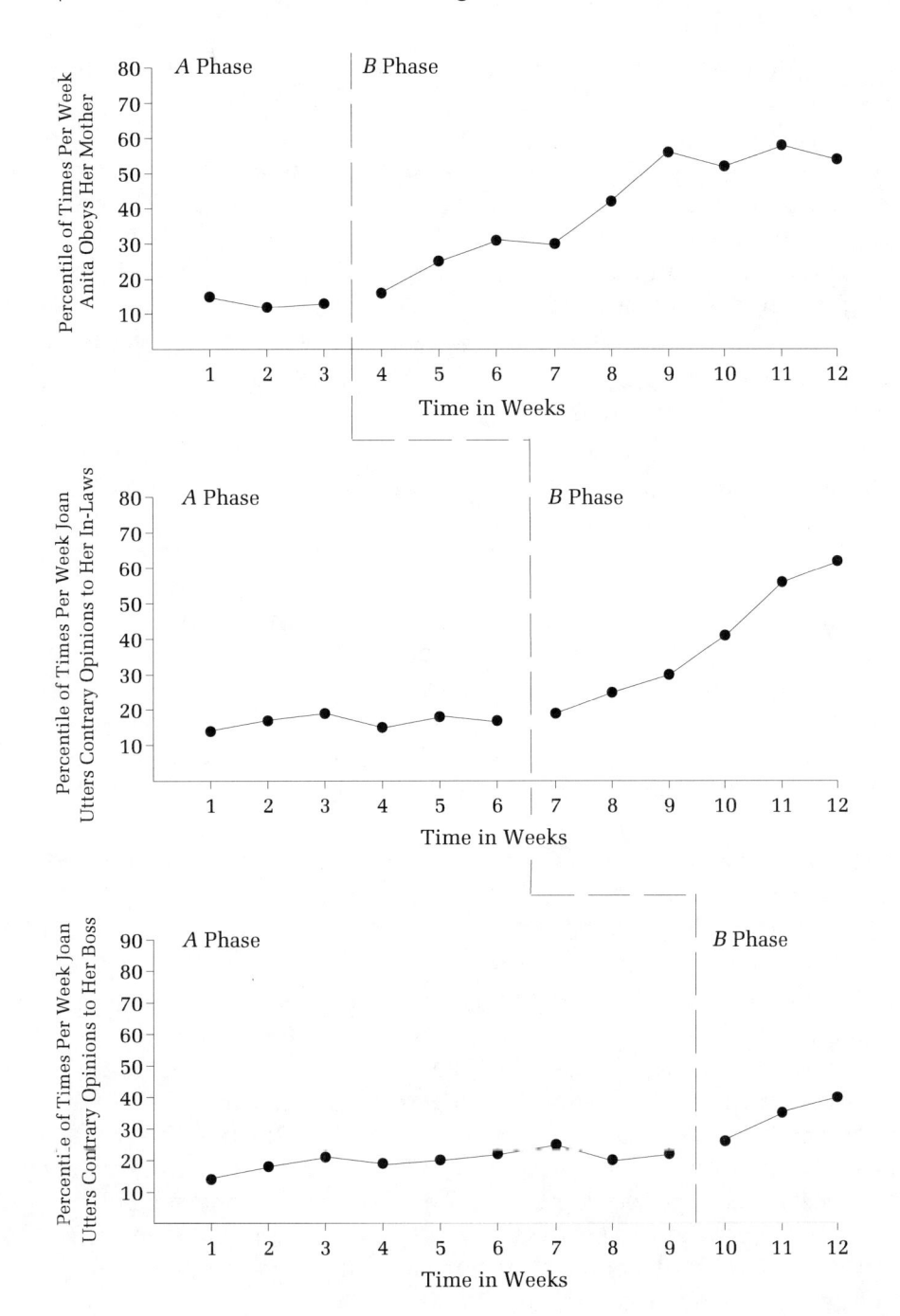

FIGURE 9.10
Multiple-Baseline Design Across Client Problems:
Magnitude of Three Client Target Problem Areas for One Client

After exploration, a worker may believe that all these troubles stem from her lack of assertiveness. Thus, the intervention would be assertiveness training. Progress with Anita might be measured by the number of times each day she is flagrantly disobedient. Progress can be measured with Joan's in-laws by the number of times she is able to utter a contrary opinion, and so on. Since the number of occasions on which Joan has an opportunity to be assertive will vary, these figures might best be expressed in percentiles.

Figure 9.10 illustrates an example of a multiple-baseline design that was used to assess the effectiveness of Joan's assertiveness training in three problem areas.

Whether it is a reversal design or a multiple-baseline design, an *ABAB* explanatory design involves establishing a baseline level for the client's target problem. This will not be possible if the need for intervention is acute, and sometimes the very thought of an *A*-type design will have to be abandoned. It is sometimes possible, however, to construct a retrospective baseline—that is, to determine what the level of the problem was before an intervention is implemented.

The best retrospective baselines are those that do not depend on the client's memory. If the target problem occurs rarely, memories may be accurate. For example, Tai, a teenager, and his family may remember quite well how many times he ran away from home during the past month. They may not remember nearly so well if the family members were asked how often he behaved defiantly. Depending on the target problem, it may be possible to construct a baseline from archival data: that is, from written records, such as school attendance sheets, probation orders, employment interview forms, and so forth.

Although establishing a baseline usually involves making at least three measurements before implementing an intervention, it is also acceptable to establish a baseline of zero, or no occurrences of a desired event. A target problem, for example, might focus upon the client's reluctance to enter a drug treatment program. The baseline measurement would then be that the client did not go (zero occurrences) and the desired change would be that the client did go (one occurrence). A social worker who has successfully used the same tactics to persuade a number of clients to enter a drug treatment program has conducted a multiple-baseline design across clients.

As previously discussed, a usable baseline should show either that the client's problem level is stable or that it is growing worse. Sometimes an *A*-type design can be used even though the baseline indicates a slight improvement in the target problem. The justification must be that the intervention is expected to lead to an improvement that will exceed the anticipated improvement if the baseline trend continues.

Perhaps a child's temper tantrums are decreasing by one or two a week, for example, but the total number per week is still 18 to 20. If a worker thought the tantrums could be reduced to four or five a week, or they could be stopped altogether, the worker would be justified in implementing an intervention even though the client's target problem was improving slowly by itself.

In a similar way, a worker may be able to implement an *A*-type design if the client's baseline is unstable, provided that the intervention is expected to exceed the largest of the baseline fluctuations. Perhaps the child's temper tantrums are fluctuating between 12 and 20 per week in the baseline period and it is hoped to bring them down to less than 10 per week.

Nevertheless, there are some occasions when a baseline cannot be established or is not usable, such as when a client's behaviors involve self-injurious ones. Also, sometimes the establishment of a baseline is totally inappropriate.

SUMMARY

Exploratory case-level research designs are used when little is known about the field of study and data are gathered in an effort to find out "what's out there." These ideas are then used to generate hypotheses that can be verified using more rigorous designs. No design is inherently inferior or superior to the others. Each has advantages and disadvantages in terms of time, cost, and the data that can be obtained.

REFERENCES AND FURTHER READING

Bloom, M., Fischer, J., & Orme, J. (2009). *Evaluating practice: Guidelines for the accountable professional* (6th ed.). Englewood Cliffs, NJ: Prentice-Hall.

Blythe, B.J., & Tripodi, T. (1989). *Measurement in direct practice.* Newbury Park, CA: Sage.

Fox, M., Martin, P., & Green, G. (2007). *Doing practitioner research.* Thousand Oaks, CA: Sage.

Morgan, D.L., & Morgan, R.K. (2008). *Single-case research methods for the behavioral and health sciences.* Thousand Oaks, CA: Sage.

Thyer, B.A., & Myers, L.L. (2007). *A social worker's guide to evaluating practice outcomes.* Alexandria, VA: Council on Social Work Education.

Williams, M., Grinnell, R.M., Jr., & Unrau, Y.A. (2008). Case-level designs. In R.M. Grinnell, Jr., & Y.A. Unrau (Eds.), *Social work research and evaluation: Foundations of evidence-based practice* (8th ed., pp. 157–178). New York: Oxford University Press.

Chapter 10

Group-Level Research Designs

CHARACTERISTICS OF "IDEAL" EXPERIMENTS
    Controlling the Time Order of Variables
    Manipulating the Independent Variable
    Establishing Relationships Between Variables
    Controlling Rival Hypotheses
    Holding Extraneous Variables Constant
    Using Correlated Variation
    Using Analysis of Covariance
    Using a Control Group
    Randomly Assigning Research Participants to Groups
    Matched Pairs
INTERNAL AND EXTERNAL VALIDITY
GROUP RESEARCH DESIGNS
    Exploratory Designs
    Descriptive Designs
    Explanatory Designs
SUMMARY
REFERENCES AND FURTHER READING

10

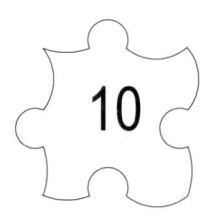

Now that we know how to draw samples for qualitative and quantitative research studies, we turn our attention to the various group-level designs that research studies can take. The two most important factors in determining what design to use in a specific study are: (1) what the research question is, and (2) how much knowledge about the problem area is available.

If there is already a substantial knowledge base in the area, we will be in a position to address very specific research questions, the answers to which could add to the explanation of previously gathered data. If less is known about the problem area, our research questions will have to be of a more general, descriptive nature. If very little is known about the problem area, our questions will have to be even more general, at an exploratory level.

Research knowledge levels are arrayed along a continuum, from exploratory at the lowest end to explanatory at the highest. Because research knowledge levels are viewed this way, the assignment of the level of knowledge accumulated in a problem area prior to a research study, as well as the level that might be attained by the research study, is totally arbitrary.

There are, however, specific designs that can be used to provide us with knowledge at a certain level. At the highest level are the explanatory designs, also called experimental designs or "ideal" experiments. These designs have the largest number of requirements (examined in the following section).

They are best used in confirmatory research studies where the area under study is well developed, theories abound, and testable hypotheses can be formulated on the basis of previous work or existing theory. These designs seek to establish causal relationships between the independent and dependent variables. In the middle range are the descriptive designs, sometimes referred to as quasi-experimental. A quasi-experiment resembles an "ideal" experiment in some aspects but lacks at least one of the necessary requirements.

At the lowest level are the exploratory designs, also called pre-experimental or non-experimental, which explore only the research question or problem area. These designs do not produce statistically sound data or conclusive results; they are not intended to. Their purpose is to build a foundation of general ideas and tentative theories, which can be explored later with more precise and hence more complex research designs, and their corresponding data gathering techniques. The research designs that allow us to acquire knowledge at each of the three levels are described in a later section of this chapter. Before considering them, however, it is necessary to establish the characteristics that differentiate an "ideal" experiment, which leads to explanatory knowledge, from other studies that lead to lower levels of knowledge.

CHARACTERISTICS OF "IDEAL" EXPERIMENTS

An "ideal" experiment is one in which a research study most closely approaches certainty about the relationship between the independent and dependent variables. The purpose of doing an "ideal" experiment is to ascertain whether it can be concluded from the study's findings that the independent variable is, or is not, the only cause of change in the dependent variable.

As pointed out in previous chapters, some social work research studies have no independent variable—for example, those studies that just want to find out how many people in a certain community wish to establish a community-based halfway house for people who are addicted to drugs.

The concept of an "ideal" experiment is introduced with the word "ideal" in quotes because such an experiment is rarely achieved in social work research situations. On a general level, in order to achieve this high degree of certainty and qualify as an "ideal" experiment, an explanatory research design must meet six conditions:

1. The time order of the independent variable must be established.
2. The independent variable must be manipulated.
3. The relationship between the independent and dependent variables must be established.
4. The research design must control for rival hypotheses.
5. At least one control group should be used.
6. Random assignment procedures (and if possible, random sampling from a population) must be employed in assigning research participants (or objects) to groups.

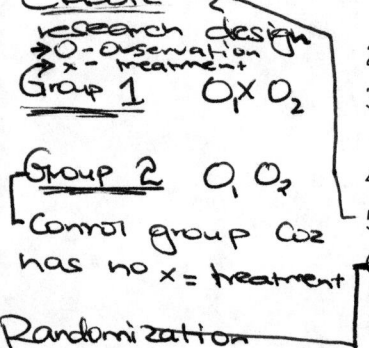

Controlling the Time Order of Variables

In an "ideal" experiment, the independent variable must precede the dependent variable in time. Time order is crucial if our research study is to show that one variable causes another, because something that occurs later cannot be the cause of something that occurred earlier. Sup-

Chapter 10: Group-Level Research Designs

pose we want to study the relationship between adolescent substance abuse and gang-related behavior. The following hypothesis is formulated after some thought:

Adolescent substance abuse causes gang-related behaviors

In the preceding hypothesis, the independent variable is adolescent substance abuse, and the dependent variable is gang-related behavior. The substance abuse must come *before* gang-related behavior because the hypothesis states that adolescent drug use causes gang-related behavior. We could also come up with the following hypothesis, however:

Adolescent gang-related behavior causes substance abuse

In this hypothesis, adolescent gang-related behavior is the independent variable, and substance abuse is the dependent variable. According to this hypothesis, gang-related behavior must come *before* the substance abuse.

Manipulating the Independent Variable

Manipulation of the independent variable means that we must do something with the independent variable in terms of at least one of the research participants in the study. In the general form of the hypothesis, if X occurs then Y will result, the independent variable (X) must be manipulated in order to effect a variation in the dependent variable (Y). There are essentially three ways in which independent variables can be manipulated:

1. *X present versus X absent.* If the effectiveness of a specific treatment intervention is being evaluated, an experimental group and a control group could be used. The experimental group would be given the intervention, the control group would not.
2. *A small amount of X versus a larger amount of X.* If the effect of treatment time on client's outcomes is being studied, two experimental groups could be used, one of which would be treated for a longer period of time.
3. *X versus something else.* If the effectiveness of two different treatment interventions is being studied, Intervention X_1 could be used with Experimental Group 1 and Intervention X_2 with Experimental Group 2.

There are certain variables, such as the gender or race of our research participants, that obviously cannot be manipulated because they are fixed. They do not vary, so they are called constants, not variables, as was pointed out in Chapter 6. Other constants, such as socioeconomic status or IQ, may vary for research participants over their life spans, but they are fixed quantities at the beginning of the study, probably will not change during the study, and are not subject to alteration by the one doing the study.

Any variable we can alter (e.g., treatment time) can be considered an independent variable. At least one independent variable must be manipulated in a research study if it is to be considered an "ideal" experiment.

Establishing Relationships Between Variables

The relationship between the independent and dependent variables must be established in order to infer a cause-effect relationship at the explanatory knowledge level. If the independent variable is considered to be the cause of the dependent variable, there must be some pattern in the relationship between these two variables. An example is the hypothesis: The more time clients spend in treatment (independent variable), the better their progress (dependent variable).

Controlling Rival Hypotheses

Rival hypotheses must be identified and eliminated in an "ideal" experiment. The logic of this requirement is extremely important, because this is what makes a cause-effect statement possible.

The prime question to ask when trying to identify a rival hypothesis is, "What other extraneous variables might affect the dependent variable?" (What else might affect the client's outcome besides treatment time?) At the risk of sounding redundant, "What else besides X might affect Y?"

Perhaps the client's motivation for treatment, in addition to the time spent in treatment, might affect the client's outcome. If so, motivation for treatment is an extraneous variable that could be used as the independent variable in the rival hypothesis, "The higher the clients' motivation for treatment, the better their progress."

Perhaps the social worker's attitude toward the client might have an effect on the client's outcome, or the client might win the state lottery and ascend abruptly from depression to ecstasy. These extraneous variables could potentially be independent variables in other rival hypotheses. They must all be considered and eliminated before it can be said with reasonable certainty that a client's outcome resulted from the length of treatment time and not from any other extraneous variables.

Control over rival hypotheses refers to efforts on our part to identify and, if at all possible, to eliminate the extraneous variables in these alternative hypotheses. Of the many ways to deal with rival hypotheses, three of the most frequently used are to keep the extraneous variables constant, use correlated variation, or use analysis of covariance.

Holding Extraneous Variables Constant — Rival hypothesis variables.

The most direct way to deal with rival hypotheses is to keep constant the critical extraneous variables that might affect the dependent variable. As we know, a constant cannot affect or be affected by any other variable. If an extraneous variable can be made into a constant, then it cannot affect either the study's real independent variable or the dependent variable.

Let us take an example to illustrate the above point. Suppose, for example, that a social worker who is providing counseling to anxious clients wants to relate client outcome to length of treatment time, but most of the clients are also being treated by a consulting psychiatrist with antidepressant medication. Because medication may also affect the clients' outcomes, it is a potential independent variable that could be used in a rival hypothesis. However, if the study included only clients who have been taking medication for some time before the treatment intervention began, and who continue to take the same medicine in the same way throughout treatment, then medication can be considered a constant (in this study, anyway).

Any change in the clients' anxiety levels after the intervention will, therefore, be due to the intervention with the help of the medication. The extraneous variable of medication, which might form a rival hypothesis, has been eliminated by holding it constant. In short, this study started out with one independent variable, the intervention, then added the variable of medication to it, so the final independent variable is the intervention plus the medication.

This is all very well in theory. In reality, however, a client's drug regime is usually controlled by the psychiatrist and may well be altered at any time. Even if the regime is not altered, the effects of the drugs might not become apparent until the study is under way.

In addition, the client's level of anxiety might be affected by a host of other extraneous variables over which the social worker has no control at all: for example, living arrangements, relationships with other people, the condition of the stock market, or an unexpected visit from an IRS agent. These kinds of pragmatic difficulties tend to occur frequently in social work practice and research. It is often impossible to identify all rival hypotheses, let alone eliminate them by keeping them constant.

Using Correlated Variation

Rival hypotheses can also be controlled with correlated variation of the independent variables. Suppose, for example, that we are concerned that income has an effect on a client's compulsive behavior. The client's income, which in this case is subject to variation due to seasonal employment, is identified as an independent variable. The client's living conditions—in a hotel room rented by the week—are then identified as the second independent variable that might well affect the client's level of compulsive behavior. These two variables, however, are correlated since living conditions are highly dependent on income.

Correlated variation exists if one potential independent variable can be correlated with another. Then only one of them has to be dealt with in the research study.

Using Analysis of Covariance

In conducting an "ideal" experiment, we must always aim to use two or more groups that are as equivalent as possible on all important variables. Sometimes this goal is not feasible, however. Perhaps we are obliged to use existing groups that are not as equivalent as we would like. Or, perhaps during the course of the study we discover inequivalencies between the groups that were not apparent at the beginning.

A statistical method called *analysis of covariance* can be used to compensate for these differences. The mathematics of the method is far beyond the scope of this text, but an explanation can be found in most advanced statistics texts.

Using a Control Group

An "ideal" experiment should use at least one control group in addition to the experimental group. The experimental group may receive an intervention that is withheld from the control group, or equivalent groups may receive different interventions or no interventions at all.

A social worker who initiates a treatment intervention is often interested in knowing what would have happened if the intervention had not been used or had some different intervention been substituted. Would members of a support group for alcoholics have recovered anyway, without the social worker's efforts? Would they have recovered faster or more completely had family counseling been used instead of the support group approach?

Chapter 10: Group-Level Research Designs

The answer to these questions will never be known if only the support group is studied. But, what if another group of alcoholics is included in the research design? In a typical design with a control group, two equivalent groups, 1 and 2, would be formed, and both would be administered the same pretest to determine the initial level of the dependent variable (e.g., degree of alcoholism).

Then an intervention would be initiated with Group 1 but not with Group 2. The group treated—Group 1 or the experimental group—would receive the independent variable (the intervention). The group not treated—Group 2 or the control group—would not receive it.

At the conclusion of the intervention, both groups would be given a posttest (the same measure as the pretest). The pretest and posttest consist of the use of some sort of data gathering procedure, such as a survey or self-report measure, to measure the dependent variable before and after the introduction of the independent variable. There are many types of group research designs and there are many ways to graphically display them. In general, group designs can be written in symbols as follows:

R_s = Random selection from a population
R_a = Random assignment to a group
O = First measurement of the dependent variable
X = Independent variable, or intervention
O_2 = Second measurement of the dependent variable

The Rs in the research design on the top of the following page indicate that the research participants were randomly assigned to each group. The symbol X, which, as usual, stands for the independent variable, indicates that an intervention is to be given to the experimental group after the pretest (O_1) and before the posttest (O_2). The absence of X for the control group indicates that the intervention is not to be given to the control group. This design is called a classical experimental design because it comes closest to having all the characteristics necessary for an "ideal" experiment.

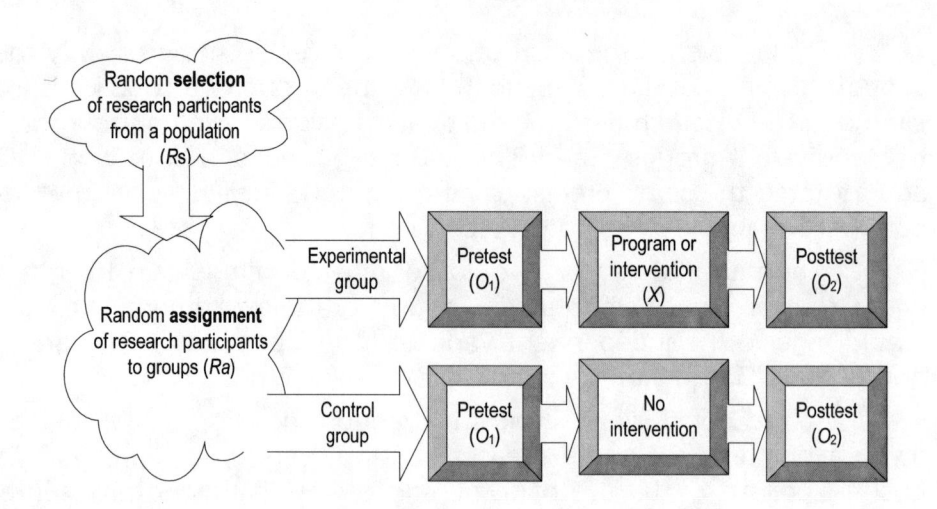

Randomly Assigning Research Participants to Groups

Once a sample has been selected (see previous chapter), the individuals (or objects or events) in it are randomly assigned to either an experimental or a control group in such a way that the two groups are equivalent. This procedure is known as random assignment or randomization. In random assignment, the word *equivalent* means equal in terms of the variables that are important to the study, such as the clients' motivation for treatment, or problem severity.

If the effect of treatment time on clients' outcomes is being studied, for example, the research design might use one experimental group that is treated for a comparatively longer time, a second experimental group that is treated for a shorter time, and a control group that is not treated at all. If we are concerned that the clients' motivation for treatment might also affect their outcomes, the research participants can be assigned so that all the groups are equivalent (on the average) in terms of their motivation for treatment. The process of random sampling from a population followed by random assignment of the sample to groups is illustrated on the top of the following page.

Let us say that the research design calls for a sample size of one-tenth of the population. From a population of 10,000, therefore, a random sampling procedure is used to select a sample of 1,000 individuals. Then random assignment procedures are used to place the sample of 1,000 into two equivalent groups of 500 individuals each. In theory, Group A will be equivalent to Group B, which will be equivalent to the random sample, which will be equivalent to the population in respect to all important variables contained within the research study.

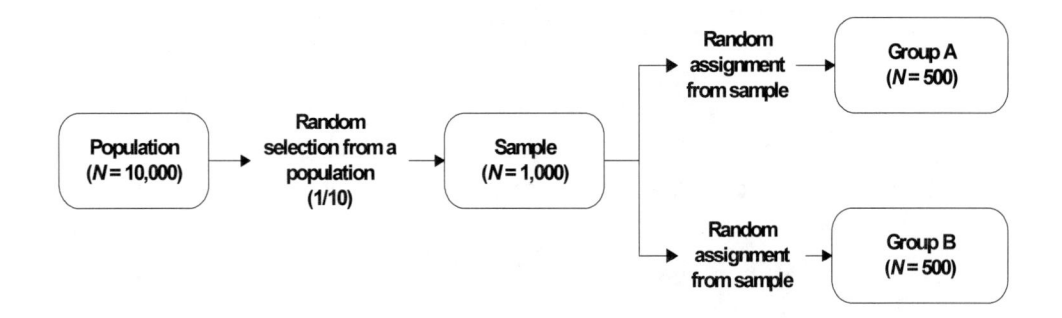

Matched Pairs

Besides randomization, another, more deliberate method of assigning people or other units to groups involves matching. The matched pairs method is suitable when the composition of each group consists of variables with a range of characteristics. One of the disadvantages of matching is that some individuals cannot be matched and so cannot participate in the study.

Suppose a new training program for teaching parenting skills to foster mothers is being evaluated, and it is important that the experimental and control groups have an equal number of highly skilled and less skilled foster parents before the training program is introduced. The women chosen for the sample would be matched in pairs according to their parenting skill level; the two most skilled foster mothers are matched, then the next two, and so on. One person in each pair of approximately equally skilled foster parents is then randomly assigned to the experimental group and the other is placed in the control group.

Let us suppose that in order to compare the foster mothers exposed to the new training program with women who were not, a standardized measuring instrument that measures parenting skill level (the dependent variable) is administered to a sample of ten women. The scores can range from 100 (excellent parenting skills) to zero (poor parenting skills). Then their scores are rank-ordered from the highest to the lowest, and out of the foster mothers with the two highest scores, one is selected to be assigned to either the experimental group or the control group.

It does not make any difference which group our first research participant is randomly assigned to, as long as there is an equal chance that she will go to either the control group or the experimental group. In this example the first person is randomly chosen to go to the experimental group, as illustrated on the following page:

Rank Order of Parenting Skills Scores (in parentheses)

First Pair:
(99) Randomly assigned to the experimental group
(98) Assigned to the control group

Second Pair:
(97) Assigned to the control group
(96) Assigned to the experimental group

Third Pair:
(95) Assigned to the experimental group
(94) Assigned to the control group

Fourth Pair:
(93) Assigned to the control group
(92) Assigned to the experimental group

Fifth Pair:
(91) Assigned to the experimental group
(90) Assigned to the control group

The foster parent with the highest score (99) is randomly assigned to the experimental group, and this person's "match," with a score of 98, is assigned to the control group. This process is reversed with the next matched pair, where the first person is assigned to the control group and the match is assigned to the experimental group. If the assignment of research participants according to scores is not reversed for every other pair, one group will be higher than the other on the variable being matched.

To illustrate this point, suppose the first participant (highest score) in each match is always assigned to the experimental group. The experimental group's average score would be 95 (99 + 97 + 95 + 93 + 91 = 475/5 = 95), and the control group's average score would be 94 (98 + 96 + 94 + 92 + 90 = 470/5 = 94).

If every other matched pair is reversed, however, as in the example, the average scores of the two groups are closer together; 94.6 for the experimental group (99 + 96 + 95 + 92 + 91 = 473/5 = 94.6) and 94.4 for the control group (98 + 97 + 94 + 93 + 90 = 472/5 = 94.4). In short, 94.6 and 94.4 (difference of 0.2) are closer together than 95 and 94 (difference of 1).

Chapter 10: Group-Level Research Designs

INTERNAL AND EXTERNAL VALIDITY

We must remember that the research design we finally select should always be evaluated on how close it comes to an "ideal" experiment in reference to the characteristics presented at the beginning of this chapter. As stressed throughout this book, most research designs used in social work do not closely resemble an "ideal" experiment.

The research design finally selected needs to be evaluated on how well it meets its primary objective—to adequately answer a research question or to test a hypothesis. In short, a research design will be evaluated on how well it controls for:

- Internal validity (Box 10.1)—the ways in which the research design ensures that the introduction of the independent variable (if any) can be identified as the *sole cause* of change in the dependent variable.
- External validity (Box 10.2)—the extent to which the research design allows for generalization of the findings of the study to other groups and other situations.

BOX 10.1
Threats to Internal Validity

In any explanatory research study, we should be able to conclude from our findings that the independent variable is, or is not, the only cause of change in the dependent variable. If our study does not have internal validity, such a conclusion is not possible, and the study's findings can be misleading.

Internal validity is concerned with one of the requirements for an "ideal" experiment—the control of rival hypotheses, or alternative explanations for what might bring about a change in the dependent variable. The higher the internal validity of any research study, the greater the extent to which rival hypotheses can be controlled; the lower the internal validity, the less they can be controlled. Thus, we must be prepared to rule out the effects of factors other than the independent variable that could influence the dependent variable.

HISTORY

The first threat to internal validity, history, refers to any outside event, either public or private, that may affect the dependent variable and was not taken into account in our research design. Many times, it refers to events occurring between the first and second measurement of the dependent variable (the pretest and the posttest). If events occur that have the potential to alter the second measurement, there would be no way of knowing how much (if any) of the observed change in the dependent variable is a function of the independent variable and how much is attributable to these events.

Suppose, for example, we are investigating the effects of an educational program on racial tolerance. We may decide to measure the dependent variable, racial tolerance in the community, before introducing the independent variable, the educational program.

The educational program is then implemented. Since it is the independent variable, it is represented by X. Finally, racial tolerance is measured again, after the program has run its course. This final measurement yields a posttest score, represented by O_2. The one-group pretest-posttest study design is presented in Figure 10.11.

The difference between the values O_2 and O_1 represent the difference in racial tolerance in the community before and after the educational program. If the study is internally valid, $O_2 - O_1$ will be a crude measure of the effect of the educational program on racial tolerance; and this is what we were trying to discover. Suppose before the posttest could be administered, an outbreak of racial violence, such as the type that occurred in Los Angeles in the summer of 1992, occurred in the community.

Violence can be expected to have a negative effect on racial tolerance, and the posttest scores may, therefore, show a lower level of tolerance than if the violence had not occurred. The effect, $O_2 - O_1$, will now be the combined effects of the educational program *and* the violence, not the effect of the program alone, as we intended.

Racial violence is an extraneous variable that we could not have anticipated and did not control for when designing the study. Other examples might include an earthquake, an election, illness, divorce, or marriage—any event, public or private that could affect the dependent variable. Any such variable that is unanticipated and uncontrolled for is an example of history.

MATURATION

Maturation, the second threat to internal validity, refers to changes, both physical and psychological, that take place in our research participants over time and can affect the dependent variable. Suppose that we are evaluating an interventive strategy designed to improve the behavior of adolescents who engage in delinquent behavior. Since the behavior of adolescents changes naturally as they mature, the observed changed behavior may have been due as much to their natural development as it was to the intervention strategy.

Maturation refers not only to physical or mental growth, however. Over time, people grow older, more or less anxious, more or less bored, and more or less motivated to take part in a research study. All these factors and many more can affect the way in which people respond when the dependent variable is measured a second or third time.

TESTING

The third threat to internal validity, testing, is sometimes referred to as the initial measurement effect. Thus, the pretests that are the starting point for many research designs are another potential threat to internal validity. One of the most utilized research designs involves three steps: measuring some dependent variable, such as learning behavior in school or attitudes toward work; initiating a program to change that variable (the independent variable); then measuring the dependent variable again at the conclusion of the program.

The testing effect is the effect that taking a pretest might have on posttest scores. Suppose that Roberto, a research participant, takes a pretest to measure his initial level of racial tolerance before being exposed to a racial tolerance educational program. He might remember some of the questions on the pretest, think about them later, and change his views on racial issues before taking part in the educational program.

After the program, his posttest score will reveal his changed opinions, and we may incorrectly assume that the program was responsible, whereas the true cause was his experience with the pretest.

Sometimes, a pretest induces anxiety in a research participant, so that Roberto receives a worse score on the posttest than he should have; or boredom with the same questions repeated again may be a factor. In order to avoid the testing effect, we may wish to use a design that does not require a pretest. If a pretest is essential, we then must consider the length of time that elapses between the pretest and posttest measurements.

A pretest is far more likely to affect the posttest when the time between the two is short. The nature of the pretest is another factor. Questions dealing with factual matters, such as knowledge levels, may have a larger testing effect because they tend to be more easily recalled.

INSTRUMENTATION ERROR

The fourth threat to internal validity is instrumentation error, which refers to all the troubles that can afflict the measurement process. The instrument may be unreliable or invalid, as presented in Chapters 5 and 6. It may be a mechanical instrument, such as an electroencephalogram (EEG), that has malfunctioned. Occasionally, the term *instrumentation error* is used to refer to an observer whose observations are inconsistent; or to measuring instruments, such as the ones presented in Chapter 6, that are reliable in themselves, but not administered properly.

"Administration," with respect to a measuring instrument, means the circumstances under which the measurement is made: where, when, how, and by whom. A mother being asked about her attitudes toward her children, for example, may respond in one way in the social worker's office and in a different way at home when her children are screaming around her feet.

A mother's verbal response may differ from her written response; or she may respond differently in the morning than she would in the evening, or differently alone than she would in a group. These variations in situational responses do not indicate a true change in the feelings, attitudes, or behaviors being measured, but are only examples of instrumentation error.

STATISTICAL REGRESSION

The fifth threat to internal validity, statistical regression, refers to the tendency of extremely low and extremely high scores to regress, or move toward the average score for everyone in the research study. Suppose that a student, named Maryanna, has to take a multiple-choice exam on a subject she knows nothing about. There are many questions, and each question has five possible answers.

Since, for each question, Maryanna has a 20 percent (one in five) chance of guessing correctly, she might expect to score 20 percent on the exam just by guessing. If she guesses badly, she will score a lot lower; if well, a lot higher. The other members of the class take the same exam and, since they are all equally uninformed, the average score for the class is 50 percent.

Now suppose that the instructor separates the low scorers from the high scorers and tries to even out the level of the class by giving the low scorers special instruction. In order to determine if the special instruction has been effective, the entire class then takes another multiple-choice exam. The result of the exam is that the low scorers (as a group) do better than they did the first time, and the high scorers (as a group) worse. The instructor believes that this has occurred because the low scorers received special instruction and the high scorers did not.

According to the logic of statistical regression, however, both the average score of the low scorers (as a group) and the average score of the high scorers (as a group) would move toward the total average score for both groups (i.e., high and low). Even without any special instruction and still in their state of ignorance, the low scorers (as a group) would be expected to have a higher average score than they did before. Likewise, the high scorers (as a group) would be expected to have a lower average score than they did before.

It would be easy for the research instructor to assume that the low scores had increased because of the special instruction and the high scores had decreased because of the lack of it. Not necessarily so, however; the instruction may have had nothing to do with it. It may all be due to statistical regression.

DIFFERENTIAL SELECTION OF RESEARCH PARTICIPANTS

The sixth threat to internal validity is differential selection of research participants. To some extent, the participants selected for a research study are different from one another to begin with. "Ideal" experiments, however, require random sampling from a population (if at all possible) and random assignment to groups.

This assures that the results of a study will be generalizable to a larger population, thus addressing threats to external validity. In respect to differential selection as a threat to internal validity, "ideal" experiments control for this since equivalency among the groups at pretest is assumed through the randomization process.

This threat is, however, present when we are working with preformed groups or groups that already exist, such as classes of students, self-help groups, or community groups. In terms of the external validity of such designs, because there is no way of knowing whether the preformed groups are representative of any larger population, it is not possible to generalize the study's results beyond the people (or objects or events) that were actually studied.

The use of preformed groups also affects the internal validity of a study, though. It is probable that different preformed groups will not be equivalent with respect to relevant variables, and that these initial differences will invalidate the results of the posttest.

A child abuse prevention educational program for children in schools might be evaluated by comparing the prevention skills of one group of children who have experienced the educational program with the skills of a second group who have not. In order to make a valid comparison, the two groups must be as similar as possible, with respect to age, gender, intelligence, socioeconomic status, and anything else that might affect the acquisition of child abuse prevention skills.

We would have to make every effort to form or select equivalent groups, but the groups are sometimes not as equivalent as might be hoped—especially if we are obliged to work with preformed groups, such as classes of students or community groups. If the two groups are different before the intervention was introduced, there is not much point in comparing them at the end.

Accordingly, preformed groups should be avoided whenever possible. If it is not feasible to do this, rigorous pre-testing must be done to determine in what ways the groups are (or are not) equivalent, and differences must be compensated for with the use of statistical methods.

MORTALITY

The seventh threat to internal validity is mortality, which simply means that individual research participants may drop out before the end of the study. Their absence will probably have a significant effect on the study's findings because people who drop out are likely to be different in some ways from the other participants who stay in the study. People who drop out may be less motivated to participate in the intervention than people who stay in, for example.

Since dropouts often have such characteristics in common, it cannot be assumed that the attrition occurred in a random manner. If considerably more people drop out of one group than out of the other, the result will be two groups that are no longer equivalent and cannot be usefully compared. We cannot know at the beginning of the study how many people will drop out, but we can watch to see how many do. Mortality is never problematic if dropout rates are five percent or less *and* if the dropout rates are similar for the various groups.

REACTIVE EFFECTS OF RESEARCH PARTICIPANTS

The eighth threat to internal validity is reactive effects. Changes in the behaviors or feelings of research participants may be caused by their reaction to the novelty of the situation or the knowledge that they are participating in a research study. A mother practicing communication skills with her child, for example, may try especially hard when she knows the social worker is watching. We may wrongly believe that such reactive effects are due to the intervention.

The classic example of reactive effects was found in a series of studies carried out at the Hawthorne plant of the Western Electric Company in Chicago many years ago. Researchers were investigating the relationship between working conditions and productivity. When they increased the level of lighting in one section of the plant, productivity increased; a further increase in the lighting was followed by an additional increase in productivity.

When the lighting was then decreased, however, production levels did not fall accordingly but continued to rise. The conclusion was that the workers were increasing their productivity not because of the lighting level but because of the attention they were receiving as research participants in the study.

The term *Hawthorne effect* is still used to describe any situation in which the research participants' behaviors are influenced not by the intervention but by the knowledge that they are taking part in a research project. Another example of such a reactive effect is the placebo given to patients, which produces beneficial results because they believe it is medication.

Reactive effects can be controlled by ensuring that all participants in a research study, in both the experimental and control groups, appear to be treated equally. If one group is to be shown an educational film, for example, the other group should also be shown a film—some film carefully chosen to bear no relationship to the variable being investigated. If the study involves a change in the participants' routine, this in itself may be enough to change behavior, and care must be taken to continue the study until novelty has ceased to be a factor.

INTERACTION EFFECTS

Interaction among the various threats to internal validity can have an effect of its own. Any of the factors already described as threats may interact with one another, but the most common interactive effect involves differential selection and maturation. Let us say we are studying two groups of clients who are being treated for depression. The intention was for these groups to be equivalent, in terms of both their motivation for treatment and their levels of depression.

It turns out that Group A is more generally depressed than Group B, however. Whereas both groups may grow less motivated over time, it is likely that Group A, whose members were more depressed to begin with, will lose motivation more completely and more quickly than Group B. Inequivalent groups thus grow less equivalent over time as a result of the interaction between differential selection and maturation.

RELATIONS BETWEEN EXPERIMENTAL AND CONTROL GROUPS

The final group of threats to internal validity has to do with the effects of the use of experimental and control groups that receive different interventions. These effects include: (1) diffusion of treatments, (2) compensatory equalization, (3) compensatory rivalry, and (4) demoralization.

Diffusion of Treatments

Diffusion, or imitation of treatments, may occur when the experimental and control groups talk to each other about the study. Suppose a study is designed that presents a new relaxation exercise to the experimental group and nothing at all to the control group. There is always the possibility that one of the participants in the experimental group will explain the exercise to a friend who happens to be in the control group. The friend explains it to another friend, and so on. This might be beneficial for the control group, but it invalidates the study's findings.

Compensatory Equalization

Compensatory equalization of treatment occurs when the person doing the study and/or the staff member administering the intervention to the experimental group feels sorry for people in the control group who are not receiving it and attempts to compensate them.

A social worker might take a control group member aside and covertly demonstrate the relaxation exercise, for example. On the other hand, if our study has been ethically designed, there should be no need for guilt on the part of the social worker because some people are not being taught to relax. They can be taught to relax when our study is "officially" over.

Compensatory Rivalry

Compensatory rivalry is an effect that occurs when the control group becomes motivated to compete with the experimental group. For example, a control group in a program to encourage parental involvement in school activities might get wind that something is up and make a determined effort to participate too, on the basis that "anything they can do, we can do better." There is no direct communication between groups, as in the diffusion of treatment effect—only rumors and suggestions of rumors are often enough to threaten the internal validity of a study.

Demoralization

In direct contrast with compensatory rivalry, demoralization refers to feelings of deprivation among the control group that may cause them to give up and drop out of the study, in which case this effect would be referred to as mortality. The people in the control group may also get angry.

BOX 10.2
Threats to External Validity

External validity is the degree to which the results of our research study are generalizable to a larger population or to settings outside the research situation or setting.

PRETEST-TREATMENT INTERACTION

The first threat to external validity, pretest-treatment interaction, is similar to the testing threat to internal validity. The nature of a pretest can alter the way research participants respond to the experimental treatment, as well as to the posttest. Suppose, for example, that an educational program on racial tolerance is being evaluated.

A pretest that measures the level of tolerance could well alert the participants to the fact that they are going to be educated into loving all their neighbors, but many people do not want to be "educated" into anything. They are satisfied with the way they feel and will resist the instruction. This will affect the level of racial tolerance registered on the posttest.

SELECTION-TREATMENT INTERACTION

The second threat to external validity is selection-treatment interaction. This threat commonly occurs when a research design cannot provide for random selection of participants from a population. Suppose we wanted to study the effectiveness of a family service agency staff, for example. If our research proposal was turned down by 50 agencies before it was accepted by the 51st, it is very likely that the accepting agency differs in certain important aspects from the other 50.

It may accept the proposal because its social workers are more highly motivated, more secure, more satisfied with their jobs, or more interested in the practical application of the study than the average agency staff member.

As a result, we would be assessing the research participants on the very factors for which they were unwittingly (and by default) selected—motivation, job satisfaction, and so on. The study may be internally valid, but, since it will not be possible to generalize the results to other family service agencies, it would have little external validity.

SPECIFICITY OF VARIABLES

Specificity of variables has to do with the fact that a research project conducted with a specific group of people at a specific time and in a specific setting may not always be generalizable to other people at a different time and in a different setting.

For example, a measuring instrument developed to measure the IQ levels of upper-socioeconomic level, Caucasian, suburban children does not provide an equally accurate measure of IQ when it is applied to lower-socioeconomic level children of racial minorities in the inner city.

REACTIVE EFFECTS

The fourth threat to external validity is reactive effects which, as with internal validity, occur when the attitudes or behaviors of our research participants are affected to some degree by the very act of taking a pretest. Thus, they are no longer exactly equivalent to the population from which they were randomly selected, and it may not be possible to generalize our study's results to that population. Because pretests affect research participants to some degree, our results may be valid only for those who were pretested.

MULTIPLE-TREATMENT INTERFERENCE

The fifth threat to external validity, multiple-treatment interference, occurs if a research participant is given two or more interventions in succession, so that the results of the first intervention may affect the results of the second one. A client attending treatment sessions, for example, may not seem to benefit from one therapeutic technique, so another is tried.

In fact, however, the client may have benefitted from the first technique but the benefit does not become apparent until the second technique has been tried. As a result, the effects of both techniques become commingled, or the results may be erroneously ascribed to the second technique alone.

Because of this threat, interventions should be given separately if possible. If our research design does not allow this, sufficient time should be allowed to elapse between the two interventions in an effort to minimize the possibility of multiple-treatment interference.

RESEARCHER BIAS

The final threat to external validity is researcher bias. Researchers, like people in general, tend to see what they want to see or expect to see. Unconsciously and without any thought of deceit, they may manipulate a study so that the actual results agree with the anticipated results. A practitioner may favor an intervention so strongly that the research study is structured to support it, or the results are interpreted favorably.

If we know which individuals are in the experimental group and which are in the control group, this knowledge alone might affect the study's results. Students who an instructor believes to be bright, for example, often are given higher grades than their performance warrants, while students believed to be dull are given lower grades.

> The way to control for such researcher bias is to perform a double-blind experiment in which neither the research participants nor the researcher knows who is in the experimental or control group or who is receiving a specific treatment intervention.

Both internal and external validity are achieved in a research design by taking into account various threats that are inherent in all research efforts. A design for a study with both types of validity will recognize and attempt to control for potential factors that could affect our study's outcome or findings. An "ideal" experiment tries to control as many threats to internal and external validity as possible.

GROUP RESEARCH DESIGNS

While, in a particular case, a group research design may need to be complex to accomplish the purpose of the study, a design that is unnecessarily complex costs more, takes more time, and probably will not serve its purpose nearly as well as a simpler one. In choosing a research design (whether a single case [see last chapter] or group), therefore, the principle of parsimony must be applied: The simplest and most economical route to the objective is the best choice.

Exploratory Designs

At the lowest level of the continuum of knowledge that can be derived from research studies are exploratory group research designs. An exploratory study explores a research question about which little is already known, in order to uncover generalizations and develop hypotheses that can be investigated and tested later with more precise and, hence, more complex designs and data gathering techniques.

The examples of exploratory designs given in this section do not use pretests; they simply measure the dependent variable only after the independent variable has been introduced. Therefore, they cannot be used to determine whether changes took place in the study's research participants; these designs simply describe the state of the research participants after they had received the independent variable (if any—see Box 10.3).

> **BOX 10.3**
>
> **Treatment: A Variable or a Constant?**
>
> For instructional purposes, group designs are displayed using symbols where X is the independent variable (treatment) and O is the measure of the dependent variable. This presentation is accurate when studies are designed with two or more groups. When one-group designs are used, however, this interpretation does not hold. In one-group designs, the treatment, or program, cannot truly vary because all research participants have experienced the same event; that is, they all have experienced the program. Without a comparison or control group, treatment is considered a constant because it is a quality shared by all members in the research study. In short, *time* is the independent variable.
>
> There does not necessarily have to be an independent variable in a study, however; we may just want to measure some variable in a particular population such as the number of people who receive a certain type of social service intervention over a 10-year period. In this situation, there is no independent or dependent variable.

One-Group Posttest-Only Design

The first exploratory group level design is the one-group posttest-only design. It is sometimes called the one-shot case study or cross-sectional case study design. It is the simplest of all the group research designs. Suppose in a particular community, Rome, Wisconsin, there are numerous parents who are physically abusive toward their children. The city decides to hire a school social worker, Antonia, to implement a program that is supposed to reduce the number of parents who physically abuse their children.

She conceptualizes a 12-week child abuse prevention program (the intervention) and offers it to parents who have children in her school who wish to participate on a voluntary basis. A simple research study is then conducted to answer the question, "Did the parents who completed the program stop physically abusing their children?" The answer to this question will determine the success of the intervention.

There are many different ways in which this program can be evaluated. For now, and to make matters as simple as possible, we are going to evaluate it by simply counting how many parents stopped physically abusing their children after they attended the program. At the simplest level, the program could be evaluated with a one-group posttest-only design. The basic elements of this design can be written as illustrated in Figure 10.1.

```
        Not an independent variable                    Dependent variable
        (see Box 10.3)
```

```
  No random sample  →  Program or  →  Posttest
  (accidental sample)  intervention   (O₁)
                       (X)
```

```
                                            Percent of parents who
                                            stopped abusing their children
        Results are not generalizable beyond the
        people who participated in the program
```

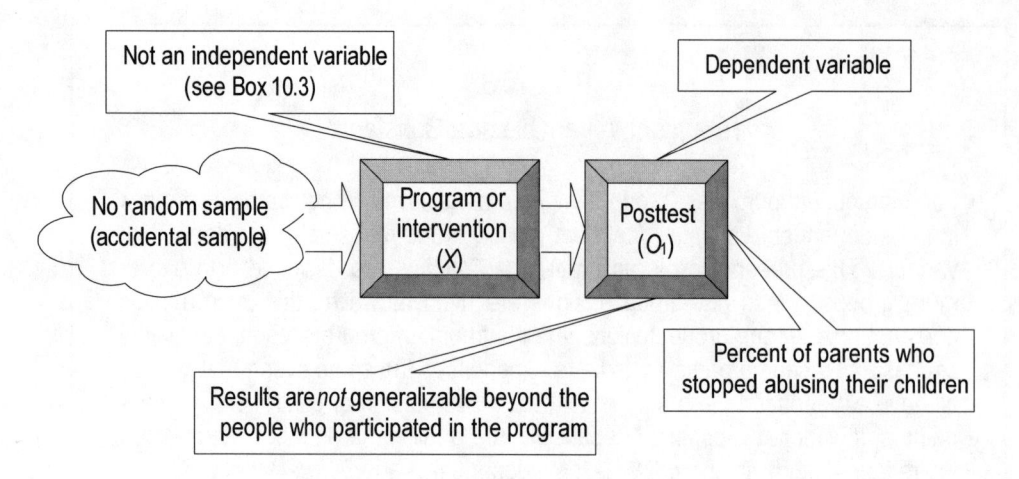

FIGURE 10.1
One-Group Post-Test Only Design

Where:

X = Child Abuse Prevention Program, or the intervention (see Box 10.3)

O_1 = First and only measurement of the dependent variable (percentage of parents who stopped physically abusing their children, the program's outcome, or program objective)

All that this design provides is a single measure (O_1) of what happens when one group of people is subjected to one treatment or experience (X). The program's participants were not randomly selected from any particular population, and, thus, the results of the findings cannot be generalized to any other group or population. It is safe to assume that all the members within the program had physically abused their children before they enrolled, since people who do not have this problem would not enroll in such a program.

But, even if the value of O_1 indicates that some of the parents did stop being violent with their children after the program, it cannot be determined whether they quit because of the intervention (the program) or because of some other rival hypothesis.

Perhaps a law was passed that made it mandatory for the police to arrest anyone who behaves violently toward his or her child, or perhaps the local television station started to report such incidents on the nightly news, complete with pictures of the abusive parent. These other extraneous variables may have been more important in persuading the parents to cease their abusive behavior toward their children than their voluntary participation in the program.

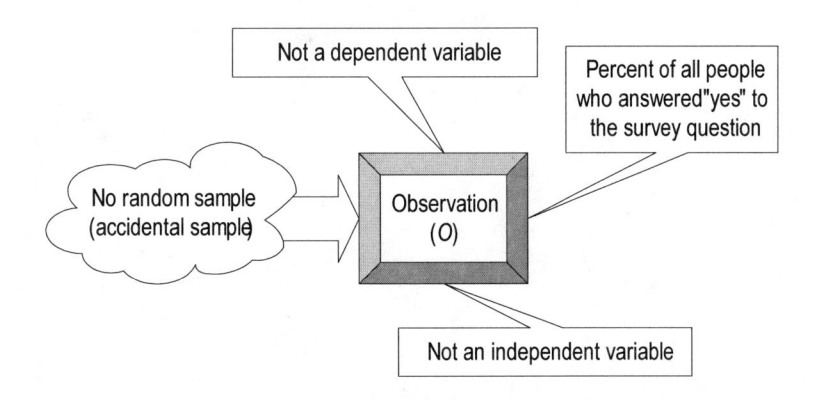

FIGURE 10.2
Cross-Sectional Survey Design

In sum, the one-group posttest only design (Figure 10.1) does not control for many of the threats to either internal or external validity. In terms of internal validity, the threats that are not controlled for in this design are history, maturation, differential selection, and mortality.

Cross-Sectional Survey Design

The second exploratory group level design is the cross-sectional survey design as illustrated in Figure 10.2. Let us take another example of a one-group posttest-only design that *does not* have an intervention. In survey research, this kind of a group research design is called a cross-sectional survey design.

In doing a cross-sectional survey, we survey *only once* a cross-section of some particular population. In addition to Antonia's child abuse prevention program geared for abusive parents, she may also want to start another program geared for all the children in the school (whether they come from abusive families or not)—a child abuse educational program taught to children in the school.

Before Antonia starts her educational program geared for the children, however, she wants to know what parents think about the idea. She may send out questionnaires to all the parents or she may decide to personally telephone every second parent, or every fifth or tenth, depending on how much time and money she has. The results of her survey constitute a single measurement, or observation, of the parents' opinions of her second proposed program (the one for the children) and may be written depicted in Figure 10.2.

The symbol *O* represents the entire cross-sectional survey design since such a design involves making only a single observation, or measurement, at one time period.

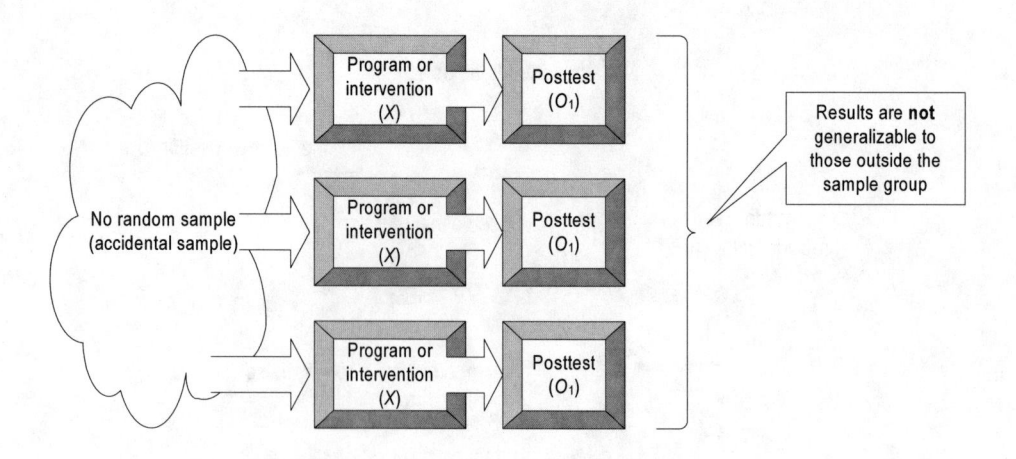

[Figure 10.3 diagram: accidental sample branching into three groups, each receiving Program or intervention (X) then Posttest (O₁); annotation: "Results are **not** generalizable to those outside the sample group"]

FIGURE 10.3
Multi-Group Posttest-Only Design

Note that there is no X in Figure 10.2, as there is really no independent variable. Antonia only wants to ascertain the parents' attitudes toward her proposed program—nothing more, nothing less.

Multi-Group Posttest-Only Design

The third exploratory group level design is the multi-group posttest-only design as illustrated in Figure 10.3. The multi-group posttest-only design is an elaboration of the one-group posttest-only design in which more than one group is used. To check a bit further into the effectiveness of Antonia's program for parents who have been physically abusive toward their children, for example, she might decide to locate several more groups of parents who had completed her program and see how many of them had stopped abusing their children—and so on, with any number of groups.

The multi-group posttest-only design can be diagramed as in Figure 10.3 and written in symbols:

Where:

X = Child Abuse Prevention Program, or the intervention (see Box 10.3)

O_1 = First and only measurement of the dependent variable (percentage of parents who stopped physically abusing their children, the program's outcome, or program objective)

Chapter 10: Group-Level Research Designs **211**

> Since the sample was not randomly drawn from a population, the results from O_1, O_2, O... are **not** generalizable beyond the people who participated in the study

No random sample (accidental sample) → Program or intervention (X) → Observation (O_1) [3-months] → Observation (O_2) [6-months] → Observation ($O...$) [9-months]

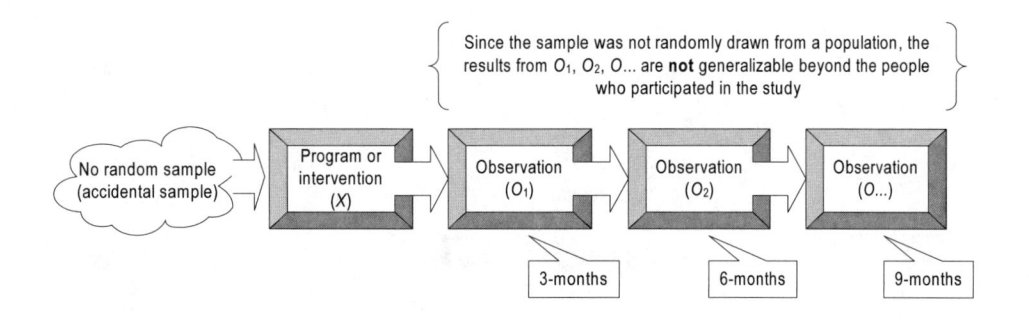

FIGURE 10.4
Longitudinal One-Group Only Posttest Design

Where:

X = Child Abuse Prevention Program, or the intervention (see Box 10.3)

O_1 = First measurement of the dependent variable (percentage of parents who stopped physically abusing their children, the program's outcome, or program objective)

O_2 = Second measurement of the dependent variable (percentage of parents who stopped physically abusing their children, the program's outcome, or program objective)

O_3 = Third measurement of the dependent variable (percentage of parents who stopped physically abusing their children, the program's outcome, or program objective)

With the multi-group design it cannot be assumed that all three Xs are equivalent because the three programs might not be exactly the same; one group might have had a different facilitator, the program might have been presented differently, or the material could have varied in important respects.

In addition, nothing is known about whether any of the research participants would have stopped being violent anyway, even without the program. It certainly cannot be assumed that any of the groups were representative of the larger population. Thus, as in the case of the one-group posttest-only design, the same threats to the internal and the external validity of the study might influence the results of the multi-group posttest design.

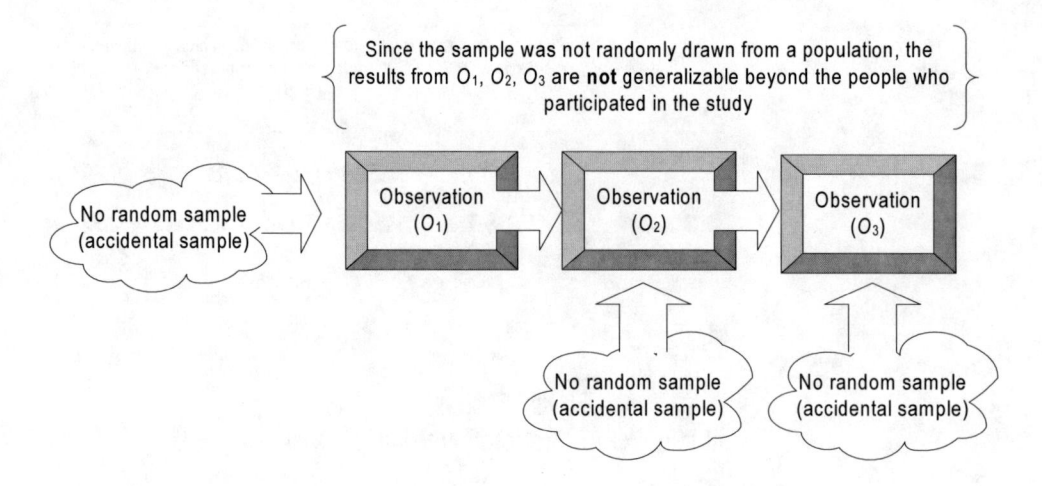

FIGURE 10.5
Antonia's Trend Research Study

Where:

O_1 = First measurement of parents' attitudes toward the school offering abuse prevention education to second-grade students in Sample 1

O_2 = Second measurement of parents' attitudes toward the school offering abuse prevention education to second-grade students in Sample 2

O_3 = Third measurement of parents' attitudes toward the school offering abuse prevention education to second-grade students in Sample 3

Longitudinal Case Study Design

The fourth exploratory group level design is the longitudinal case study design. The longitudinal case study design is exactly like the one-group posttest-only design, except that it provides for more measurements of the dependent variable (Os). This design can be written as illustrated in Figure 10.4.

Suppose that Antonia is interested in the long-term effects of the child abuse prevention program. Perhaps the program was effective in helping some people to stop physically abusing their children, but will they continue to refrain from abusing their children?

One simple way to find out is to calculate the percentage of parents who physically abuse their children at intervals—say at the end of the program, the first three months after the program, then the next three months after that, and every three months for the next two years.

Chapter 10: Group-Level Research Designs   **213**

This design can be used to monitor the effectiveness of treatment interventions over time and can be applied not just to groups but also to single-client systems, as described in Chapter 8. However, all of the same threats to the internal and external validity that were described in relation to the previous two exploratory designs also apply to this design.

Longitudinal Survey Design

The fifth exploratory group level design is the longitudinal survey design. Unlike cross-sectional surveys, where the variable of interest is measured at one point in time, longitudinal surveys provide data at various points in time so that changes can be monitored over time. They can be down into three types: (1) trend studies, (2) cohort studies, and (3) panel studies.

1. Trend Studies. A trend study is used to find out how a population, or sample, changes over time as illustrated in Figure 10.5. Antonia, the school social worker mentioned previously in this chapter, may want to know if parents of young children enrolled in her school are becoming more receptive to the idea of the school teaching their children child abuse prevention education in the second grade (Williams, Tutty, & Grinnell, 1995). She may survey all the parents of Grade 2 children this year, all the parents of the new complement of Grade 2 children next year, and so on until she thinks she has sufficient data.

Each year the parents surveyed will be different, but they will all be parents of Grade 2 children. In this way, Antonia will be able to determine whether parents are becoming more receptive to the idea of introducing child abuse prevention material to their children as early as Grade 2. In other words, she will be able to measure any attitudinal trend that is, or is not, occurring. The research design can still be written as illustrated in Figure 10.5.

2. Cohort Studies. Cohort studies are used over time to follow a single group of people who have shared a similar experience—for example, AIDS survivors, sexual abuse survivors, or parents of grade-school children. Perhaps Antonia is interested in knowing whether parents' attitudes toward the school offering abuse prevention education to second-grade students change as their children grow older.

She may survey a random sample of the Grade 2 parents who attend a Parent Night this year, and survey a different random sample of parents who attend a similar meeting for the same parents next year, when their children are in Grade 3.

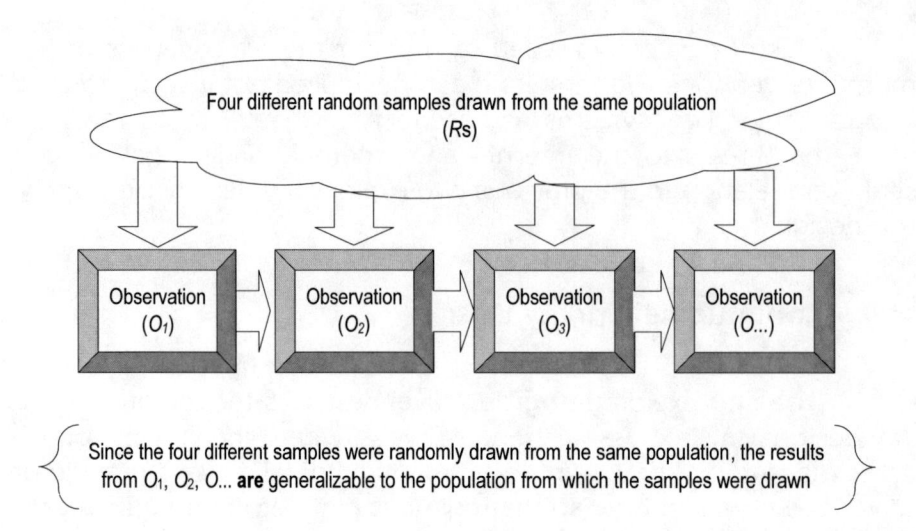

[Diagram: Four different random samples drawn from the same population (Rs) flowing down to four Observation boxes: Observation (O_1) → Observation (O_2) → Observation (O_3) → Observation ($O...$)]

Since the four different samples were randomly drawn from the same population, the results from O_1, O_2, $O...$ **are** generalizable to the population from which the samples were drawn

FIGURE 10.6
Antonia's Cohort Research Study

The following year, when the children are in Grade 4, she will take another, different random sample of those parents who attend Parent Night. Although different parents are being surveyed every year, they all belong to the same population of parents whose children are progressing through the grades together. Her research design can be written as depicted in Figure 10.6.

Where:

O_1 = First measurement of parents' attitudes toward the school offering abuse prevention education to second-grade students for a random sample drawn from some population

O_2 = Second measurement of parents' attitudes toward the school offering abuse prevention education to second-grade students for a different random sample drawn from the same population one year later

O_3 = Third measurement of parents' attitudes toward the school offering abuse prevention education to second-grade students for a still different random sample, drawn from the same population after two years

3. Panel Studies. In a panel study, the *same individuals* are followed over a period of time. Antonia might select one random sample of parents, for example, and measure their attitudes toward child abuse prevention education in successive years. Again, the design can be illustrated as Figure 10.7 and written:

Chapter 10: Group-Level Research Designs   **215**

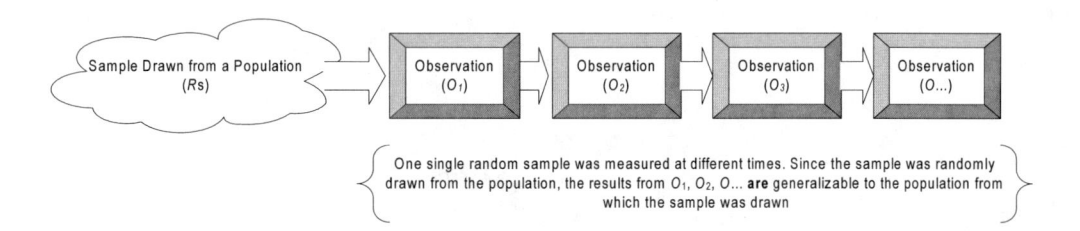

```
Sample Drawn from a Population (Rs) → Observation (O₁) → Observation (O₂) → Observation (O₃) → Observation (O...)
```

{ One single random sample was measured at different times. Since the sample was randomly drawn from the population, the results from O_1, O_2, O... **are** generalizable to the population from which the sample was drawn }

FIGURE 10.7
Antonia's Panel Research Study

Where:

O_1 = First measurement of attitudes toward child abuse prevention education for a sample of individuals

O_2 = Second measurement of attitudes toward child abuse prevention education for the same sample of individuals one year later

O_3 = Third measurement of attitudes toward child abuse prevention education for the same sample of individuals after two years

In summary, a trend study is interested in broad trends over time, whereas a cohort study provides data about people who have shared similar experiences. In neither case do we know anything about *individual* contributions to the changes that are being measured. A panel study provides data that we can use to look at change over time as experienced by particular individuals.

Descriptive Designs

At the midpoint of the knowledge continuum are descriptive designs, which have some but not all of the requirements of an "ideal" experiment. They usually require specification of the time order of variables, manipulation of the independent variable, and establishment of the relationship between the independent and dependent variables.

They may also control for rival hypotheses and use a second group as a comparison (not as a control). The requirement that descriptive designs lack most frequently is the random assignment of research participants to two or more groups. We are seldom in a position to randomly assign research participants to either an experimental or control group. Sometimes the groups to be studied are already in existence; sometimes ethical issues are involved.

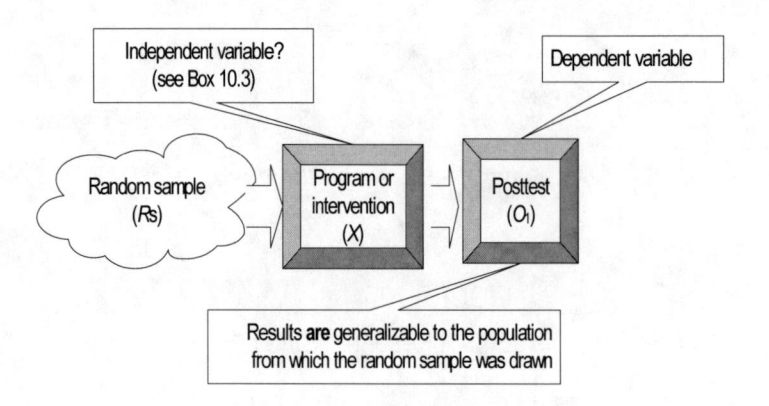

FIGURE 10.8
Randomized One-Group Posttest-Only Design

Where:

R = Random selection from a population
X = Program, or the intervention (see Box 10.3)
O_1 = First and only measurement of the dependent variable

It would be unethical, for example, to assign clients who need immediate help to two random groups, only one of which is to receive the intervention. Since a lack of random assignment will affect the internal and external validities of the study, the descriptive research design must try to compensate for this.

Randomized One-Group Posttest-Only Design

The distinguishing feature of the randomized one-group posttest-only design is that members of the group are randomly selected for it. Otherwise, this design is identical to the exploratory one-group posttest-only design. The randomized one-group posttest-only design is written as depicted in Figure 10.8.

In the example of the child abuse prevention program, the difference in this design is that the group does not accidentally assemble itself by including anyone who happened to be interested in volunteering for the program.

Instead, group members are randomly selected from a population, say, of all the 400 parents who were reported to child welfare authorities for having physically abused a child and who wish to receive voluntary treatment in Rome, Wisconsin, in 2010. These 400 parents comprise the population of all the physically abusive parents who wish to receive treatment in Rome, Wisconsin.

Chapter 10: Group-Level Research Designs **217**

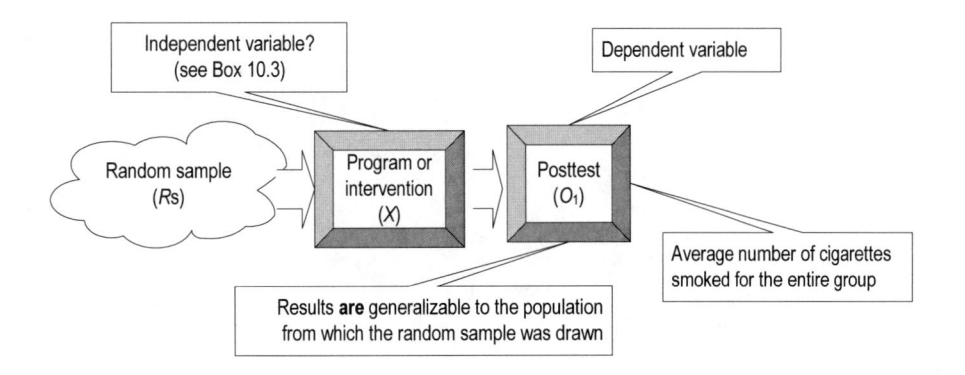

FIGURE 10.9
Randomized One-Group Posttest-Only Design
for Physically Abusive Parents in Roam Wisconsin

Where:

R = Random selection of 40 people from the population of physically abusive parents who voluntarily wish to receive treatment in Rome, Wisconsin

X = Child Abuse Prevention Program, or the intervention (see Box 10.3)

O_1 = Percentage of parents in the program who stopped being physically abusive to their children

The sampling frame of 400 people is used to select a simple random sample of 40 physically abusive parents who voluntarily wish to receive treatment. The program (X) is administered to these 40 people, and the number of parents who stopped being abusive toward their children after the program is determined (O_1). The design can be written as illustrated in Figure 10.9.

Say that the program fails to have the desired effect, and 80 percent of the people continue to physically harm their children after participating in the program. Because the program was ineffective for the sample and the sample was randomly selected, it can be concluded that the program would be ineffective for the physically abusive parent population of Rome, Wisconsin—the other 360 who did not go through the program.

In other words, because a representative random sample was selected, it is possible to generalize the program's results to the population from which the sample was drawn. Since no change in the dependent variable occurred, we don't need to consider the control of rival hypotheses. Antonia need not wonder what might have caused the miniscule change—X, her program, or an alternative explanation.

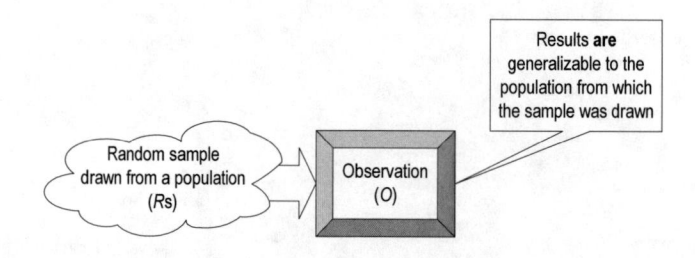

FIGURE 10.10
Randomized Cross-Sectional Survey Design

If Anita's program had been successful, however, it would not be possible to ascribe her success *solely* to the program.

Randomized Cross-Sectional Survey Design

As discussed earlier, a cross-sectional survey obtains data only once from a sample of a particular population. If the sample is a random sample—that is, if it represents the population from which it was drawn—then the data obtained from the sample can be generalized to the entire population. A cross-sectional survey design using a random sample (e.g., Figure 10.10) can be written:

Where:

R = Random sample drawn from a population
O_1 = First and only measurement of the dependent variable

Explanatory surveys look for associations between variables. Often, the suspected reason for the relationship is that one variable caused the other. In Antonia's case, she has two studies going on: the child abuse prevention program for parents who have physically abused their children, and her survey of parental attitudes toward the school that is teaching second-grade children child abuse prevention strategies.

The success of the child abuse prevention program (her program) may have caused parents to adopt more positive attitudes toward the school in teaching their children child abuse prevention (her survey). In this situation, the two variables, the program and survey, become commingled.

Demonstrating causality is a frustrating business at the best of times because it is so difficult to show that nothing apart from the independent variable could have caused the observed change in the dependent variable.

Chapter 10: Group-Level Research Designs

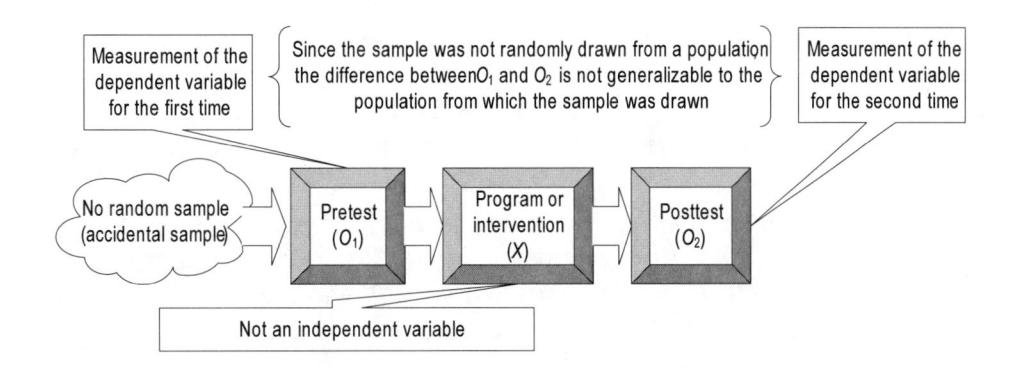

FIGURE 10.11
One-Group Pretest-Posttest Design

Even supposing that this problem is solved, it is impossible to demonstrate causality unless data are obtained from random samples and are generalizable to entire populations.

One-Group Pretest-Posttest Design

The one-group pretest-posttest design is also referred to as a before-after design because it includes a pretest of the dependent variable, which can be used as a basis of comparison with the posttest results. It is illustrated in Figure 10.11 and written as:

Where:

O_1 = First measurement of the dependent variable
X = Program, or the intervention (see Box 10.3)
O_2 = Second measurement of the dependent variable

The one-group pretest-posttest design, in which a pretest precedes the introduction of the intervention and a posttest follows it, can be used to determine precisely how the intervention affects a particular group. The design is used often in social work decision making—far too often, in fact, because it does not control for many rival hypotheses. The difference between O_1 and O_2, on which these decisions are based, therefore, could be due to many other factors rather than the intervention.

Let us take another indicator of how Antonia's child abuse prevention program could be evaluated. Besides calculating the percentage of parents who stopped physically abusing their children as the only indicator of the program's success, she could have a second outcome indicator such as reducing the parents' risk for abusive and neglecting parenting behaviors. This dependent variable could be easily measured by an instrument that measures their attitudes of physical punishment of children. Let us say that Antonia had the parents complete the instrument *before* the child abuse prevention program (O_1) and *after* it (O_2).

In this example, history would be a rival hypothesis or threat to internal validity because all kinds of things could have happened between O_1 and O_2 to affect the participants' behaviors and feelings—such as the television station deciding to publicize the names of parents who are abusive to their children.

Testing also could be a problem. Just the experience of taking the pretest could motivate some participants to stop being abusive toward their children. Maturation—in this example, the children becoming more mature with age so that they became less difficult to discipline—would be a further threat. This design controls for the threat of differential selection, since the participants are the same for both pretest and posttest.

Second, mortality would not affect the outcome, because it is the differential drop-out between groups that causes this threat and, in this example, there is only one group (Williams, Tutty, & Grinnell, 1995).

Comparison Group Posttest-Only Design

The comparison group posttest-only design improves on the exploratory one-group and multi-group posttest-only designs by introducing a comparison group that does not receive the independent variable, but is subject to the same posttest as those who do (the experimental group). The group used for purposes of comparison is usually referred to as a comparison group in an exploratory or descriptive design and as a control group in an explanatory design. While a control group is always randomly assigned, a comparison group is not.

The basic elements of the comparison group posttest-only design are as follows (Figure 10.12):

Chapter 10: Group-Level Research Designs

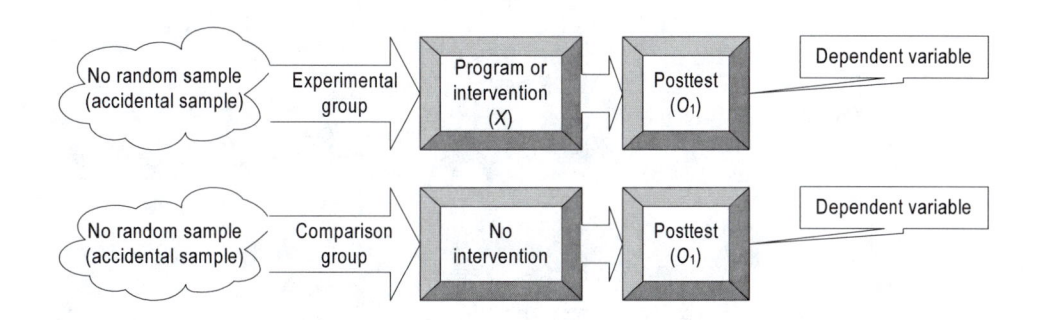

FIGURE 10.12
Comparison Group Posttest-Only Design

Where:

X = Independent variable, or the intervention

O_1 = First and only measurement of the dependent variable

In Antonia's child abuse prevention program, if the January, April, and August sections are scheduled but the August sessions are canceled for some reason, those who would have been participants in that section could be used as a comparison group. If the values of O_1 on the measuring instrument were similar for the experimental and comparison groups, it could be concluded that the program was of little use, since those who had experienced it (those receiving X) were not much better or worse off than those who had not.

A problem with drawing this conclusion, however, is that there is no evidence that the groups were equivalent to begin with. Selection, mortality, and the interaction of selection and other threats to internal validity are, thus, the major difficulties with this design. The comparison group does, however, control for such threats as history, testing, and instrumentation.

Comparison Group Pretest-Posttest Design

The comparison group pretest-posttest design (Figure 10.13) elaborates on the one-group pretest-posttest design by adding a comparison group. This second group receives both the pretest (O_1) and the posttest (O_2) at the same time as the experimental group, but it does not receive the independent variable. This design is written as follows:

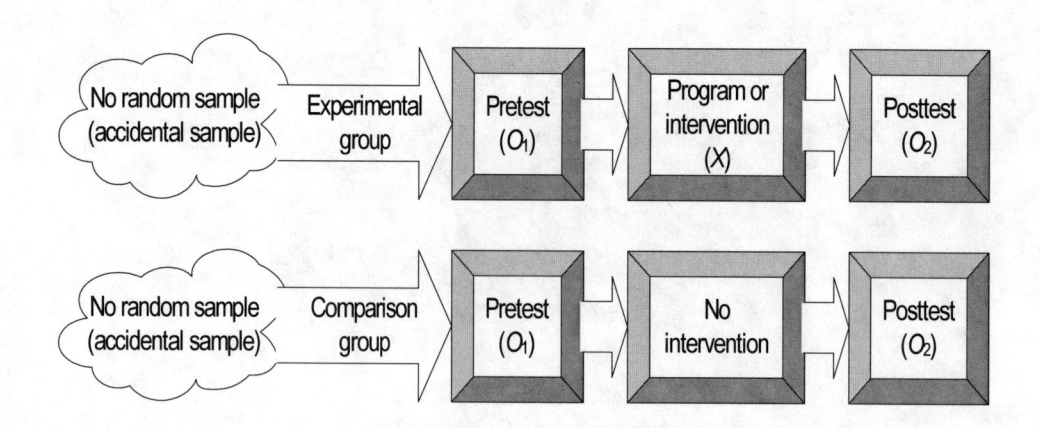

FIGURE 10.13
Comparison Group Pretest-Posttest Design

Where:

O_1 = First measurement of the dependent variable, the parents' scores on the measuring instrument

X = Independent variable, or the intervention

O_2 = Second measurement of the dependent variable, the parents' scores on the measuring instrument

The experimental and comparison groups formed under this design will probably not be equivalent, because members are not randomly assigned to them. The pretest scores, however, will indicate the extent of their differences. If the differences are not statistically significant, but are still large enough to affect the posttest, the statistical technique of analysis of covariance can be used to compensate for this. As long as the groups are equivalent at pretest, then, this design controls for nearly all of the threats to internal validity. But, because random selection and assignment were not used, the external validity threats remain.

Interrupted Time-Series Design

In the interrupted time-series design, a series of pretests and posttests are conducted on a group of research participants over time, both before and after the independent variable is introduced. The basic elements of this design are illustrated in Figure 10.14 and is considered a single-case design as described in the previous chapter.

Chapter 10: Group-Level Research Designs

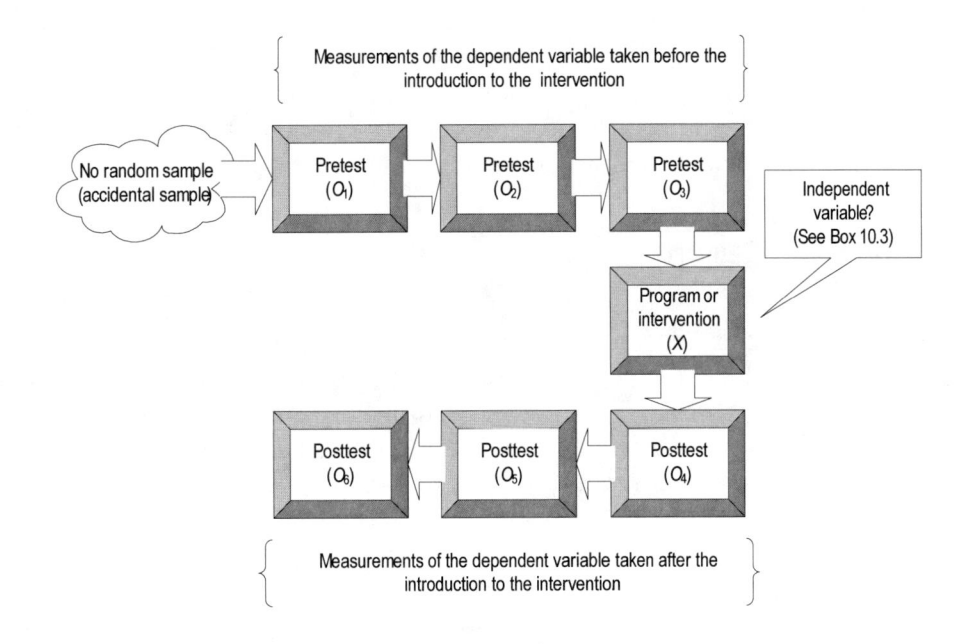

223

FIGURE 10.14
Interrupted Time-Series Design

Where:

Os = Measurements of the dependent variable
X = Independent variable, or the intervention

The interrupted time-series design (basically a case-level design as presented in the previous chapter) takes care of the major weakness in the descriptive one-group pretest-posttest design, which does not control for many rival hypotheses.

Suppose, for example, that a new policy is to be introduced into an agency whereby all promotions and raises are to be tied to the number of educational credits acquired by social workers. Since there is a strong feeling among some workers that years of experience should count for more than educational credits, the agency's management decides to examine the effect of the new policy on morale.

Because agency morale is affected by many things and varies normally from month to month, it is necessary to ensure that these normal fluctuations are not confused with the results of the new policy. Therefore, a baseline is first established for morale by conducting a number of pretests over, say, a six-month period before the policy is introduced.

Then, a similar number of posttests is conducted over the six months following the introduction of the policy. The same type of time-series design can be used to evaluate the result of a treatment intervention with a client or client system, as in case-level designs described in the previous chapter. Again, without randomization, threats to external validity still could affect the study's generalizability, but most of the threats to internal validity are addressed.

Explanatory Designs

Explanatory group research designs approach the "ideal" experiment most closely. They are at the highest level of the knowledge continuum, have the most rigid requirements, and are most able to produce results that can be generalized to other people and situations. Explanatory designs, therefore, are most able to provide valid and reliable research results that can serve as additions to our professions' knowledge base.

The purpose of an explanatory design is to establish a causal connection between the independent and dependent variable. The value of the dependent variable could always result from chance rather than from the influence of the independent variable, but there are statistical techniques for calculating the probability that this will occur.

Classical Experimental Design

The classical experimental design is the basis for all the experimental designs. It involves an experimental group and a control group, both created by a random assignment method (and if possible, random selection from a population). Both groups take a pretest (O_1) at the same time, after which the independent variable (X) is given only to the experimental group, and then both groups take the posttest (O_2). Figure 10.15 illustrates the classical experimental design, or commonly referred to as the pretest posttest control group design.

Because the experimental and control groups have been randomly assigned, they are equivalent with respect to all important variables. This group equivalence in the design helps control for rival hypotheses, because both groups would be affected by them in the same way.

Chapter 10: Group-Level Research Designs **225**

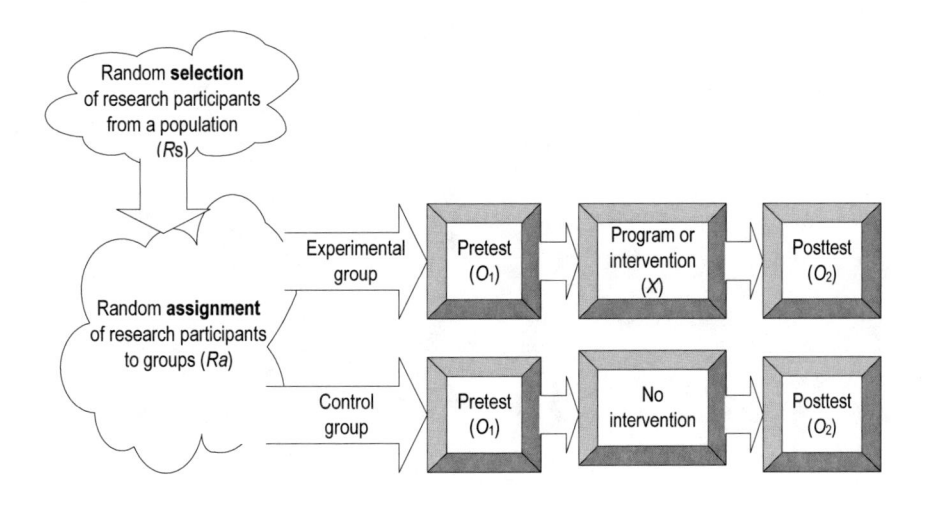

FIGURE 10.15
Pretest-Posttest Control Group Design
(Classical Experimental Design)

Where:

R = Random selection from a population and random assignment to group
O_1 = First measurement of the dependent variable
X = Independent variable, or the intervention
O_2 = Second measurement of the dependent variable

Randomized Posttest-Only Control Group Design

The randomized posttest-only control group research design is identical to the descriptive comparison group posttest-only design, except that the research participants are randomly assigned to two groups. This design, therefore, has a control group rather than a comparison group.

The randomized posttest-only control group research design usually involves only two groups, one experimental and one control. There are no pretests. The experimental group receives the independent variable and takes the posttest; the control group only takes the posttest. This design can be diagramed as in Figure 10.16 and written as follows:

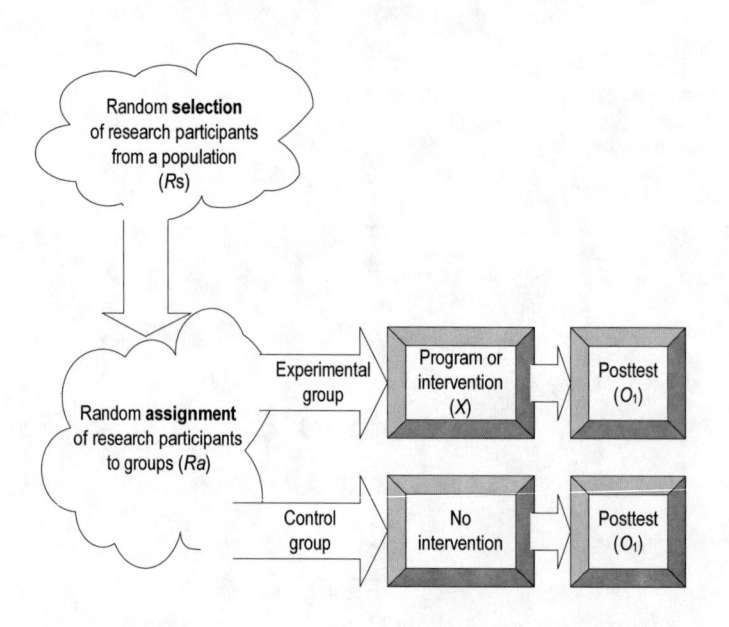

FIGURE 10.16
Randomized Posttest-Only Control Group Design

Where:

R = Random selection from a population and random assignment to group

X = Independent variable, or the intervention

O_1 = First and only measurement of the dependent variable

SUMMARY

Group research designs are conducted with groups of cases rather than on a case-by-case basis. They cover the entire range of research questions and provide designs that can be used to gain knowledge on the exploratory, descriptive, and explanatory levels.

A group research study is said to be internally valid if any changes in the dependent variable, Y, result only from the introduction of an independent variable, X. In order to demonstrate internal validity, we must first document the time order of events. Next, we must identify and eliminate extraneous variables.

Finally, we must control for the factors that threaten internal validity. In summary, threats to the internal validity of a research design address the assumption that changes in the dependent variable are solely because of the independent variable. "Ideal" experimental designs account for virtually all threats to internal validity—a rarity in social work research studies.

External validity is the degree to which the results of a research study are generalizable to a larger population or to settings other than the research setting. If a research study is to be externally valid, we must be able to demonstrate conclusively that the sample we selected was representative of the population from which it was drawn. If two or more groups are used in the study, we must be able to show that the two groups were equivalent at the beginning of the study.

Most importantly, we must be able to demonstrate that nothing happened during the course of the study, except for the introduction of the independent variable, to change either the representativeness of the sample or the equivalence of the groups.

The degree of control we try to exert over threats to internal and external validity varies according to the research design. Threats to internal and external validity may be more or less problematic depending on what particular research design we select. When we design a study, we must be aware of which threats will turn into real problems and what can be done to prevent or at least to minimize them. When doing an exploratory study, for example, we will not be much concerned about threats to external validity because an exploratory study is not expected to have any external validity anyway. Nor do we attempt to control very rigorously for threats to internal validity.

When we use a descriptive research design, we might be trying to determine whether two or more variables are associated. Often, descriptive designs are employed when we are unable, for practical reasons, to use the more rigorous explanatory designs. We do our best to control for threats to internal validity because, unless we can demonstrate internal validity, we cannot show that the variables are associated.

When using explanatory designs, we are attempting to show causation; that is, we are trying to show that changes in one variable cause changes in another. We try hard to control threats to internal validity because, if the study is not internally valid, we cannot demonstrate causation. We would also like the results of the study to be as generally applicable as possible and, to this end, we do our best to control for threats to external validity.

Exploratory designs are used when little is known about the field of study and data are gathered in an effort to find out "what's out there." These ideas are then used to generate hypotheses that can be verified using more rigorous research designs. Descriptive designs are one step closer to determining causality. Explanatory designs are useful when considerable preexisting knowledge is available about the research question under study and a testable hypothesis can be formulated on the basis of previous work.

They have more internal and external validity than exploratory and descriptive designs, so they can help establish a causal connection between two variables. No one group research design is inherently inferior or superior to the others. Each has advantages and disadvantages. Those of us who are familiar with all three categories of group research designs will be equipped to select the one that is most appropriate to a particular research question.

REFERENCES AND FURTHER READING

Babbie, E., Halley, F., & Zaino, J. (2007). *Adventures in social research* (6th ed.). Thousand Oaks, CA: Sage.

Chambliss, D.F., & Schutt, R.K. (2006). *Making sense of the social world: Methods of investigation* (2nd ed.). Thousand Oaks, CA: Sage.

Grinnell, R.M., Jr., Unrau, Y.A., & Williams, M. (2008). Group-level designs. In R.M. Grinnell, Jr., & Y.A. Unrau (Eds.), *Social work research and evaluation: Foundations of evidence-based practice* (8th ed., pp. 179–204). New York: Oxford University Press.

Kumar, R. (2005). *Research methodology: A step-by-step guide for beginners* (2nd ed.). Thousand Oaks, CA: Sage.

Osborne, J.W. (Ed.). (2007). *Best practices in quantitative methods.* Thousand Oaks, CA: Sage.

Singh, K. (2007). *Quantitative social research methods.* Thousand Oaks, CA: Sage.

Williams, M., Tutty, L.M., & Grinnell, R.M., Jr. (1995). *Research in social work: An introduction* (2nd ed.). Itasca, IL: F.E. Peacock.

Williams, M., Unrau, Y.A., & Grinnell, R.M., Jr. (1998). *Introduction to social work research.* Itasca, IL: F.E. Peacock.

Check out our Website for useful links and chapters (PDF) on:
- introductory quantitative research concepts
- Web-based links to quantitative research sites

Go to: www.pairbondpublications.com
Click on: Student Resources
    Chapter-by-Chapter Resources
    Chapter 10

Part V

Collecting Data

Chapter 11: Collecting Quantitative Data   230

Chapter 12: Collecting Qualitative Data   250

Chapter 13: Selecting a Data Collection Method   266

Chapter 11

Collecting Quantitative Data

DATA COLLECTION METHODS VERSUS DATA SOURCES
TYPES OF DATA
SURVEY QUESTIONNAIRES
    Establishing Procedures to Collect Survey Data
    Recording Survey Data
    Establishing the Reliability and Validity of Survey Data
    Advantages and Disadvantages of Survey Data
STRUCTURED OBSERVATION
    Establishing Procedures to Collect Structured Observational Data
    Recording Structured Observational Data
    Establishing the Reliability and Validity of Structured Observational Data
    Advantages and Disadvantages of Structured Observational Data
SECONDARY DATA
    Establishing Procedures to Collect Secondary Data
    Recording Secondary Data
    Establishing the Reliability
    and Validity of Secondary Data
    Advantages and Disadvantages of Secondary Data
EXISTING STATISTICS
    Establishing Procedures to Collect Existing Statistics
    Recording Existing Statistical Data
    Establishing the Reliability and Validity of Existing Statistics
    Advantages and Disadvantages of Using Existing Statistics
SUMMARY
REFERENCES AND FURTHER READING

11

DATA COLLECTION IS THE HEARTBEAT of all research studies—quantitative or qualitative. Our goal is to collect good data—qualitative data (i.e., words) for interpretive studies and quantitative data (i.e., numbers) for positivist studies—with a steady rhythm and in a systematic manner. When data collection becomes erratic or stops prematurely, all research studies are in grave danger.

With the above in mind, this chapter briefly presents a variety of data collection methods that can produce data for quantitative research studies. Likewise, the following chapter presents various data collection options that can be used with qualitative studies.

DATA COLLECTION METHODS VERSUS DATA SOURCES

There is a critical distinction between a data collection method and a data source. This distinction must be clearly understood before we collect any data whatsoever. A *data collection method* consists of a detailed plan of procedures that aims to gather data for a specific purpose; that is, to answer a research question or to test a hypothesis.

Any data collection method can tap into a variety of *data sources*. The primary source of data in most social work research studies is people. Rarely do we use machines (e.g., biofeedback) to monitor change in people's attitudes, knowledge, or behaviors. Rather, we tend to collect data about people from the people themselves.

Data collected directly from people can be first- or second-hand data. First-hand data are obtained from people who are closest to the problem we are studying. Single mothers participating in a parent support group, for example, can easily provide first-hand data to describe their own stresses as single mothers.

Second-hand data may come from other people who are indirectly connected to the primary problem (i.e., stress of single mothers) we are studying. A parent support group facilitator, for example, can record second-hand observations of "stress" behaviors displayed by parents in the parent support group. The facilitator can also collect data from family members. Grandparents, for example, can provide data that reflect their perceptions of how "stressed out" their granddaughters may be.

Regardless of the data collection method or the data source, *all* data are eventually collected, analyzed, and interpreted as part of the research process. We will discuss, shortly, the basic data collection methods that commonly produce quantitative data for positivist studies. Before we do this, however, we need to understand the various types of data.

TYPES OF DATA

What exactly are data? They are recorded units of information that are used as the basis for reasoning, calculation, and discussion (a single unit of information is called a *datum*). We can collect *original data* and/or *existing data*. The distinction between the two is a simple one. We collect original data (for the first time) during the course of our study to fulfill a specific purpose; that is, to answer our research question or test our hypothesis. Unlike original data, we can use existing data that have been previously collected and stored, either manually or in a computer, before our study was even fully conceptualized.

We can also distinguish between types of data by the research approach used. In a positivist study, for example, we analyze *numerical data* using mathematical computations and calculations (Chapter 14). On the other hand, an interpretive study typically analyzes words or *text data* (Chapter 15).

We analyze text data by reading and rereading; our task is to look for common and differentiating characteristics and themes within the words. A single research study can easily incorporate both quantitative and qualitative types of data. In fact, interpretive and positivist data collection methods can produce complimentary data.

As you know, quantitative data are used in research studies that utilize the positivist research approach. Chapters 2 and 4 discussed how this approach aims to reduce research topics into concepts and clearly defined variables. One of the major steps in a positivist study is "Focusing the Question," which requires that we develop operational definitions for all the variables in our study. Let us revisit our two-variable research question that has been used throughout this book:

Research Question:
Do people who come from ethnic minority backgrounds have difficulty in accessing social services?

As we know, a positivist research study requires that we operationally define "ethnicity" and "difficulty in accessing social services" in such a way that we can measure each variable. As you know, we could easily measure both of our study's variables using two categories each:

What is your ethnicity *(Circle one category below)*?
- Ethnic Minority
- Ethnic Majority

Did you have difficulty in accessing any form of social services over the last 12-month period *(Circle one category below)*?
- Yes
- No

In Chapter 2, however, we learned that our two variables (i.e., ethnicity and difficulty in accessing social services) can be operationally defined in a variety of ways. Let us consider the above two-variable research question for the various data collection methods that are most often associated with producing quantitative data: (1) survey questionnaires, (2) structured observation, (3) secondary data, and (4) existing statistics.

SURVEY QUESTIONNAIRES

Survey research, or surveys, in itself is a method for researching social problems. The aim of a survey is to collect data from a population, or a sample of research participants, in order to describe them as a group (McMurtry, 2005). One form that a survey can take is a questionnaire, which is a carefully selected set of questions that relate to the variables in our research question.

When data are collected using survey questionnaires, we get our research participants' perceptions about the variable being measured. Think about a survey questionnaire that you have filled out recently—a marketing survey, a consumer satisfaction questionnaire, or a teaching evaluation of your instructor. Your answers reflected your perceptions and were probably different from the perceptions of others who answered the same survey.

A survey questionnaire is one data collection method that we can use to collect data in order to answer our two-variable research question above. It would have to include items that measure our two variables—ethnicity and difficulty in accessing social services. We would ask clients within our study to "self-report" what they perceive their ethnicity to be and whether they have experienced difficulty in accessing social services over the last 12-month period.

There are two basic types of survey questionnaires. Nonstandardized survey questionnaires are one type. Our two questions posed above are a very basic example of a nonstandardized survey questionnaire. The two-question survey is nonstandardized because the items have not been tested for reliability or validity. We have no way of knowing whether our research participants will respond to our two questions in a uniform way.

At best, we can comment on our survey questionnaire's face validity; that is, do its items "look" like they are measuring the variables we are interested in? Nonstandardized questionnaires are often developed when standardized questionnaires are not available or suited to our specific research question.

Most standardized survey questionnaires are scientifically constructed to measure one specific variable. Let us say we have searched the library and computer data bases for a standardized survey questionnaire that measures our dependent variable, difficulty in accessing social services.

Suppose the closest standardized questionnaire we find is *Reid-Gundlach Social Service Satisfaction Scale (R-GSSSS)*, which contains 34-items and measures clients' satisfaction with social services they have received (see Figure 7.1). Clients would read each item and rate how much they agree with each item. A five-point scale is used: 1 strongly agree, 2 agree, 3 undecided, 4 disagree, and 5 strongly disagree.

The *R-GSSSS* specifically measures satisfaction with social services, a concept that closely relates to our access variable but is not an exact measure of it. In other words, satisfaction with social services is not the same as difficulty in accessing social services but it could be used as a "proxy measure" to answer our research question. As a proxy measure, we would make a humongous leap to say that clients who are most dissatisfied with the social services are the same clients that will have the most difficulty in accessing new social services.

How we proceed in collecting survey questionnaire data varies depending upon whether we collect them via the mail, telephone interviews, or face-to-face interviews. When a questionnaire is sent through the mail, our research participants must also be provided with sufficient instructions to "self-administer" the questionnaire. This means that they must have all the necessary instructions to successfully complete the survey questionnaire alone.

When telephone or face-to-face interviews are used, the interviewer is available to assist the research participant with any difficulties in completing the questionnaire. Regardless of how our questionnaires are administered, there are some basic procedures that we can use to increase the likelihood of obtaining accurate and complete data.

Establishing Procedures to Collect Survey Data

There are four basic procedures that must be considered before any survey questionnaire is administered to potential research participants. First, it is essential that straightforward and simple instructions accom-

Chapter 11: Collecting Quantitative Data

pany the questionnaire. Our objective is to have each research participant complete the questionnaire in exactly the same way, which means they must have the same understanding of each question. Suppose for a moment, that we ask one of our two questions posed earlier—Did you have difficulty in accessing any form of social services over the last 12-month period? It is clear that we want our research participants to answer our question with respect to the past 12 months and not any other time frame.

A second procedure that must be established is how informed consent will be obtained from the research participants. This is a task that can be accomplished by writing a cover letter explaining the purpose of the research study and questionnaire, who the researchers are, that participation is entirely voluntary, how data will be used, and the steps taken to guarantee confidentiality. A cover letter is usually sent out with mailed questionnaires, read aloud over the telephone, and presented in face-to-face interviews.

To decrease the likelihood of any misunderstanding, elementary language should be used in the cover letter (the same as in the questionnaire). It may even be possible to have the cover letter (and questionnaire) translated into our research participants' native language(s). In our study, we are asking about ethnicity, which suggests that English could be a second language for many of our research participants. Depending on the geographic area in which we are conducting our study, we may wish to translate our cover letter and questionnaire into the dominant language of the geographic area.

If our survey questionnaire is to be administered over the telephone or face-to-face by an interviewer, training of an interviewer will be a necessary third procedure. In the same way that we want research participants to have the same understanding of the questionnaire's items, we want interviewers to handle potential queries in a consistent manner. Interviewers must be trained to ask questions within our questionnaire in such a way that they do not influence research participants' answers.

Basically, interviewers should refrain from discussing individual questions with research participants, from varying their tone of voice when asking questions, and from commenting on research participants' answers. The interviewer's task is to keep the research participant focused on answering the questions at hand, to answer questions for clarification, and to record the answers.

To save on time and resources, it may be possible to administer survey questionnaires in a group format. This usually means that we meet a group of research participants in their own environment. Suppose, for example, we were specifically interested in the accessibility of social services among Native people. We would need to travel to the reserve and administer our questionnaire at a central location on the reserve that would be convenient to our research participants.

Recording Survey Data

The data we gather from questionnaires usually are in numerical form. Our question that asks whether clients had difficulty accessing any form of social services over the last 12 months will produce a number of "yes" responses and a number of "no" responses. We can add up the total sum of responses for each category and calculate the number of clients who had difficulty accessing social services broken down by ethnicity (i.e., ethnic minority and ethnic majority).

Establishing the Reliability and Validity of Survey Data

Reliability and validity are critical to determining the credibility of any data collection instrument. Standardized questionnaires have associated with them values of both. Reliability values tell us how confident we can be in using our questionnaire over two or more time periods, across different people, and across different places. Validity values give us information about how good our questionnaire is at measuring what it purports to measure. The validity value for the *R-GSSSS*, for example, is reasonably high at .95 (the highest possible value is 1.0).

Reports on the *R-GSSSS* also note that the standardized questionnaire is useful for measuring differences in race. Specifically, African Americans and Mexican Americans are reported to predictably have lower scores. We can also assess the face validity of the *R-GSSSS* by simply looking at the 34 items contained within it (see Figure 6.1).

A quick glance tells us that the items reflect the "idea" of client satisfaction with social services. When using unstandardized questionnaires, we do not have the luxury of reporting reliability or validity. In this instance, we must ensure that our questionnaire, at the very least, has face validity.

Advantages and Disadvantages of Survey Data

Overall, the major advantage of survey questionnaires is that they offer a relatively inexpensive way to collect new data. In addition, they can: easily reach a large number of people, provide specific data, collect data efficiently, and be input into a computer easily. An additional advantage of standardized questionnaires is that the data they generate can be compared with the data collected in other studies that used the same standardized questionnaire.

The overall disadvantages of survey questionnaires include: research participants must be literate, surveys usually get a low response rate, the research participants must have a mailing address or telephone to be reached, in-depth or open-end responses are missed, and simple questions are utilized to provide answers to complex research questions.

STRUCTURED OBSERVATION

The term "structured observation" describes itself as a data collection method. A trained observer records the interaction of others in a specific place over an agreed upon amount of time using specific procedures and measurements. Structured observation aims to observe the natural interactions between and among people that occur either in natural or in artificial settings.

If we want to observe how people of different ethnic backgrounds interact with their social worker, for example, we could set up structured observations at the social worker's offices (natural environment), or we could artificially set up interviews behind one-way mirrors (artificial setting).

The data collected in structured observation reflect the trained observers' perceptions of the interactions. That is, the observers do not interact with the people they are observing. They simply watch the interactions and record the presence or absence of certain behaviors. Observers could, for example, count the number of times clients (comparing ethnic minority and ethnic majority groups) mention a barrier or obstacle to accessing a social service, or monitor whether or not social workers engage in behaviors that specifically assist clients in accessing new services. These observations, of course, would take place during "normal" meetings between social workers and their clients.

Structured observation requires that the variables being measured are specifically defined. We must "micro-define" our variables. In our two-variable research question, for example, we have already operationally defined "ethnic" as having two categories: ethnic minority and non–ethnic minority. In measuring our difficulty in accessing social services variable, we could measure it by the number of social services referred by a social worker in a 60-minute interview.

We could then compare the results of our difficulty in accessing social services variable across our two ethnicity groups. Would we find that social workers are more or less likely to offer social service referrals to clients who are from an ethnic minority or a non–ethnic minority group?

There are three general types of recording that can be used with structured observation. Observations can be measured for frequency (e.g., count a behavior every time it occurs), for duration (e.g., how long does the behavior last each time it occurs), and for magnitude (e.g., the varying intensity of a behavior such as mild, moderate, or severe).

Whether we use frequency, duration, or magnitude (or some combination), depends on the picture we want to develop to answer our research question. Do we want to describe difficulty in accessing social services by how many times services are offered to clients (frequency), by how much time social workers spend explaining how services can be accessed (duration), or by the degree of sincerity that the workers display when offering service referrals (magnitude)?

When we record the presence or absence of behaviors during an interaction between and among people, our observations can be structured in such a way as to help the observers record the behaviors accurately. Suppose for a moment, social worker interviews with clients take an average of one hour. We could set recording intervals every minute so that we have 60 recording intervals in one hour.

The observers would sit with a stop-watch and record whether or not the social workers offered any social services to their clients, in the first minute, the second minute, and each minute thereafter until the interviews end. Interval recording assists the observers in making continuous observations throughout the interaction. Spot-check recording is another way to schedule observation recordings.

Unlike interval recording, which is continuous, spot-check recording is a way to record intermittent observations. It may be possible to observe and record the presence or absence of a behavior for one minute every five minutes, for example.

Establishing Procedures to Collect Structured Observational Data

The procedures established for structured observation are firmly decided upon before actual observations take place. That is, the observers are specifically trained to watch for the presence or absence of specific behaviors and the recording method is set. The types of data to be recorded are selected (i.e., frequency, duration, or magnitude) and the nature of the recording is decided (i.e., continuous, interval, or spot-check).

Data generally are recorded by observers and not the persons engaging in the behaviors. As such, it is essential that observers be trained to "see" behaviors in a reliable way. Imagine that you are one of the observers and your main task is to record magnitude ratings of the degree of sincerity that social workers display when they offer social service referrals. You must decide if the social workers' gestures are mildly, moderately, or extremely sincere. How would you know to rate a behavior that you observed as "mildly sincere" versus "moderately sincere?"

Training of observers requires that all variables be unequivocally clear. Thus, observers are usually trained using "mock" trials. They are selected for their ability to collect data according to the "rules of the research study," not for their unique views or creative observations. In a nutshell, the observers are an instrument for data collection and an effort is made to calibrate all instruments of the research in the same way. To allow for several different observers to view the same situation, we may choose to videotape interactions.

A decision must also be made about who will be the best observers. Should only professionally trained outside observers be selected? This would include people who are completely unfamiliar with the context of the study but are skilled observers. How about indigenous observers—people who are familiar with the nature of the interaction to be observed?

In our example, indigenous observers could include people such as other social workers, social work supervisors, other clients, or staff from other social service offices. Who we choose to observe may depend upon availability, expense, or how important we feel it is for the observers to be familiar or unfamiliar with the situation being observed. Measurement that reflects cultural variables, for example, may require indigenous observers.

Recording Structured Observational Data

The recording instrument for structured observation generally takes the form of a grid or checklist. The behaviors being observed and the method of recording are identified. The simplest recording form to construct is one that records frequency. The recording form would simply identify the period of observation and the number of times a specific behavior occurs:

Observation Period	Frequency
3:00 P.M. to 4:00 P.M.	✓✓✓✓

For duration recording it is necessary to add the duration in minutes at each occurrence of the behavior:

Frequency and Observation Period	Duration (in minutes)
3:00 P.M. to 4:00 P.M.	1(5), 2(11), 3(15)

For magnitude recording, we record the time of occurrence and give a corresponding rating for the behavior. To simplify recording, the magnitude ratings are generally coded by assigning a number to each category. For example, "1" is mildly sincere, "2" is moderately sincere, and "3" is extremely sincere.

Observation Time	Magnitude
3:17 P.M.	1
3:25 P.M.	1
3:29 P.M.	2

Establishing the Reliability and Validity of Structured Observational Data

As we know from Chapters 6 and 7, validity refers to whether we are measuring what we think we are measuring. In other words, are we really measuring or observing "difficulty in accessing social services?" Validity, in this instance, should make us think about whether our criteria for observation (i.e., how we operationalized our variable) are reasonable representations of the variable.

Reliability, on the other hand, can be assessed in more concrete ways. Once we have established an operational definition for each variable and the procedures for observation and recording, we can test their reliability. A simple method for determining reliability is to use two independent raters who observe the exact same situation. How well do their recordings match? Do the two observers produce the same frequencies?

The same duration periods? The same magnitude ratings? Do they agree 100 percent or only 50 percent? One hundred percent agreement suggests that the measure is a reliable one, compared to 50 percent, which suggests we should go back to the drawing board to come up with more precise operational definitions for both variables or more exact recording procedures.

Advantages and Disadvantages of Structured Observational Data

Structured observation helps us to collect precise, valid, and reliable data within complex interactions. We are able to tease out important behaviors that are direct (or indirect) measures of the variable we are interested in. The data produced are objective observations of behaviors and thus are not tainted by our individual self-perceptions. Think back to the survey questionnaire method of data collection that we discussed at the beginning of this chapter.

The data collected using a survey questionnaire reflect research participants' own perceptions. Structured observation, on the other hand, gives a factual account of what actually took place. It can explain one component of a social interaction with objectivity, precision, and detail.

The major disadvantage of structured observation is the time and resources needed to train skilled observers. Think about the amount of time it would take, for example, to get you and, say, two other observers to agree on what constitutes "mild," "moderate," or "extremely" sincere behavior of social workers. Another disadvantage of structured observation is that it is a microscopic approach to dealing with complex social interactions. By focusing on one or two specific details (variables) of an interaction, we may miss out on many other important aspects of the interactions.

SECONDARY DATA

When existing data are used to answer a newly developed research question (or to test a hypothesis), the data are considered secondary. In other words, the data are being used for some purpose other than the original one for which they were collected. Unlike survey questionnaires or structured observation mentioned previously, collecting secondary data is unobtrusive.

Since data already exist, it is not necessary to ask people questions (as in survey questionnaires) or to be observed (as in structured observations). Furthermore, secondary data can exist in numerical form (quantitative) or text form (qualitative). Our discussion in this section focuses on existing quantitative data.

Let us go back to our two-variable research question that asks about the relationship between clients' ethnicity and whether they have difficulty in accessing social services. So far, we have discussed data collection methods within a social service context and have focused on the interaction between social workers and their clients. Within a social service program there are many client records that could provide meaningful data to answer our simple research question. It is likely, for example, that client intake forms collect data about clients' ethnicity.

We can be assured that an "ethnicity" question on an existing client intake form was not "thought up" to answer our specific research question. Rather, the program likely had other reasons for collecting the data such as needing to report the percentage of clients who come from an ethnic minority that they see in a given fiscal period.

The social service program may also have records that contain data that we could use for our difficulty in accessing social services variable. Social workers, for example, may be required to record each service referral made for each client. By reading each client file, we would be able to count the number of service referrals made for each client.

Secondary data can also be accessed from existing data bases around the world. Census data bases are a common example. With the advances of computer technology, data bases are becoming easier to access. The Inter-University Consortium for Political and Social Research (ICPSR) is the largest data archive in the world; it holds more than 17,000 files of data from more than 130 countries (Krysik, 2005).

We could use the ICPSR data, for example, to compare the accessibility of social services for clients from different ethnic groups across various countries. Of course, we could only do this if meaningful data already exist within the data base to answer our research question.

Establishing Procedures to Collect Secondary Data

Given that secondary data already exist, there is no need to collect them. Rather, our focus shifts to evaluating the data set's worth with respect to answering our research question. The presence of data sets have important influences on how we formulate our research questions or test our hypotheses. When original data are collected, we design our research study and tailor our data collection procedures to gather the "best" data possible. When data exist, we can only develop a research question that is as good as the data that we have available.

Since secondary data influence how we formulate our research questions, we must firmly settle on a research question before analyzing them. Data sets can have a vertigo effect, leaving us to feel dizzy about the relationships between and among variables contained within them. It may be that we begin with a general research question, such as our two-variable question about ethnicity and difficulty in accessing social services and move to a more specific hypothesis.

Suppose our existing data set had within it data about client ethnicity (e.g., Asians, African Americans, Caucasians, Native Americans) and data about difficulty in accessing social services (e.g., yes or no). We could return to the literature for studies that might help us formulate a directional (one-tailed) hypothesis.

A directional hypothesis would suggest that one or more of these four ethnic groups would have more (or less) difficulty accessing social services than the remaining groups. If no such literature exists, however, we would pose a nondirectional (two-tailed) hypothesis, which indicates that we have no basis to suggest that one of the four ethnic groups would have more or less difficulty accessing social services than any of the others.

Another influence of an existing data set is on how we operationally define the variables in our study. Simply put, our study's variables will have already been operationally defined for us. The definition used to collect the original data will be the definition we use in our study as well. It is possible to create a new variable by recasting one or more of the original variables in an existing data set.

For instance, suppose we were interested to know whether or not our clients were parents. Existing program records, however, only list the number of children in each family. We could recast these data by categorizing those clients that have one or more children versus those that have no children, thus creating a two-category variable. Clients with one or more children would be categorized as "parents" versus clients without any children who would be classified as "not parents."

One final consideration for using existing secondary data is that of informed consent. Just because data exist does not mean that we have free rein to use them in future studies. It may be that clients have provided information on intake forms because they believed it was necessary to do so in order to receive services. As we have seen in Chapter 3, when data are originally collected from clients, they must be informed of the present and future uses of the data and give their explicit consent.

Recording Secondary Data

Recording secondary data bases are a simple matter because data have already been collected, organized, and checked for accuracy. Our task is to work with them and determine the best possible procedures for data analyses. One feature of existing data bases is that they generally include a large number of cases and variables. Thus, some advanced knowledge of statistical software packages is required to extract the variables of interest and conduct the data analyses.

Establishing the Reliability and Validity of Secondary Data

Once our research question is formulated and our variables have been operationally defined, it is important to establish the data set's credibility. We must remember that "all data sets are not created equal," particularly with respect to their validity and reliability. Most importantly, we want to check out the data source. If ethnic status is recorded on a social service program's client intake form, for example, we would want to know how the data were obtained.

The data would be considered unreliable and, even invalid, if the workers simply looked at their clients and checked one of several ethnicity categories. A more reliable and valid procedure would be if they asked their clients what ethnic category they come from and, at the same time, showed them the categories to be selected. Data sets that are accompanied by clear data collection procedures are generally more valid and reliable than those data sets that are not.

Advantages and Disadvantages of Secondary Data

The advancement of computer technology increases the likelihood that secondary data will be used more often in future social work research studies. It is a reasonably inexpensive way to gather and analyze data to answer a research question or to test a hypothesis.

Given that data sets are developing around the world, researchers can fairly easily compare data sets from different countries and across different time spans. Data sets are at "our fingertips," provided that we have a research question or hypothesis to match. One major disadvantage of using an existing data set is that our research question is limited by the possibilities of the data set. We must make the best use of the data available to us, with the understanding that the "best" may not be good enough to answer our research question or test our hypothesis.

EXISTING STATISTICS

Existing statistics are a special form of secondary data. They *exist* in a variety of places. Box 11.1 provides an example of the many places that Fran, a clinical director of a private agency, accessed to gather existing statistics for her research question. One difference between existing statistics and secondary data is that statistics summarize data in aggregate form.

Another unique feature of existing statistics is that they exist only in numerical form. We might have statistics, for example, that report: 20 percent of clients served by social services were Native, 30 percent were Hispanic, and 50 percent were Caucasian.

BOX 11.1

Using Governmental and Private-Agency Statistics

Fran is clinical director of a private, nonprofit agency that provides foster care, group-home care, and residential treatment for children in a small city in

Arizona. She has noticed that there is a higher number of ethnic minority children in her agency's caseload than would be expected based on the proportion of ethnic minority children in the general population. She decides to do a study of this issue using existing statistics already gathered by various sources as her data collection method.

Fran first talks with the information specialist at the local library, asking for help to conduct a computer search through the library's existing data bases. The search reveals a series of reports, titled Characteristics of Children in Substitute and Adoptive Care, that were sponsored by the American Public Welfare Association and that provide several years of data reported by states on their populations of children in various kinds of foster and adoptive care.

She also locates an annual publication produced by the Children's Defense Fund, and this provides a variety of background data on the well-being of children in the United States. Next, Fran checks in the library's government documents section. She locates recent census data on the distribution of persons under the age of 18 across different ethnic groups in her county.

Fran now quickly turns to state-level resources, where a quick check of government listings in the telephone book reveals two agencies that appear likely to have relevant data. One is the Foster Care Review Board, which is comprised of citizen volunteers who assist juvenile courts by reviewing the progress of children in foster care statewide. A call to the Board reveals that they produce an annual report that lists a variety of statistics. These include the number of foster children in the state, where they are placed, and how long they have been in care.

Fran also learns that the Board's annual reports from previous years contain similar data, thus a visit to the Board's office provides her with the historical data needed to identify trends in the statistics she is using. Finally, she discovers that two years earlier the Board produced a special issue of its annual report that was dedicated to the topic of ethnic minority children in foster care, and this issue offers additional statistics not normally recorded in most annual reports.

Another state agency is the Administration for Children, Youth, and Families, a division of the state's social services department. A call to the division connects her with a staff member who informs her that a special review of foster children was conducted by the agency only a few months before. Data from this review confirm her perception that minority children are overrepresented in foster care in the state, and it provides a range of other data that may be helpful in determining the causes of this problem.

From these sources Fran now has the data she needs to paint a detailed picture of minority foster children at the national and state levels. She also has the ability to examine the problem in terms of both point-in-time circumstances and longitudinal trends, and the latter suggest that the problem of overrepresentation has grown worse. There is also evidence to indicate that the problem is more severe in her state than nationally.

Finally, corollary data on related variables, together with the more intensive work done in the special studies by the Foster Care Review Board and the Administration for Children, Youth, and Families, gives Fran a basis for beginning to understand the causes of the problem and the type of research study that must be done to investigate solutions.

In the same way that secondary data are used for a purpose other than what was originally stated, so it is for existing statistics. With existing statistics, however, we are one step further removed from the original data. A statistic, for example, can be computed from 3,000 cases or from 30 cases—each produces only a single value.

There are two main types of statistics for us to be concerned with. The first is descriptive statistics, which simply describe the sample or population being studied. Statistics used for descriptive purposes include things like percentages, percentiles, means, standard deviations, medians, ranges, and so on.

The second type of statistics, inferential statistics, include test statistics such as chi-square, *t*-test, ANOVA, regression, and correlation. Do not panic; we will give an introductory discussion on statistics in Chapter 14. For now, we need only to know that descriptive statistics tell us about the characteristics of a sample or a population. Inferential statistics tell us about the likelihood of a relationship between and among variables within a population.

When using existing statistics as our "data" we proceed in much the same way as when we use secondary data. Both are unobtrusive methods of data collection in that no persons will be asked to provide data about themselves—the data already exist and have been used to calculate a statistic. Therefore, we must establish their credibility.

Establishing Procedures to Collect Existing Statistics

Using existing statistics in a research study influences our research question (and hypotheses) to a greater degree than when using secondary data. If existing statistics are to assist in developing a research question, then advanced knowledge of statistics is required. We would need to understand the purpose of each statistic, the assumptions behind it, and how it was calculated.

We would not want to gather existing statistics without also collecting information about how the original data (used to calculate our existing statistics) were collected and analyzed. This information is important to assessing the credibility of the statistics that we would use for our research study.

Recording Existing Statistical Data

Recording existing statistics is different from recording secondary data. In the case of secondary data, the data are already in a recorded form. When working with existing statistics, it is usually necessary to extract the statistics from their original sources and reconfigure them in a way that permits us to conduct an appropriate data analysis. It may be, for example, that we need to extract percentage figures from a paragraph within a published article, or mean and standard deviation scores from an already existing table in a research report.

Establishing the Reliability and Validity of Existing Statistics

Checking the reliability and validity of existing statistics requires us to look for a few basic things. First, we want to be sure that the studies or reports from which we are extracting the statistics use comparable conceptual and operational definitions. If we were to compare statistical results about the accessibility of social services to clients who are from different ethnic groups across two different studies, we want to be certain that both studies had reasonably similar operational definitions for the two variables that we are interested in.

We can also examine previously conducted studies for their own assessment of their validity and reliability. Were sound procedures used to collect the original data? Were the original data analyzed in an appropriate way? What were the limitations of each study? By asking these types of questions about previous studies, we can get clues as to the credibility of the statistics that we are about to use in our own research study.

Advantages and Disadvantages of Using Existing Statistics

The advantages of using existing statistics are many. Given that existing data are used, via the form of statistics, it is a relatively inexpensive approach and is unobtrusive. By using existing statistics, we can push the knowledge envelope of what we already know and ask more complex research questions. The use of existing statistics also provides us with the opportunity to compare the results of several research studies in an empirical way.

One disadvantage of using existing statistics is that we are not provided with data for individual cases. We can only answer research questions and test hypotheses about groups of people. Because data are already presented in a summary (i.e., statistical) form, it is difficult to assess how the data were collected and analyzed. Since the data have been handled by different individuals, there is an increased chance that human error has occurred somewhere along the way.

SUMMARY

In this chapter we have described different methods for collecting quantitative data. We have demonstrated that one research question can be answered using any data collection method. Different data collection methods produce different types of data. Data collection methods that are more commonly associated with quantitative or numerical data are: survey questionnaires, structured observation, secondary data, and existing statistics.

In the next chapter, we will look at the various data collection methods that can be used to collect qualitative data.

REFERENCES AND FURTHER READING

- *On Structured Observation*

    Poslter, R.A., & Collins, D. (2008). Structured observation. In R.M. Grinnell, Jr., & Y.A. Unrau (Eds.), *Social work research and evaluation: Foundations of evidence-based practice* (8th ed., pp. 207–221). New York: Oxford University Press.

- *On Surveys*

    Bulmer, M., Sturgis, P.L., & Allum, N. (2009). *The secondary analysis of survey data* (4 vols.). Thousand, Oaks, CA: Sage.

    de Vaus, D. (Ed.). (2007). *Social surveys* (4 vols.). Thousand Oaks, CA: Sage.

    Engel, R.J., & Schutt, R.K. (2008). Survey research. In R.M. Grinnell, Jr., & Y.A. Unrau (Eds.), *Social work research and evaluation: Foundations of evidence-based practice* (8th ed., pp. 265–304). New York: Oxford University Press.

    Fink, A. (2002). *The survey kit* (2nd ed.). Thousand Oaks, CA: Sage.

    Lavrakas, P.J. (Ed.). (2008). *Encyclopedia of survey research methods.* Thousand Oaks, CA: Sage.

- *On Existing Statistics*

    Sieppert, J.D., McMurtry, S.L., & McClelland, R.W. (2005). Utilizing existing statistics. In R.M. Grinnell, Jr., & Y.A. Unrau (Eds.), *Social work research and evaluation: Quantitative and qualitative approaches* (7th ed., pp. 324–336). New York: Oxford University Press.

- *On Content Analysis*

    LeCroy, C.W., & Soloman, G. (2008). Content analysis. In R.M. Grinnell, Jr., & Y.A. Unrau (Eds.), *Social work research and evaluation: Foundations of evidence-based practice* (8th ed., pp. 315–325). New York: Oxford University Press.

    Neuendorf, K.A. (2002). *The content analysis guidebook.* Thousand Oaks, CA: Sage.

- *On Secondary Analysis*

    Bulmer, M., Sturgis, P.L., & Allum, N. (2009). *The secondary analysis of survey data* (4 vols.). Thousand, Oaks, CA: Sage.

    Krysik. J.L. (2005). Secondary analysis. In R.M. Grinnell, Jr., & Y.A. Unrau (Eds.), *Social work research and evaluation: Quantitative and qualitative approaches* (7th ed., pp. 291–301). New York: Oxford University Press.

    Rubin, A. (2008). Secondary analysis. In R.M. Grinnell, Jr., & Y.A. Unrau (Eds.), *Social work research and evaluation: Foundations of evidence-based practice* (8th ed., pp. 305–314). New York: Oxford University Press.

- *On Selecting a Data-Collection Method*

    Sapsford, R., & Jupp, V. (2006). *Data collection and analysis* (2nd ed.). Thousand Oaks, CA: Sage.

    Unrau, Y.A. (2008). Selecting a data collection method and data source. In R.M. Grinnell, Jr., & Y.A. Unrau (Eds.), *Social work research and evaluation: Foundations of evidence-based practice* (8th ed., pp. 327–341). New York. Oxford University Press.

Chapter 12

Collecting Qualitative Data

NARRATIVE INTERVIEWING
    Establishing Data Collection Procedures for Narrative Interviewing
    Recording Narrative Data
    Trustworthiness and Truth Value of Narrative Data
    Advantages and Disadvantages of Narrative Data
PARTICIPANT OBSERVATION
    Establishing Procedures to Collect
    Participant Observational Data
    Recording Participant Observational Data
    Trustworthiness and Truth Value of Participant Observational Data
    Advantages and Disadvantages of Participant Observational Data
SECONDARY CONTENT DATA
    Establishing Procedures to Collect Secondary Content Data
    Recording Secondary Content Data
    Establishing the Trustworthiness and Truth Value of Content Data
    Determining the Advantages and Disadvantages of Content Data
HISTORICAL DATA
    Establishing Procedures to Collect Historical Data
    Recording Historical Data
    Establishing the Trustworthiness and Truth Value of Historical Data
    Determining the Advantages and Disadvantages of Historical Data
SUMMARY
REFERENCES AND FURTHER READING

12

AS WE KNOW FROM THE LAST CHAPTER, any research study requires data—regardless of whether a positivistic or interpretive approach is used. We have already discussed that quantitative data are represented numerically, while qualitative data are expressed using words. A second key difference between these two types of data is that qualitative data are collected to "build" a story or understanding of a concept or variable, compared to quantitative data, which narrows in on a select few variables. A third difference worth noting is that qualitative data are generally "bulkier" than quantitative data. As we will see, qualitative data can be represented by a single word, a sentence, or even an entire page of text.

The exact point in time when data are collected in the research process is another difference between quantitative and qualitative data. We have stressed that quantitative data collection can *only occur* once variables have been completely operationally defined—in fact, the reliability and validity of our measurements depend on it.

Qualitative data, on the other hand, can be collected during many different steps in the research process. We may collect data near the beginning of our study to help us focus our research question and identify key variables to be explored later on in the research process. Qualitative data collected in later steps in our study can be used to check out any assumptions we may have and any new ideas that emerge (see Chapter 5).

When collecting quantitative data, the rules for data collection are tried and tested *before* collecting any data—procedures can be outlined in a checklist format, where the data collector checks off each procedure as it is completed. For qualitative data, however, explicit procedures of data collection are not necessarily known before the data collection process starts. Rather, the procedures used are documented *as they happen*. It is only after data collection is complete that a detailed description of the procedure can be articulated.

This is not to say that in qualitative data collection we can "willy nilly" change our minds or that "anything goes." Research is still research, which means that a systematic approach to inquiry is used. The big difference for interpretive research is that the systematic nature of data collection applies to how we record and monitor the data collection process.

Any changes in data collection (e.g., change in questions, research participants, or literature review) are based on data already collected. More will be explained about how qualitative data collection procedures are monitored in Chapter 15. For now, let us return to our research question from the preceding chapter.

Research Question:

Do people who come from ethnic minority backgrounds have difficulty in accessing social services?

We have already discussed in the last chapter four different methods of collecting quantitative data to answer the above research question. Now we will turn to four data collection methods that are more commonly associated with producing qualitative data to answer the same research question: (1) narrative interviewing, (2) participant observation, (3) secondary content data, and (4) historical data.

NARRATIVE INTERVIEWING

We have seen in the previous chapter how interviewing is used to collect quantitative data via survey questionnaires, in which interviewers are trained to ask questions of research participants in a uniform way and interaction between interviewers and their research participants is minimized. Interviewing for qualitative data, on the other hand, has a completely different tone.

The aim of narrative interviewing is to have research participants tell stories in their own words. Their stories usually begin when we ask our identified research question. For example, we could begin an interview by saying, "Could you tell me about your experiences as a Native client in terms of accessing social services?"

Qualitative interviewers are not bound by strict rules and procedures. Rather, they make every effort to engage research participants in meaningful discussions. In fact, the purpose is to have research participants tell their own stories in their own ways. The direction the interviews take may deviate from the original research problem being investigated depending on what research participants choose to tell interviewers. Data collected in one research interview can be used to develop or revise further research questions in subsequent research interviews.

The narrative interview is commonly used in case study research where numerous interviews are usually conducted to learn more about a "case," which could be defined as a person, a group of people, an event, or an organization. The narrative interview is also a basic component of other research pursuits, such as feminist research and participatory action research.

Establishing Data Collection Procedures for Narrative Interviewing

Data collection and data analysis are intertwined in an interpretive research study. Thus, the steps we take in collecting data from a narrative interview will evolve with our study. Nevertheless, we must be clear about our starting point and the procedures that we will use to "check and balance" the decisions we make along the way.

One of the aims of an interpretive research approach is to tell a story about a problem without the cobwebs of existing theories, labels, and interpretations. As such, we may choose to limit the amount of literature we review before proceeding with data collection. The amount of literature we review is guided by the interpretive research approach we use. A grounded theory approach, for example, suggests that we review some literature at the beginning of our study in order to help frame our research question. We would then review more literature later on, taking direction from the new data that we collect.

It is necessary for us to decide when and how to approach potential research participants before data collection occurs. Think about our simple research question for a moment—Do clients who come from an ethnic minority have difficulty in accessing social services? Who would we interview first? Would we interview someone receiving multiple social services or someone not receiving any? Is there a particular ethnic group that we would want represented in our early interviews?

Remember that the data we collect in our beginning interviews will directly influence how we proceed in later interviews. It is possible for the professional literature, or our expert knowledge, to point us to a particular starting place. Regardless of where we begin, it is critical for us to document our early steps in the data collection process. The notes we take will be used later on to recall the steps we took, and more importantly, why we took them.

Let us say, for example, that our interest in our research question stems from the fact that many ethnic minority people we know in our personal lives (not our professional lives) have complained about access to social services. We may choose to begin interviewing an individual which we already know.

On the other hand, we may decide after a cursory review of the literature that most of the existing research on our research problem is based on interviews with people who, in fact, are receiving services. We may then choose to begin interviewing clients who are from an ethnic minority and who are *not* currently receiving any social services but who are named on social service client lists.

We must also establish how the interview will be structured. Interviewing for qualitative data can range from informal casual conversations to more formal guided discussions. An informal interview is unplanned. The interviewer begins, for example, by asking Native clients a general question, such as the one presented earlier—could you tell me about your experiences in accessing social services?

The discussion that would follow would be based on the natural and spontaneous interaction between the interviewer and the research participant. The interviewer has no way of knowing at the outset what direction the interview will take or where it will end.

A guided interview, on the other hand, has more structure. The exact amount of structure varies from one guided interview to another. In any case, the interviewer has an interview schedule, which essentially amounts to an outline of questions and/or concepts, to guide the interview with the research participant.

A loosely structured interview schedule may simply identify key concepts or ideas to include in the interview at the appropriate time. In our study, for example, we might want the interviewer to specifically ask each research participant about language barriers and social isolation, if these concepts do not naturally emerge in the interviews. A highly structured interview schedule would list each question to be asked during the interviews. The interviewer would read the question out to the research participants and record their answers.

Recording Narrative Data

The most reliable way to record interview data is through the use of audiotape. Videotape is also a possibility but many people are uncomfortable with being filmed. The audiotape is an excellent record of an interview because it gives us verbatim statements made by our research participants. The tone, pace, and "atmosphere" of the interview can be recalled by simply replaying the tape.

To prepare data for analysis, however, it is necessary to transcribe every word of the interview. As we will see in Chapter 15, the transcription provides text data, which will be read (and reread) and analyzed. Because audio data are transformed into text data, verbatim statements are used. In addition, any pauses, sighs, and gestures are noted in the text. The following is an example of an excerpt of text data based on our question asked earlier.

Research Participant Number 6: Being a Native in social services, huh...well, it's not so good, eh people don't know...they don't see the reserve or the kids playing...even when they see they don't see. The kids you know, they could have some help, especially the older ones (sighs and pauses for a few seconds). My oldest you know, he's 12 and he goes to school in the city. He comes home and doesn't know what to do. He gets into trouble. Then the social services people come with their briefcases and tell you what to do (voice gets softer). They put in stupid programs and the kids don't want to go you know, they just want to play, get into trouble stuff (laughs)....

Interviewer: Hmmm. And how would you describe your own experiences with social services?

Because the interviewing approach to data collection is based on the interaction of the interviewer and the research participant, both parts of the discussion are included in the transcript. But data collection and recording do not stop here. The interviewer must also keep notes on the interview to record impressions, thoughts, perspectives, and any data that will shed light on the transcript during analysis. An example of the interviewer's notes from the above excerpt is:

September 12 (Research Participant Number 6): I was feeling somewhat frustrated because the research participant kept talking about his children. I felt compelled to get him to talk about his own experiences but that seemed sooooo much less important to him.

As we will see in Chapter 15, the verbatim transcript and the interviewer's notes are key pieces of data that will be jointly considered when text data are analyzed.

Trustworthiness and Truth Value of Narrative Data

Regardless of the type of interview or the data recording strategy that we select, we must have some way to assess the credibility of the data we collect. With quantitative data, we do this by determining the reliability and validity of the data—usually by calculating a numerical value. In contrast, with qualitative data we are less likely to generate numerical values and more likely to document our own personal observations and procedures.

An important question to assess the trustworthiness and truth value of the text data (and our interpretation of it) is whether we understand what our research participants were telling us from their points of view. Two common ways to check the credibility of interview data are triangulation and member checking. These procedures will be discussed in more detail in Chapter 15.

Briefly, triangulation involves comparing data from multiple perspectives. It may be that we interview several people (i.e., data sources) on one topic, or even that we compare quantitative and qualitative data for the same variable. Member checking simply involves getting feedback from our research participants about our interpretation of what they said.

Advantages and Disadvantages of Narrative Data

The major advantage of narrative interviewing is the richness of data that are generated. Narrative interviewing allows us to remain open to learning new information or new perspectives about old ideas. Because the interviewer is usually the same person who is the researcher, there is a genuine interest expressed in the interviews.

The major disadvantage associated with narrative interviewing is that it is time consuming. Not only does it take time to conduct the interviews, but considerable time also must be allotted for transcribing the text data. Narrative interviews produce reams and reams of text, which make it difficult to conduct an analysis. Researchers can tire easily and are subject to imposing their own biases and perspectives on what the research participants say.

PARTICIPANT OBSERVATION

Participant observation is a way for us to be a part of something and study it at the same time. This method of data collection requires us to establish and maintain ongoing relationships with research participants in the field setting. We not only listen, observe, and record data, but also participate in events and activities as they happen. Our role in participant observation can be described on a continuum (see Figure 12.1), with one end emphasizing the *participant role* and the other end emphasizing the *observer role*.

An "observer-participant" has the dominant role of observation, while the "participant-observer" is predominantly a participant. Suppose we were to use participant observation as a data collection method for our study. An example of an observer-participant is a researcher who joins a group of people, say Natives living on their reserve, for events related to our research study. We may travel with a community social worker, for example, and participate in community meetings on the reserve to discuss access to social services.

An example of a participant-observer could be when a social worker has the dual responsibility of conducting the meeting and observing the interactions of community members. A participant-observer could also be identified within his or her unique community.

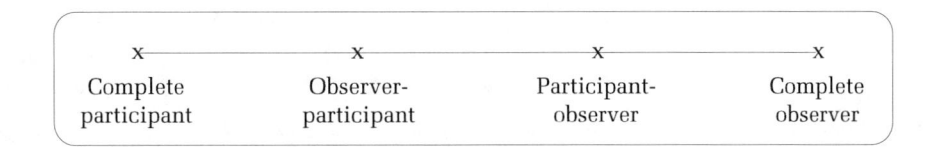

```
x─────────x─────────x─────────x
Complete   Observer-  Participant-  Complete
participant participant  observer    observer
```

FIGURE 12.1
Continuum of Participant Observation

A Native resident of the reserve, for example, could participate in the meeting as a resident and as an observer. In any case, participant observation always requires that one person assumes a dual role. For ethical reasons, the role must be made explicit to all persons participating in the research study. Whether the researcher is a social worker or a Native resident of the reserve, he or she must declare his or her dual role before the community meeting (i.e., data collection) gets underway.

Establishing Procedures to Collect Participant Observational Data

Given the dual role of researchers in participant observation, formulating steps for data collection can be tricky. It is essential for the researcher to keep a balance between "participation" and "observation." Suppose our Native resident who is a participant-observer gets so immersed in the issues of the community meeting that she forgets to look around or ask others questions related to the research study.

Someone who completely participates will not be very effective in noting important detail with respect to how other people participated in the meeting. On the other hand, if our Native researcher leans too far into her observation role, others might see her as an outsider and express their views differently than if they believed she had a vested interest within the community.

An important consideration for a participant observation study is how to gain access to the group of people being studied. Imagine who might be welcomed into a community meeting on a Native reserve and who might be rejected. Chances are that the Native people on a reserve would be more accepting of a Native person, a person known to them, or a person whom they trust, as compared to a non-Native person, a stranger, or a person whom they know but do not trust.

In participant observation we would need to understand the culture of the reserve in order to know who to seek permission from. In participant observation, entry and access is a process "more analogous to peeling away the layers of an onion than it is to opening a door" (Rogers & Bouey, 2008). Forming relationships with people is a critical part of data collection. The more meaningful relationships we can establish, the more meaningful data we will collect. This is not to say, however, that we somehow trade relationship "tokens" for pieces of information. Rather, research participants should feel a partnership with the researcher that is characterized by mutual interest, reciprocity, trust, and cooperation.

Recording Participant Observational Data

The data generated from participant observation come from observing, interviewing, using existing documents and artifacts, and reflecting on personal experiences. Below are a few examples:

> *Observation (September 12, 9:30 A.M. Meeting Start Time):* When the social worker announced the purpose of the meeting was to discuss access to social services for people living on the reserve, there was a lot of agreement from a crowd of 42 people. Some cheered, some nodded their heads, and others made comments like, "it's about time."

> *Research participant's verbatim comment (September 12, 10:15 A.M.):* "Native people don't use social services because they are not offered on the reserve."

> *Existing document:* A report entitled, "*The Social Service Needs of a Reserve Community*" reports that there have been three failed attempts by state social workers to keep a program up and running on the reserve. Worker turnover is identified as the main contributing factor of program failures.

> *Researcher's personal note (September 12, 10:44 A.M.):* I feel amazed at the amount of positive energy in the room. I can sense peoples' frustrations (including my own), yet I am in awe of the hopeful and positive outlook everyone has to want to develop something that works for our community. I am proud to be a part of this initiative to improve services in the community.

Overall, participant observers produce "participant observational" data through the use of detailed and rich notes. Several strategies can be used for documenting notes. We can take notes on an ongoing basis, for example, by being careful to observe the time and context for all written entries. A more efficient method, however, would be to carry a Dictaphone, which gives us the flexibility to state a thought, to record a conversation, and to make summary comments about what we observe. Collecting data on an ongoing basis is preferable because it increases the likelihood of producing accurate notes in addition to remembering key events. In some instances, however, participation in a process that prevents us from recording any data, which leaves us to record our notes after the event we are participating in has ended.

Trustworthiness and Truth Value of Participant Observational Data

Given that multiple data sources are possible in participant observation, it is possible to check the credibility of the data through triangulation. By reviewing the four data recording examples above, we can be reasonably confident that there is agreement about the state of social services on the reserve. The four separate data entries seem to agree that social services on the reserve are inadequate in terms of the community's needs.

The flexibility of participant observation allows us to seek out opportunities to check out our personal assumptions and ideas. After hearing the general response of Native community members at the beginning of the meeting, for example, we may choose to ask related questions to specific individuals. We can also check our perceptions along the way by sharing our own thoughts, asking people to comment on data summaries, or asking people how well they think two data sources fit together.

Advantages and Disadvantages of Participant Observational Data

A major advantage of participant observation is that we can collect multiple sources of data and check the credibility of the data as we go. Because of the participant role, any observations made are grounded within the context in which they were generated. A "complete participant" is more likely to pick up subtle messages, for example, than would be a "complete observer."

The disadvantages of participant observation are related to time considerations. As with narrative interviewing, a considerable amount of time must be allotted to data recording and transcription. The researcher also runs the risk of becoming too immersed as a participant or too distant as an observer—both are situations that will compromise the data collected.

SECONDARY CONTENT DATA

Secondary content data are existing text data. In the same way that existing quantitative data can be used to answer newly developed research questions, so can existing qualitative (or content) data. In this case, the text data were recorded at some other time and for some other purpose than the research question that we have posed. Let us return to our research question—Do clients who come from an ethnic minority have difficulty in accessing social services?

With the use of content data, we could answer our question using context data that already exist. It may be, for example, that social workers are mandated to record any discussion they had about barriers to accessing social services for their clients on their client files. Case notes, for example, may read "client does not have reliable transportation" or "client has expressed a strong wish for a worker who has the same ethnic background."

These types of comments can be counted and categorized according to meaningful themes. In this instance, we are zeroing in on text data for specific examples of recorded behaviors; much like the microscopic approach of structured observation.

We must be aware that content data can be first- or second-hand. In our example, first-hand data would be generated when the social workers document their own behaviors in relation to offering services to their clients. When social workers record their own impressions about how their clients respond to offers of services, second-hand data are produced.

Establishing Procedures to Collect Secondary Content Data

Content data are restricted to what is available. That is, if 68 client files exist, then we have exactly 68 client files to work from. We do not have the luxury of collecting more data. Like secondary quantitative data, we are in a position only to compile the existing data, not to collect new data. It is possible, however, that content data are available from several sources. Client journals may be on file, or perhaps the agency recently conducted an open-ended client satisfaction survey, which has handwritten responses to satisfaction-type questions from clients.

A major consideration of content data is whether or not a sufficient amount exists to adequately answer our research question. If there are not enough data, we do not want to spend precious time reading and analyzing them. We must also remember that we are reading and reviewing text data for a purpose that they were not originally intended for.

Suppose, for example, you kept a daily journal of your personal experience in your research course. Would you write down the same thoughts if you knew your journal would be private, compared to if you were to share your writings with your research instructor? The original intent of writing is essential to remember because it provides the context in which the writing occurred.

Because we are examining confidential client case files, we must establish a coding procedure to ensure that our clients' confidentiality and anonymity are maintained. In fact, it is not at all necessary for us, as researchers, to know the names of the client files we are reading. Given our research question, we need only to have access to those portions of clients' files that contain information about client ethnicity and the difficulty clients have in accessing social services.

Recording Secondary Content Data

As we know, content data exist in text form. If they exist in handwritten form, it is necessary to type them out into transcript form. It is useful to make several copies of the transcripts to facilitate data analysis. More will be said about preparing transcripts in Chapter 15.

Establishing the Trustworthiness and Truth Value of Content Data

Clearly, first-hand data are more credible than second-hand data when dealing with existing text data. Because the data are secondary, questions of credibility center more around data analysis than data collection. In our example, it may be possible to triangulate data by comparing files of different workers or by comparing different client files with the same worker. It also may be possible to member check (Chapter 15) if workers are available and consent to it.

Determining the Advantages and Disadvantages of Content Data

A major advantage of content data is that they already exist. Thus, time and money are saved in the data collection process. The task of the researcher is simply to compile data that are readily available and accessible. The disadvantages of content data are that they are limited in scope. The data were recorded for some other purpose and may omit important detail that would be needed to answer our research question.

HISTORICAL DATA

Historical data are collected in an effort to study the past (Stuart, 2005). Like the participant observation method to data collection, historical data can come from different data sources. Data collection can be obtrusive, as in the case of interviewing people about past or historical events, or unobtrusive, as in the case when existing documents (primarily content data) are compiled. When our purpose is to study history, however, special cases of content data and interviewing are required.

Suppose we recast our research question to ask—Did clients who come from an ethnic minority have difficulty accessing social services from 1947 to 1965? Phrased this way, our question directs us to understand *what has been* rather than *what is*. In order for us to answer our new research question, we would need to dig up remains from the past that could help us to describe the relationship, if any, between ethnicity and difficulty in accessing social services for our specified time period. We would search for documents such as, related reports, memos, letters, transcription, client records, and documentaries, all dating back to the time period. Many libraries store this information in their archives.

We could also interview people (e.g., social workers and clients) who were part of social services between 1947 and 1965. People from the past might include clients, workers, supervisors, volunteers, funders, or ministers. The purpose of our interviews would be to have people remember the past—that is, to describe factual events from their memories. We are less interested in people's opinions about the past than we are about what "really" happened.

When collecting historical data, it is important to sort out first-hand data from second-hand data. First-hand data, of course, are more highly valued because the data are less likely to be distorted or altered. Diaries, autobiographies, letters, home videos, photographs, and organizational minutes are all examples of first-hand data. Second-hand data, on the other hand, include documents like biographies, history books, and published articles.

Establishing Procedures to Collect Historical Data

It is probably more accurate to say that we retrieve historical data than it is to say that we collect it. The data already exist, whether in long-forgotten written documents, dusty videotapes, or in peoples' memories. Our task is to resurface a sufficient amount of data so that we can describe, and sometimes explain, what happened (Stuart, 2005; Danto, 2008). One of the first considerations for conducting a historical research study is to be sure that the variable being investigated is one that existed in the past. Suppose, for example, we had a specific interest in the relationship among three variables: client ethnicity, difficulty in accessing social services, and computer technology.

More specifically, let us say that we were interested in knowing about the relationship among these three variables, if any, for our specified time period, 1947 to 1965. It would be impossible to determine the relationship among these three variables from a historical perspective since computers were not (or were rarely) used in the social services from 1947 to 1965. Thus, there would be no past to describe or explain—at least not where computers are concerned.

Once we have established that our research question is relevant, we can delineate a list of possible data sources and the types of data that each can produce. The total list of data sources and types must be assessed to determine if, in fact, sufficient data exist to answer our research question. It is possible, however, that data do exist but they are not accessible. Client records, for example, may be secured. Documents may be out of circulation; thus, requiring a researcher to travel to where the data are stored.

Recording Historical Data

Despite the fact that historical data, in many cases, already exist, it is usually necessary to reproduce the data. When interviews are used, interview notes are recorded and the interviews are transcribed as we have discussed in other interpretive data collection methods. When past documents are used, they ought to be duplicated to create "working" copies. Original documents should be protected and preserved so that they can be made available to other interested researchers.

Establishing the Trustworthiness and Truth Value of Historical Data

Assessing the trustworthiness and truth value of historical data, in many ways, is the process of data analysis. Do different people recall similar facts? Do independent events from the past tell the same story? Much effort goes into triangulating pieces of data. The more corroboration we have among our data, the stronger our resulting conclusions will be. When reconstructing the past, our conclusions can only be as good as the data we base them on.

Are the data authentic? How much of our data are first-hand, compared to second-hand? Do we have sufficient data to describe the entire time period of interest? Perhaps, we might have to cut back a few years. Or, if other data emerge, perhaps we can expand our time period.

Determining the Advantages and Disadvantages of Historical Data

Historical data are unique and used for a specific purpose—to describe and explain the past. It is the only way for us to research history. When secondary data are readily available and accessible, the cost is minimized. When interviews are used, they provide us with an opportunity to probe further into the past and our area of interest. The disadvantages of historical data are that they are not always easily available or accessible. There is also a risk of researcher bias such as when a letter or document is analyzed out of context. The past cannot be reconstructed, for example, if we impose present-day views, standards, and ideas.

SUMMARY

In this chapter we discussed different methods to collect qualitative data. With the previous chapter in mind we now know that that one research question can be answered using any data collection method—quantitative or qualitative. Different data collection methods produce different types of data. Data collection methods that are more commonly associated with qualitative or text data are: narrative interviewing, participant observation, secondary content data, and historical data.

REFERENCES AND FURTHER READING

- *On Interviewing*

   Gochros, H. (2008). Qualitative interviewing. In R.M. Grinnell, Jr., & Y.A. Unrau (Eds.), *Social work research and evaluation: Foundations of evidence-based practice* (8th ed., pp.239–264). New York: Oxford University Press.

   Morgan, D.L., & Krueger, R.A. (1997). *The focus group kit* (6 vols.). Thousand Oaks, CA: Sage.

   Rubin, H.J., & Rubin, I.S. (2005). *Qualitative interviewing: The art of hearing data* (2nd ed.). Thousand Oaks, CA: Sage.

- *On Participant Observation*

   Alaszewski, A. (2006). *Using diaries for social research.* Thousand Oaks, CA: Sage.

Rogers, G., & Bouey, E. (1996). Phase two: Collecting your data. In L.M. Tutty, M.A. Rothery, & R.M. Grinnell, Jr. (Eds.), *Qualitative research for social workers: Phases, steps, and tasks* (pp. 50–87). Boston: Allyn & Bacon.

Rogers, G., & Bouey, E. (2008). Participant observation. In R.M. Grinnell, Jr., & Y.A. Unrau (Eds.), *Social work research and evaluation: Foundations of evidence-based practice* (8th ed., pp. 223–237). New York: Oxford University Press.

- *On Historical Research*

Danto, E. (2008). *Historical research*. New York: Oxford University Press.

Stuart, P. (2005). Historical research. In R.M. Grinnell, Jr., & Y.A. Unrau (Eds.), *Social work research and evaluation: Quantitative and qualitative approaches* (7th ed., pp. 338–346). New York: Oxford University Press.

- *On General Qualitative Research*

Atkinson, P., & Delamont, S. (2008). *Representing ethnography: Reading, writing, and rhetoric in qualitative research* (4 vols.). Thousand Oaks, CA: Sage.

Bailey, C.A. (2007). *A guide to qualitative field research* (2nd ed.). Thousand Oaks, CA: Sage.

Bickman, L., & Rog, D.J. (Eds.). (2008). *The SAGE handbook of applied social research methods* (2nd ed.). Thousand Oaks, CA: Sage.

Bryant, A., & Charmaz, K. (Eds.). (2007). *The SAGE handbook of grounded theory*. Thousand Oaks, CA: Sage.

Fink, A. (2007). *The SAGE qualitative research kit*. Thousand Oaks, CA: Sage.

Gilbert, N. (Ed.). (2008). *Researching social life* (3rd ed.). Thousand Oaks, CA: Sage.

Given, L.M. (Ed.). (2008). *The SAGE encyclopedia of qualitative research methods*. Thousand Oaks, CA: Sage.

Tutty, L.M., Rothery, M.L., & Grinnell, R.M., Jr. (Eds.). (1996). *Qualitative research for social workers: Phases, steps, and tasks*. Boston: Allyn & Bacon.

Chapter 13

Selecting a Data Collection Method

DATA COLLECTION AND THE RESEARCH PROCESS
    Steps 1 and 2: Selecting a General Research Topic and Focusing the Topic into a Research Question
    Steps 3 and 4: Designing the Research Study and Collecting the Data
    Steps 5 and 6: Analyzing and Interpreting the Data
    Steps 7 and 8: Presentation and Dissemination of Findings
CRITERIA FOR SELECTING A DATA COLLECTION METHOD
    Size
    Scope
    Program Participation
    Worker Cooperation
    Intrusion Into the Lives of Research Participants
    Resources
    Time
    Previous Research
TRYING OUT THE SELECTED DATA COLLECTION METHOD
IMPLEMENTATION AND EVALUATION
    Implementation
    Evaluation
SUMMARY
REFERENCE AND FURTHER READING

13

IN THE LAST TWO CHAPTERS we discussed eight methods of data collection, which were divided by the type of data that each is most likely to produce—quantitative data (Chapter 11) or qualitative data (Chapter 12). This chapter examines the data collection process from the vantage point of choosing the most appropriate data collection method and data source for a given research question.

DATA COLLECTION AND THE RESEARCH PROCESS

Data collection is a critical step in the research process because it is the link between theory and practice. Our research study always begins with an idea that is molded by a conceptual framework, which uses preexisting knowledge about our study's problem area. Once our research problem and question have been refined to a researchable level, data are sought from a selected source(s) and gathered using a data collection method. The data collected are then used to support or supplant our original study's conceptions about our research problem under investigation.

The role of data collection in connecting theory and practice is understood when looking at the entire research process. As we have seen in previous chapters of this book, choosing a data collection method and data source follows the steps of selecting a research topic area, focusing the topic into a research question, and designing the research study.

Data collection comes before the steps of analyzing the data and writing the research report. Although data collection is presented in this text as a distinct phase of the research process, in reality it cannot be tackled separately or in isolation. All steps of the research process must be considered if we hope to come up with the best strategy to gather the most relevant, reliable, and valid data to answer a research question or to test a hypothesis. This section discusses the role of data collection in relation to the other steps of the generic research process.

- Steps 1 and 2—selecting a general research topic and focusing the topic into a research question
- Steps 3 and 4—designing the research study and collecting the data

- Steps 5 and 6—analyzing and interpreting the data
- Steps 7 and 8—presentation and dissemination of findings

Steps 1 and 2: Selecting a General Research Topic and Focusing the Topic into a Research Question

Our specific research question identifies the general problem area and the population to be studied. It tells us what we want to collect data about and alerts us to potential data sources. It does not necessarily specify the exact manner in which our data will be gathered, however. Let's return to our research question:

Research Question:
Do people who come from ethnic minority backgrounds have difficulty in accessing social services?

Our research question identifies a problem area (difficulty in accessing social services for people who are ethnic minorities) and a population (social service clients). It does not state how the question will be answered. We have seen in the last two chapters, that our research question, in fact, could be answered using various data collection methods. One factor that affects how our question is answered depends upon how we conceptualize and measure the variables within it. As we know, "accessing social services" could be measured in a variety of ways.

Another factor that affects how a research question is answered (or a hypothesis is tested) is the source of data; that is, who or what is providing them. If we want to get first-hand data about the accessibility of social services, for example, we could target the clients as a potential data source. If such first-hand data sources were not a viable option, second-hand data sources could be sought.

Social workers, for example, can be asked for their perceptions of how accessible social services are to clients who come from ethnic minorities. In other instances, second-hand data can be gleaned from existing reports (secondary data or content data) written about clients (or client records) that monitor client progress and social worker-client interactions.

By listing all possible data collection methods and data sources that could provide sound data to answer a research question, we develop a fuller understanding of our initial research problem. It also encourages us to think about our research problem from different perspectives, via the data sources. Because social work problems are complex, data collection is strengthened when two or more data sources are used. For example, if social workers and clients were to each report their perceptions of service accessibility in a similar way, then we could be more confident that the data (from both these sources) accurately reflect the problem being investigated.

Steps 3 and 4: Designing the Research Study and Collecting the Data

As we know, the research design flows from the research question, which flows from the problem area. A research design organizes our research question into a framework that sets the parameters and conditions of the study. As mentioned, the research question directs *what* data are collected and *who* data could be collected from. In a positivistic study, the research design refines the *what* question by operationalizing variables and the *who* question by developing a sampling strategy.

In an interpretive study, however, the research design identifies the starting point for data collection and how such procedures will be monitored and recorded along the way. In both research approaches, the research design also dictates (more or less) *when, where,* and *how* data will be collected.

The research design outlines how many data collection points our study will have and specifies the data sources. Each discrete data gathering activity constitutes a data collection point and defines *when* data are to be collected. Thus, using an exploratory one-group, post-test-only design, we will collect data only once from a single group of research participants. On the other hand, if a classical experimental design is used, data will be collected at two separate times with two different groups of research participants—for a total of four discrete data collection points.

Where the data are collected is also important to consider. If our research question is too narrow and begs for a broader issue that encompasses individuals living in various geographic locations, then mailed survey questionnaires would be more feasible than face-to-face interviews. If our research question focuses on a specific population where all research participants live in the same geographic location, however, it may be possible to use direct observations or individual or group interviews.

Because most social work studies are applied, the setting of our study usually involves clients' in their natural environments where there is little control over extraneous variables. If we want to measure the clients' perceptions about their difficulty in accessing social services, for example, do we observe clients in agency waiting rooms, observe how they interact with their social workers, or have them complete a survey questionnaire of some kind?

In short, we must always consider which method of data collection will lead to the most valid and reliable data to answer a research question or to test a hypothesis.

The combination of potential data collection methods and potential data sources is another important consideration. A research study can have one data collection source and still use multiple data collection methods. Our program's clients (data source 1) in our study, for example, can fill out a standardized questionnaire that measures their perceptions of how accessible social services are (data collection method 1) in addition to participating in face-to-face interviews (data collection method 2).

In the same vein, another study can have multiple data sources and one data collection method. In this case, we can collect data about difficulty in accessing social services through observation recordings by social workers (data source 1), administrators (data source 2), or other citizens (data source 3). The combination of data collection methods should not be too taxing on any research participant or any system, such as the social service program itself.

That is, data collection should not interfere greatly with the day-to-day activities of the persons providing (or responsible for collecting) the data. In some studies, there is no research design *per se*. Instead we can use existing data to answer our research question.

Such is the case when a secondary or content data are used. When the data already exist, we put more effort into ensuring that the research question is a good fit with the data at hand. Regardless, of what data collection method is used, once the data are collected, they are subject to analysis.

Steps 5 and 6: Analyzing and Interpreting the Data

Collecting data is a resource intensive endeavor that can be expensive and time consuming—both for interpretive and positivistic research studies. The truth of this statement is realized in the data analysis step of our research study. Without a great deal of forethought about what data to collect, data can be thrown out because they cannot be organized or analyzed in any meaningful way.

In short, data analyses should always be considered when choosing a data collection method and data source because the analysis phase must summarize, synthesize, and ultimately organize the data in an effort to have as clear-cut an answer as possible to our research question. When too much (or too little) data are collected, we can easily become bogged down (or stalled) by difficult decisions that could have been avoided with a little forethought.

After thinking through our research problem and research question and selecting a viable data collection and data source, it is worth-

Chapter 13: Selecting a Data Collection Method

while to list out the details of the type of data that will be produced. Specifically, we must think about how the data will be used in our data analysis. This exercise provides a clearer idea of the type of results we can expect.

The main variable in our research question is difficulty in accessing social services. Suppose the social worker decides to collect data about this variable by giving clients (data source) a brief standardized questionnaire (data collection method) to measure their perceptions. Many standardized questionnaires contain several subscales that, when combined, give a quantitative measure of a larger concept.

The *R-GSSSS* introduced in Chapter 7 (Page 142), for example, is made up of three subscales: (1) relevance, (2) impact, and (3) gratification. Thus, the *R-GSSSS* has four scores associated with it: a relevance score, an impact score, a gratification score, and a total score (which is a measure of satisfaction). With three separate subscales, we can choose to use any one subscale (one variable), all three subscales (three variables), or a total score (one variable).

Alternatively, if data about difficulty in accessing social services were to be collected using two different data sources such as social worker (data source 1) and a client (data source 2) observations, we must think about how the two data sources "fit" together. That is, will data from the two sources be treated as two separate variables? If so, will one variable be weighted differently in our analysis than the other? Thinking about how the data will be summarized helps us to expose any frivolous data—that is, data that are not suitable to answer our research question.

Besides collecting data about our study's variables, we must also develop a strategy to collect demographic data about the people who participate in our study. Typical demographic variables include: age, gender, education level, and family income. These data are not necessarily used in the analysis of the research question. Rather, they provide a descriptive context for our study. Some data collection methods, such as standardized questionnaires, include these types of data. Often, however, we are responsible for obtaining them as part of the data collection process.

Steps 7 and 8: Presentation and Dissemination of Findings

It is useful to think about our final research report when choosing a data collection method and data source as it forces us to visualize how our study's findings will ultimately be presented. It identifies both who the audience of the study will be and the people interested in our findings. Knowing who will read our research report and how it will be disseminated helps us to take more of an objective stance toward our study. In short, we can take a third-person look at what our study will finally look like.

Such objectivity helps us to think about our data collection method and data source with a critical eye. Will consumers of our research study agree that the clients in fact were the best data collection source? Were the data collection method and analysis sound? These are some of the practical questions that bring scrutiny to the data collection process.

CRITERIA FOR SELECTING A DATA COLLECTION METHOD

Thinking through the steps in the research process, from the vantage point of collecting data, permits us to refine the conceptualization of our study and the place of data collection within it. It also sets the context within which our data will be gathered. Clearly, there are many viable data collection methods and data sources that can be used to answer any research question.

Nevertheless, there are many practical criteria that ultimately refine the final data collection method (and sources) to fit the conditions of any given research study. These criteria are: (1) size, (2) scope, (3) program participation, (4) worker cooperation, (5) intrusion into the lives of research participants, (6) resources, (7) time, and (8) previous research findings. They all interact with one another, but for the sake of clarity each one is presented separately.

Size

The size of our study reflects just how many people, places, or systems are represented in it. As with any planning activity, the more people involved, the more complicated the process and the more difficult it is to arrive at a mutual agreement. Decisions about which data collection method and which data source to use can be stalled when several people, levels, or systems are involved. This is because individuals have different interests and opinions.

Imagine if our research question about whether ethnicity is related to difficulty in accessing social services were examined on a larger scale such that all social service programs in the country were included. Our study's complexity is dramatically increased because of such factors as the increased number of programs, clients, funders, government representatives, and social workers involved. The biases within each of these stakeholder groups make it much more difficult to agree upon the best data collection method and data source for our study.

Our study's sample size is also a consideration. This is particularly true for positivistic studies, which aim to have a sample that is representative of the population of interest. With respect to sample size, this means that we should strive for a reasonable representation of the sampling frame. When small-scale studies are conducted, such as a program evaluation in one social work program, the total number of people in the sampling frame may be in the hundreds or fewer. Thus, randomly selecting clients poses no real problem.

On the other hand, when large-scale studies are conducted, such as when the federal government chooses to examine a social service program that involves hundreds of thousands of people, sampling can become more problematic. If our sample is in the hundreds, it is unlikely that we would be able to successfully observe or interview all participants to collect data for our study. Rather, a more efficient manner of data collection—say a survey—may be more appropriate.

Scope

The scope of our research study is another matter to consider. Scope refers to how much of our *problem area* will be covered. If in our research question, for example, we are interested in gathering data on other client-related variables such as language ability, degree of social isolation, and level of assertiveness, then three different aspects of our problem area will be covered.

In short, we need to consider whether one method of data collection and one data source can be used to collect all the data. It could be that client records, for example, are used to collect data about clients' language abilities, interviews with clients are conducted to collect data about social isolation, and observation methods are used to gather data about clients' assertiveness levels.

Program Participation

Many social work research efforts are conducted in actual real-life program settings. Thus, it is essential that we gain the support of program personnel to conduct our study. Program factors that can impact the choice of our data collection methods and data sources include variables such as the program's clarity in its mandate to serve clients, its philosophical stance toward clients, and its flexibility in client record keeping. First, if a program is not able to clearly articulate a client service delivery plan, it will be difficult to distinguish between clinical activity and research activity, and to determine when the two overlap.

Second, some programs tend to base themselves on strong beliefs about a client population, which affect who can have access to clients and in what manner. A child sexual abuse investigation program, for example, may be designed specifically to avoid the problem of using multiple interviewers and multiple interviews of children in the investigation of an allegation of sexual abuse.

As a result, the program would be hesitant for us to conduct interviews with the children to gather data for "research purposes." Finally, to save time and energy there is often considerable overlap between program client records and research data collection. The degree of willingness of a program to adapt to new record-keeping techniques will affect how we might go about collecting certain types of data.

Worker Cooperation

On a general level, programs have fewer resources than they need and more clients than they can handle. Such conditions naturally lead their administrators and social workers to place intervention activity as a top priority (versus research activity). When our research study has social workers collecting data as a part of their day-to-day activities, it is highly likely that they will view data collection as additional paper work and not as a means to expedite decision-making in their work.

Getting cooperation of social workers within a program is a priority in any research study that relies directly or indirectly on their meaningful participation. Program workers will be affected by our study whether they are involved in the data collection process or not. They may be asked to schedule additional interviews with families or adjust their intervention plans to ensure that data collection occurs at the optimal time.

Given the fiscal constraints faced by programs, the workers themselves often participate as data collectors. They may end up using new client recording forms or administer questionnaires. Whatever the level of their participation, it is important for us to strive to achieve a maximum level of their cooperation.

There are three factors to consider when trying to achieve maximum cooperation from workers. First, we should make every effort to work effectively and efficiently with the program's staff. Cooperation is more likely to be achieved when workers participate in the development of our study plan from the beginning. Thus, it is worthwhile to take time to explain the purpose of our study and its intended outcomes at an early stage in the study. Furthermore, administrators and front-line workers alike can provide valuable information about which data collection method(s) may work best.

Second, we must be sensitive to the workloads of the program's staff. Data collection methods and sources should be designed to enhance the work of professionals. Client recording forms, for example, can be designed to provide focus for supervision meetings, as well as summarize facts and worker impressions about a case.

Third, a mechanism ought to be set up by which workers receive feedback based on the data they have collected. When a mechanism for feedback is put in place, for example, workers are more likely to show interest in the data collection activity. When data are reported back to the program's staff before the completion of our study, however, we must ensure that the data will not bias later measurements (if any).

Intrusion Into the Lives of Research Participants

When clients are used as a data source, client self-determination takes precedence over research activity. As we know, clients have every right to refuse participation in a research study and cannot be denied services because they are unwilling to participate. It is unethical, for example, when a member of a group-based treatment intervention has not consented to participate in the study, but participant observation is used as the data collection method.

This is unethical because the group member ends up being observed as part of the group dynamic in the data collection process after refusing to give his or her consent. The data collection method(s) we finally select must be flexible enough to allow our study to continue, even with the possibility that some clients will not participate.

Ethnic and cultural consideration must also be given to the type of data collection method used. One-to-one interviewing with Cambodian refugees, for example, may be extremely terrifying for them, given the interrogation they may have experienced in their own country. Moreover, if we, as data collectors, have different ethnic backgrounds than our research participants, it is important to ensure that interpretation of the data (e.g., their behaviors, events, or expressions) is accurate from the clients' perspectives and not our own.

We must also recognize the cultural biases of standardized measuring instruments, since most are based on testing with Caucasian groups. The problems here are twofold. First, we cannot be sure if the concept that the instrument is measuring is expressed the same way in different cultures. For instance, a standardized self-report instrument that measures family functioning may include an item such as, "We have flexible rules in our household that promote individual differences," which would likely be viewed positively by North American cultures, but negatively by many Asian cultures.

Second, because standardized measuring instruments are written in English, research participants must have a good grasp of English to ensure that the data collected from them are valid and reliable.

Another consideration comes into play when particular populations have been the subject of a considerable amount of research study already. Many aboriginal people living on reserves, for example, have been subjected to government surveys, task force inquiries, independent research projects, and perhaps even to the curiosities of social work students learning in a practicum setting.

When a population has been extensively researched, for example, it is even more important that we consider how the data collection method will affect those people participating in the study. Has the data collection method been used previously? If so, what was the nature of the data collected? Could the data be collected using less intrusive methods?

Resources

There are various costs associated with collecting data in any given research study. Materials and supplies, equipment rental, transportation costs, and training for data collectors are just a few things to consider when choosing a data collection method. In addition, once the data are collected, additional expenses can arise when the data are entered into a computer or transcribed.

An efficient data collection method is one that collects credible data to answer a research question or test a hypothesis while requiring the least amount of time and money. In our example, to ask clients about their perceptions about the difficulty in accessing social services via an open-ended interview may offer rich data, but we take the risk that clients will not fully answer our questions in the time allotted for the interview.

On the other hand, having them complete a self-report questionnaire about access to social services is a quicker and less costly way to collect data, but it gives little sense about how well the clients understood the questions being asked of them or whether the data obtained reflect their true perceptions.

Time

Time is a consideration when our study has a fixed completion date. Time constraints may be self- or externally imposed. Self-imposed time constraints are personal matters we need to consider. Is our research project a part of a thesis or dissertation? What are our personal time commitments? Externally imposed time restrictions are set by someone other than the person who is doing the study.

For instance, our research study may be limited by the fiscal year of a social service program and/or funding source. Other external pressures may be political, such as an administrator who wants research results for a funding proposal or to present at a conference.

Previous Research

Having reviewed the professional literature on our problem, we need to be well aware of other data collection methods that have been used in similar studies. We can evaluate earlier studies for the strengths and weaknesses of their data collection methods and thereby make a more informed decision as to the best data collection strategy to use in our specific situation. Further, we need to look for diversity when evaluating other data collection approaches; that is, we can triangulate results from separate studies that used different data collection methods and data sources.

TRYING OUT THE SELECTED DATA COLLECTION METHOD

Data collection is a particularly vulnerable time for a research study because it is the point where "talk" turns into "action." So far, all the considerations that have been weighed in the selection of a data collection method have been in theory. All people involved in our research endeavor have cast their suggestions and doubts on the entire process.

Once general agreement has been reached about which data collection method and data source to use, it is time to test the waters. Trying out a data collection method can occur informally by simply testing it out with available and willing research participants or, at the very least, with anyone who has not been involved with the planning of the study. The purpose of this trial run is to ensure that those who are going to provide data understand the questions and procedures in the way that they were intended.

Data collection methods might also be tested more formally, such as when a pilot study is conducted. A pilot study involves carrying out all aspects of the data collection plan on a mini-scale. That is, a small portion of our study's actual sample is selected and run through all steps of the data collection process.

In a pilot study, we are interested in the process of the data collection as well as the content. In short, we want to know whether our chosen data collection method produces the expected data. Are there any unanticipated barriers to gathering the desired data? How do research participants respond to our data collection procedures?

IMPLEMENTATION AND EVALUATION

The data collection step of a research study can go smoothly if we act proactively. That is, we should guide and monitor the entire data collection process according to the procedures and steps that were set out in the planning stage of our study and were tested in the pilot study.

Implementation

The main guiding principle to implementing the selected data collection method is that a systematic approach to data collection must be used. This means that the steps to gathering data should be methodically detailed so that there is no question about the tasks of the person(s) collecting the data—the data collector(s).

This is true whether using a positivistic or interpretive research approach. As we know, the difference between these two research approaches is that the structure of the data collection process within an interpretive research study is documented as the study progresses. On the other hand, in a positivistic research study, the data collection process is decided at the study's outset and provides much less flexibility after the study is underway.

It must be very clear from the beginning who is responsible for collecting the data. When we take on the task, there is reasonable assurance that the data collection will remain objective and be guided by our research interests. Data collection left to only one person may be a formidable task. We must determine the amount of resources available to decide what data collection method is most realistic. Regardless of the study size, we must attempt to establish clear roles and boundaries with those involved in the data collection process.

The clearer our research study is articulated, the less difficulty there will be in moving through all the steps of the study. In particular, it is critical to identify who will and will not be involved in the data collection process. To further avoid mix-up and complications, specific tasks must be spelled out for all persons involved in our study. Where will the data be stored? Who will collect them? How will the data collection process be monitored?

In many social work research studies, front-line social workers are involved in data collection activities as part of their day-to-day activities. They typically gather intake and referral data, write assessment notes, and even use standardized questionnaires as part of their assessments. Data collection in programs can easily be designed to serve the dual purposes of research *and* intervention inquiry. Thus, it is important to establish data collection protocols to avoid problems of biased data. As mentioned, everyone in a research study must agree *when* data will be collected, *where*, and in *what* manner. Agreement is more likely to occur when we have fully informed and involved everyone participating in our study.

Evaluation

The process of selecting a chosen data collection method is not complete without evaluating it. Evaluation occurs at two levels. First, the strengths and weaknesses of a data collection method and data source are evaluated, given the research context in which our study takes place. If, for example, data are gathered about clients' presenting problems by a referring social worker, it must be acknowledged that the obtained data offer a limited (or restricted) point of view about the clients' problems. The strength of this approach may be that it was the only means for collecting the data.

A second level of evaluation is monitoring the implementation of the data collection process itself. When data are gathered using several methods (or several sources), it is beneficial to develop a checklist of what data have been collected for each research participant. Developing a strategy for monitoring the data collection process is especially important when the data must be collected in a timely fashion.

If pretest data are needed before a client enters a treatment program, for example, the data collection must be complete before admission occurs. Once a client has entered the program, opportunity to collect pretest data is lost.

Another strategy for monitoring evaluation is to keep a journal of the data collection process. The journal records any questions or queries that arise in the data collection step. We may find, for example, that several research participants completing a questionnaire have difficulty understanding one particular question.

In addition, sometimes research participants have poor reading skills and require assistance with completion of some self-report standardized questionnaires. Documenting these idiosyncratic incidents accumulates important information by which to comment on our data's credibility.

SUMMARY

There are many possible data collection methods and data sources that can be used in any given research situation. We must weigh the pros and cons of both within the context of a particular research study to arrive at the best data collection method and data source. This process involves both conceptual and practical considerations.

On a conceptual level, we review the steps of the research process through a "data collection and data source lens." We think about how various data collection methods and data sources fit with each step of the research process. At the same time, considering the different data collection methods and data sources helps us to gain a fuller understanding of our problem area and research question.

On a practical level, there are many considerations when deciding upon the best data method(s) and data source(s) for a particular study. Factors such as worker cooperation, available resources, and consequences for the clients all influence our final choices.

Now that we know how to collect quantitative and qualitative data for any given research study, we will turn our attention to analyzing them. Thus, Chapter 13 presents how to analyze quantitative data and Chapter 14 discusses how to analyze qualitative data.

REFERENCE AND FURTHER READING

Unrau, Y.A. (2008). Selecting a data collection method and data source. In R.M. Grinnell, Jr., & Y.A. Unrau (Eds.), *Social work research and evaluation: Foundations of evidence-based practice* (8th ed., pp. 327–341). New York: Oxford University Press.

Check out our Website for useful links and chapters (PDF) on:
- introductory data-collection methodologies
- Web-based links to various data-collection tutorials

Go to: www.pairbondpublications.com
Click on: Student Resources
            Chapter-by-Chapter Resources
            Chapter 13

Check Out Our Website

PairBondPublications.com

√ *Student Workbook Exercises*
√ *Chapter Power Point Slides*
√ *On-line Glossaries*
√ *General Links*
√ *Specific Chapter Links*

Part VI

Analyzing Data

Chapter 14: Analyzing Quantitative Data   282

Chapter 15: Analyzing Qualitative Data   310

Chapter 14

Analyzing Quantitative Data

LEVELS OF MEASUREMENT
    Nominal Measurement
    Ordinal Measurement
    Interval Measurement
    Ratio Measurement
COMPUTER APPLICATIONS
DESCRIPTIVE STATISTICS
    Frequency Distributions
    Measures of Central Tendency
        Mode
        Median
        Mean
    Measures of Variability
        Range
        Standard Deviation
INFERENTIAL STATISTICS
    Statistics That Determine Associations
        Chi-Square
        Correlation
    Statistics That Determine Differences
        Dependent t-Tests
        Independent t-Tests
        One-Way Analysis of Variance
SUMMARY
REFERENCES AND FURTHER READING

14

AFTER QUANTITATIVE DATA ARE COLLECTED they need to be analyzed—the topic of this chapter. To be honest, a thorough understanding of quantitative statistical methods is far beyond the scope of this book. Such comprehension necessitates more in-depth study, through taking one or more statistics courses. Instead, we briefly describe a select group of basic statistical analytical methods that are used frequently in many quantitative *and* qualitative social work research studies. Our emphasis is not on providing and calculating formulas, but rather on helping the reader to understand the underlying rationale for their use.

We present two basic groups of statistical procedures. The first group is called *descriptive statistics*, which simply describe and summarize one or more variables for a sample or population. They provide information about only the group included in the study. The second group of statistical procedures is called *inferential statistics*, which determine if we can generalize findings derived from a sample to the population from which the sample was drawn.

In other words, knowing what we know about a particular sample, can we infer that the rest of the population is similar to the sample that we have studied? Before we can answer this question, however, we need to know the level of measurement for each variable being analyzed. Let us now turn to a brief discussion of the four different levels of measurement that a variable can take.

LEVELS OF MEASUREMENT

The specific statistic(s) used to analyze the data collected is dependent on the type of data that are gathered. The characteristics or qualities that describe a variable are known as its *attributes*. The variable *gender*, for example, has only two characteristics or attributes—*male* and *female*—since gender in humans is limited to male and female, and there are no other categories or ways of describing gender.

The variable *ethnicity* has a number of possible categories: *African American, Native American, Asian, Hispanic American,* and *Caucasian* are just five examples of the many attributes of the variable ethnicity. A point to note here is that the attributes of gender differ in kind from one another—male is different from female—and, in the same way, the attributes of ethnicity are also different from one another.

Now consider the variable *income*. Income can only be described in terms of amounts of money: $15,000 per year, $288.46 per week, and so forth. In whatever terms a person's income is actually described, it still comes down to a number. Since every number has its own category, as we mentioned before, the variable income can generate as many categories as there are numbers, up to the number covering the research participant who earns the most. These numbers are all attributes of income and they are all different, but they are not different in *kind*, as male and female are, or Native American and Hispanic; they are only different in *quantity*.

In other words, the attributes of income differ in that they represent more or less of the same thing, whereas the attributes of gender differ in that they represent different kinds of things. Income will, therefore, be measured in a different way from gender. When we come to measure income, we will be looking for categories that are lower or higher than each other; when we come to measure gender, we will be looking for categories that are different in kind from each other.

Mathematically, there is not much we can do with categories that are different in kind. We cannot subtract Hispanics from Caucasians, for example, whereas we can quite easily subtract one person's annual income from another and come up with a meaningful difference. As far as mathematical computations are concerned, we are obliged to work at a lower level of complexity when we measure variables like ethnicity than when we measure variables like income. Depending on the nature of their attributes, all variables can be measured at one (or more) of four measurement levels: (1) nominal, (2) ordinal, (3) interval, or (4) ratio.

Nominal Measurement

Nominal measurement is the lowest level of measurement and is used to measure variables whose attributes are different in kind. As we have seen, gender is one variable measured at a nominal level, and ethnicity is another. *Place of birth* is a third, since "born in California," for example, is different from "born in Chicago," and we cannot add "born in California" to "born in Chicago," or subtract them or divide them, or do anything statistically interesting with them at all.

Ordinal Measurement

Ordinal measurement is a higher level of measurement than nominal and is used to measure those variables whose attributes can be rank ordered: for example, socioeconomic status, sexism, racism, client satisfaction, and the like. If we intend to measure *client satisfaction*, we must first develop a list of all the possible attributes of client satisfaction: that is, we must think of all the possible categories into which answers about client satisfaction might be placed.

Some clients will be *very satisfied*—one category, at the high end of the satisfaction continuum; some will be *not at all satisfied*—a separate category, at the low end of the continuum; and others will be *generally satisfied, moderately satisfied,* or *somewhat satisfied*—three more categories, at differing points on the continuum, as illustrated below:

1. Not At All Satisfied
2. Somewhat Satisfied
3. Moderately Satisfied
4. Generally Satisfied
5. Very Satisfied

The above is a five-point scale with a brief description of the degree of satisfaction represented by the point. Of course, we may choose to express the anchors in different words, substituting *extremely satisfied* for *very satisfied,* or *fairly satisfied* for *generally satisfied.* We may select a three-point scale instead, limiting the choices to *very satisfied, moderately satisfied,* and *not at all satisfied;* or we may even use a ten-point scale if we believe that our respondents will be able to rate their satisfaction with that degree of accuracy.

Whichever particular method is selected, some sort of scale is the only measurement option available because there is no other way to categorize client satisfaction except in terms of more satisfaction or less satisfaction. As we did with nominal measurement, we might assign numbers to each of the points on the scale. If we used the five-point scale as illustrated above, we might assign a 5 to *very satisfied,* a 4 to *generally satisfied,* a 3 to *moderately satisfied,* a 2 to *somewhat satisfied,* and a 1 to *not at all satisfied.*

Here, the numbers do have some mathematical meaning. Five (*very satisfied*) is in fact better than 4 (*generally satisfied*), 4 is better than 3, 3 is better than 2, and 2 is better than 1. The numbers, however, say nothing about *how much better* any category is than any other. We cannot assume that the difference in satisfaction between *very* and *generally* is the same as the difference between *generally* and *moderately.*

In short, we cannot assume that the intervals between the anchored points on the scale are all the same length. Most definitely, we cannot assume that a client who rates a service at 4 (*generally satisfied*) is twice as satisfied as a client who rates the service at 2 (*somewhat satisfied*).

In fact, we cannot attempt any mathematical manipulation at all. We cannot add the numbers 1, 2, 3, 4, and 5, nor can we subtract, multiply, or divide them. As its name might suggest, all we can know from ordinal measurement is the order of the categories.

Interval Measurement

Some variables, such as client satisfaction, have attributes that can be rank-ordered—from *very satisfied* to *not at all satisfied*, as we have just discussed. As we saw, however, these attributes cannot be assumed to be the same distance apart if they are placed on a scale; and, in any case, the distance they are apart has no real meaning. No one can measure the distance between *very satisfied* and *moderately satisfied*; we only know that the one is better than the other.

Conversely, for some variables, the distance, or interval, separating their attributes *does* have meaning, and these variables can be measured at the interval level. An example in physical science would is the Fahrenheit or Celsius temperature scales. The difference between 80 degrees and 90 degrees is the same as the difference between 40 and 50 degrees. Eighty degrees is not twice as hot as 40 degrees; nor does zero degrees mean no heat at all.

In social work, interval measures are most commonly used in connection with standardized measuring instruments, as presented in Chapter 7. When we look at a standardized intelligence test, for example, we can say that the difference between IQ scores of 100 and 110 is the same as the difference between IQ scores of 95 and 105, based on the scores obtained by the many thousands of people who have taken the test over the years. As with the temperature scales mentioned above, a person with an IQ score of 120 is not twice as intelligent as a person with a score of 60: nor does a score of 0 mean no intelligence at all.

Ratio Measurement

The highest level of measurement, ratio measurement, is used to measure variables whose attributes are based on a true zero point. It may not be possible to have zero intelligence, but it is certainly possible to have zero children or zero money. Whenever a question about a particular variable might elicit the answer "none" or "never," that variable can be measured at the ratio level.

The question, "How many times have you seen your social worker?" might be answered, "Never." Other variables commonly measured at the ratio level include length of residence in a given place, age, number of times married, number of organizations belonged to, number of antisocial behaviors, number of case reviews, number of training sessions, number of supervisory meetings, and so forth. With a ratio level of measurement we can meaningfully interpret the comparison between two scores.

A person who is 40 years of age, for example, is twice as old as a person who is 20 and half as old as a person who is 80. Children aged 2 and 5, respectively, are the same distance apart as children aged 6 and 9. Data resulting from ratio measurement can be added, subtracted, multiplied, and divided. Averages can be calculated and other statistical analyses can be performed.

It is useful to note that, while some variables *can* be measured at a higher level, they may not need to be. The variable *income*, for example, can be measured at a ratio level because it is possible to have a zero income but, for the purposes of a particular study, we may not need to know the actual incomes of our research participants, only the range within which their incomes fall.

A person who is asked how much he or she earns may be reluctant to give a figure ("mind your own business" is a perfectly legitimate response) but may not object to checking one of a number of income categories, choosing, for example, between:

1. less than $5,000 per year
2. $5,001 to $15,000 per year
3. $15,001 to $25,000 per year
4. $25,001 to $35,000 per year
5. more than $35,000 per year

Categorizing income in this way reduces the measurement from the ratio level to the ordinal level. It will now be possible to know only that a person checking Category 1 earns less than a person checking Category 2, and so on. While we will not know *how much* less or more one person earns than another and we will not be able to perform statistical tasks such as calculating average incomes, we will be able to say, for example, that 50 percent of our sample falls into Category 1, 30 percent into Category 2, 15 percent into Category 3, and 5 percent into Category 4. If we are conducting a study to see how many people fall in each income range, this may be all we need to know.

In the same way, we might not want to know the actual ages of our sample, only the range in which they fall. For some studies, it might be enough to measure age at a nominal level—to inquire, for example, whether people were born during the depression, or whether they were born before or after 1990. When studying variables that can be measured at any level, the measurement level chosen depends on what kind of data are needed, and this in turn is determined by why the data are needed, which in turn is determined by our research question.

COMPUTER APPLICATIONS

The use of computers has revolutionized the analysis of quantitative and qualitative data. Where previous generations of researchers had to rely on hand-cranked adding machines to calculate every small step in a data analysis, today we can enter raw scores into a personal computer, and, with few complications, direct the computer program to execute just about any statistical test imaginable. Seconds later, the results are available. While the process is truly miraculous, the risk is that, even though we have conducted the correct statistical analysis, we may not understand what the results mean, a factor that will almost certainly affect how we interpret the data.

We can code data from all four levels of measurement into a computer for any given data analysis. The coding of nominal data is perhaps the most complex, because we have to create categories that correspond with certain possible responses for a variable. One type of nominal level data that is often gathered from research participants is *place of birth*. If, for the purposes of our study, we are interested in whether our research participants were born in either the Canada or the United States, we would assign only three categories to *place of birth*:

1. Canada
2. United States
9. Other

The *other* category appears routinely at the end of lists of categories and acts as a catch-all, to cover any category that may have been omitted.

When entering nominal level data into a computer, because we do not want to enter *Canada* every time the response on the questionnaire is Canada, we may assign it the code number 1, so that all we have to enter is 1. Similarly, the United States may be assigned the number 2, and "other" may be assigned the number 9.

These numbers have no mathematical meaning: We are not saying that Canada is better than the United States because it comes first, or that the United States is twice as good as Canada because the number assigned to it is twice as high. We are merely using numbers as a shorthand device to record *qualitative* differences: differences in *kind*, not in amount.

Most coding for ordinal, interval, and ratio level data is simply a matter of entering the final score, or number, from the measuring instrument that was used to measure the variable directly into the computer.

If a person scored a 45 on a standardized measuring instrument, for example, the number 45 would be entered into the computer. Although almost all data entered into computers are in the form of numbers, we need to know at what level of measurement the data exist, so that we can choose the appropriate statistic(s) to describe and compare the variables. Now that we know how to measure variables at four different measurement levels, let us turn to the first group of statistics that can be helpful for the analyses of data—descriptive statistics.

DESCRIPTIVE STATISTICS

Descriptive statistics are commonly used in most quantitative and qualitative research studies. They describe and summarize a variable(s) of interest and portray how that particular variable is distributed in the sample, or population. Before looking at descriptive statistics, however, let us examine a social work research example that will be used throughout this chapter.

Thea Black is a social worker who works in a treatment foster care program. Her program focuses on children who have behavioral problems who are placed with "treatment" foster care parents. These parents are supposed to have parenting skills that will help them provide the children with the special needs they present.

Thus, Thea's program also teaches parenting skills to these treatment foster care parents. She assumes that newly recruited foster parents are not likely to know much about parenting children who have behavioral problems. Therefore, she believes that they would benefit from a training program that teaches these skills in order to help them to deal effectively with the special needs of these children who will soon be living with them.

Thea hopes that her parenting skills training program will increase the knowledge about parental management skills for the parents who attend. She assumes that, with such training, the foster parents would be in a better position to support and provide clear limits for their foster children.

TABLE 14.1
Data Collection for Four Variables
from Foster Care Providers

Number	PSS Score	Gender	Previous Training	Years of Education
01	95	male	no	12
02	93	female	yes	15
03	93	male	no	08
04	93	female	no	12
05	90	male	yes	12
06	90	female	no	12
07	84	male	no	14
08	84	female	no	18
09	82	male	no	10
10	82	female	no	12
11	80	male	no	12
12	80	female	no	11
13	79	male	no	12
14	79	female	yes	12
15	79	female	no	16
16	79	male	no	12
17	79	female	no	11
18	72	female	no	14
19	71	male	no	15
20	55	female	yes	12

After offering the training program for several months, Thea became curious about whether the foster care providers who attended the program were, indeed, lacking in knowledge of parental management skills as she first believed (her tentative hypothesis). She was fortunate to find a valid and reliable standardized instrument that measures the knowledge of such parenting skills, the Parenting Skills Scale (*PSS*). Thea decided to find out for herself how much the newly recruited parents knew about parenting skills—clearly a descriptive research question.

At the beginning of one of her training sessions (before they were exposed to her skills training program), she handed out the *PSS*, asking the 20 individuals in attendance to complete it and also to include data about their gender, years of education, and whether they had ever participated in a parenting skills training program before. All of these three variables could be potentially extraneous ones that might influence the level of knowledge of parenting skills of the 20 participants.

For each foster care parent, Thea calculated the *PSS* score, called a *raw score* because it has not been sorted or analyzed in any way. The total score possible on the *PSS* is 100, with higher scores indicating greater knowledge of parenting skills. The scores for the *PSS*

TABLE 14.2
Frequency Distribution

PSS Score	Absolute Frequency
95	1
93	3
90	2
84	2
82	2
80	2
79	5
72	1
71	1
55	1

scale, as well as the other data collected from the 20 parents, are listed in Table 14.1.

At this point, Thea stopped to consider how she could best utilize the data that she had collected. She had data at three different levels of measurement. At the nominal level, Thea had collected data on gender (3rd column), and whether the parents had any previous parenting skills training (4th column). Each of these variables can be categorized into two responses.

The scores on the *PSS* (2nd column) are ordinal because, although the data are sequenced from highest to lowest, the differences between units cannot be placed on an equally spaced continuum. Nevertheless, many measures in the social sciences are treated as if they are at an interval level, even though equal distances between scale points cannot be proved. This assumption is important because it allows for the use of inferential statistics on such data.

Finally, the data on years of formal education (5th column) that were collected by Thea are clearly at the ratio level of measurement, because there are equally distributed points and the scale has an absolute zero.

In sum, it seemed to Thea that the data could be used in at least two ways. First, the data collected about each variable could be described to provide a picture of the characteristics of the group of foster care parents. This would call for descriptive statistics. Secondly, she might look for relationships between some of the variables about which she had collected data, procedures that would utilize inferential statistics. For now let us begin by looking at how the first type of descriptive statistic can be utilized with Thea's data set.

Frequency Distributions

One of the simplest procedures that Thea can employ is to develop a frequency distribution of her data. Constructing a frequency distribution involves counting the occurrences of each value, or category, of the variable and ordering them in some fashion. This *absolute* or *simple frequency distribution* allows us to see quickly how certain values of a variable are distributed in our sample.

The *mode*, or the most commonly occurring score, can be easily spotted in a simple frequency distribution (see Table 14.2). In this example, the mode is 79, a score obtained by five parents on the *PSS* scale. The highest and the lowest score are also quickly identifiable. The top score was 95, while the foster care parent who performed the least well on the *PSS* scored 55.

There are several other ways to present frequency data. A commonly used method that can be easily integrated into a simple frequency distribution table is the *cumulative frequency distribution*, shown in Table 14.3.

In Thea's data set, the highest *PSS* score, 95, was obtained by only one individual. The group of individuals who scored 93 or above on the *PSS* measure includes four foster care parents. If we want to know how many scored 80 or above, if we look at the number across from 80 in the cumulative frequency column, we can quickly see that 12 of the parents scored 80 or better.

Other tables utilize percentages rather than frequencies, sometimes referred to as *percentage distributions*, shown in the right-hand column in Table 14.3. Each of these numbers represents the percentage of participants who obtained each *PSS* value. Five individuals, for example, scored 79 on the *PSS*. Since there was a total of 20 foster care parents, 5 out of the 20, or one-quarter of the total, obtained a score of 79. This corresponds to 25 percent of the participants.

Finally, *grouped frequency distributions* are used to simplify a table by grouping the variable into equal-sized ranges, as is shown in Table 14.4. Both absolute and cumulative frequencies and percentages can also be displayed using this format. Each is calculated in the same way that was previously described for non-grouped data, and the interpretation is identical.

TABLE 14.3
Cumulative Frequency and Percentage
Distribution of Parental Skill Scores

PSS Score	Absolute Frequency	Cumulative Frequency	Percentage Distribution
95	1	1	5
93	3	4	15
90	2	6	10
84	2	8	10
82	2	10	10
80	2	12	10
79	5	17	25
72	1	18	5
71	1	19	5
55	1	20	5
Totals...	20		100

Looking at the absolute frequency column, for example, we can quickly identify the fact that seven of the foster care parents scored in the 70–79 range on the *PSS*. By looking at the cumulative frequency column, we can see that 12 of 20 parents scored 80 or better on the *PSS*. Further, from the absolute percentage column, it is clear that 30 percent of the foster parents scored in the 80–89 range on the knowledge of parenting skills scale.

Note that each of the other variables in Thea's data set could also be displayed in frequency distributions. Displaying years of education in a frequency distribution, for example, would provide a snapshot of how this variable is distributed in Thea's sample of foster care parents. With two category nominal variables, such as gender (male, female) and previous parent skills training (yes, no), however, cumulative frequencies become less meaningful and the data are better described as percentages.

Thea noted that 55 percent of the foster care parents who attended the training workshop were women (obviously the other 45 percent were men) and that 20 percent of the parents had already received some form of parenting skills training (while a further 80 percent had not been trained).

Measures of Central Tendency

We can also display the values obtained on the *PSS* in the form of a graph. A *frequency polygon* is one of the simplest ways of charting frequencies.

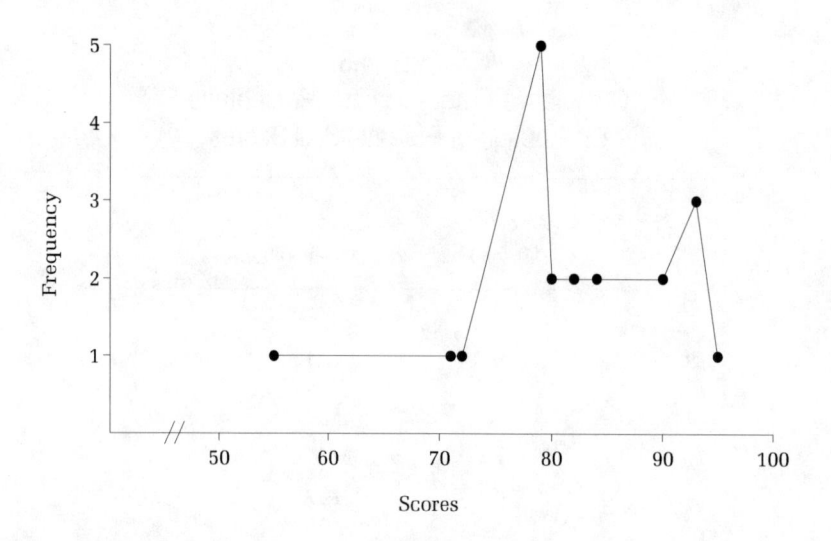

FIGURE 14.1
Frequency Polygon of Parental Skill Scores
(from Table 14.2)

The graph in Figure 14.1 displays the data that we had previously put in Table 14.2. The *PSS* score is plotted in terms of how many of the foster care parents obtained each score.

As can be seen from Table 14.2 and Figure 14.1, most of the scores fall between 79 and 93. The one extremely low score of 55 is also quickly noticeable in such a graph, because it is so far removed from the rest of the values.

A frequency polygon allows us to make a quick analysis of how closely the distribution fits the shape of a normal curve.

TABLE 14.4
Grouped Frequency Distribution
of Parental Skill Scores

PSS Scores	Absolute Frequency	Cumulative Frequency	Absolute Percentage
90 – 100	6	6	30
80 – 89	6	12	30
70 – 79	7	19	35
60 – 69	0	19	0
50 – 59	1	20	5

Chapter 14: Analyzing Quantitative Data

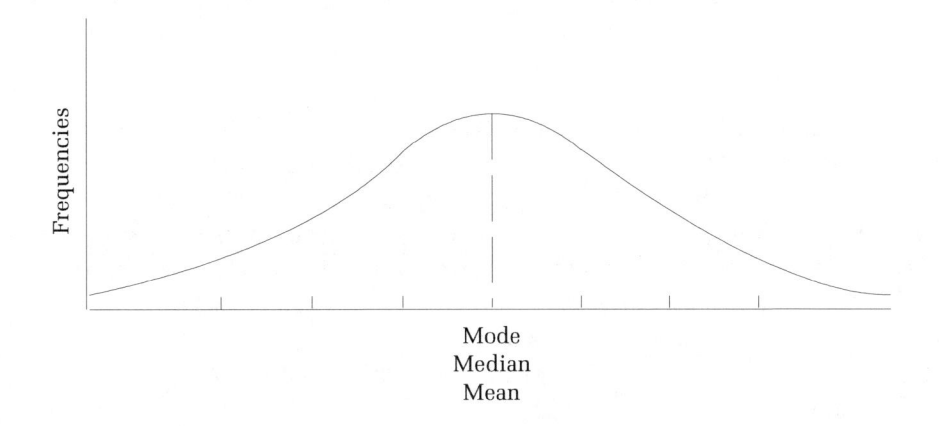

FIGURE 14.2
The Normal Distribution

A *normal curve*, also known as a *bell-shaped distribution* or a *normal distribution*, is a frequency polygon in which the greatest number of responses fall in the middle of the distribution and fewer scores appear at the extremes of either very high or very low scores (see Figure 14.2).

Many variables in the social sciences are assumed to be distributed in the shape of a normal curve. Low intelligence, for example, is thought to be relatively rare as compared to the number of individuals with average intelligence. On the other end of the continuum, extremely gifted individuals are also relatively uncommon.

Of course, not all variables are distributed in the shape of a normal curve. Some are such that a large number of people do very well (as Thea found in her sample of foster care parents and their parenting skill levels). Other variables, such as juggling ability, for example, would be charted showing a fairly substantial number of people performing poorly.

Frequency distributions of still other variables would show that some people do well, and some people do poorly, but not many fall in between. What is important to remember about distributions is that, although all different sorts are possible, most statistical procedures assume that there is a normal distribution of the variable in question in the population.

When looking at how variables are distributed in samples and populations it is common to use measures of *central tendency*, such as the mode, median, and mean, which help us to identify where the typical or the average score can be found. These measures are utilized so often because, not only do they provide a useful summary of the data, they also provide a common denominator for comparing groups to each other.

Mode

As mentioned earlier, the mode is the score, or value, that occurs the most often—the value with the highest frequency. In Thea's data set of parental skills scores the mode is 79, with five foster care parents obtaining this value. The mode is particularly useful for nominal level data. Knowing what score occurred the most often, however, provides little information about the other scores and how they are distributed in the sample or population. Because the mode is the least precise of all the measures of central tendency, the median and the mean are better descriptors of ordinal level data and above. We now turn our attention to the second measure of central tendency, the median.

Median

The median is the score that divides a distribution into two equal parts or portions. In order to do this, we must rank-order the scores, so at least an ordinal level of measurement is required. In Thea's sample of 20 *PSS* scores, the median would be the score above which the top ten scores lie and below which the bottom ten fall. As can be seen in Table 14.2, the top ten scores finish at 82, and the bottom ten scores start at 80. In this example, the median is 81, since it falls between 82 and 80.

Mean

The mean is the most sophisticated measure of central tendency and is useful for interval or ratio levels of measurement. It is also one of the most commonly utilized statistics. A mean is calculated by summing the individual values and dividing by the total number of values. The mean of Thea's sample is 95 + 93 + 93 + 93 + 90 + 90 + ... 72 + 71 + 55/20 = 81.95. In this example, the obtained mean of 82 (we rounded off for the sake of clarity) is larger than the mode of 79 or the median of 81.

The mean is one of the previously mentioned statistical procedures that assumes that a variable will be distributed normally throughout a population. If this is not an accurate assumption, then the median might be a better descriptor. The mean is also best used with relatively large sample sizes where extreme scores (such as the lowest score of 55 in Thea's sample) have less influence.

Chapter 14: Analyzing Quantitative Data

Measures of Variability

While measures of central tendency provide valuable information about a set of scores, we are also interested in knowing how the scores scatter themselves around the center. A mean does not give a sense of how widely distributed the scores may be: This is provided by measures of variability such as the range and the standard deviation.

Range

The range is simply the distance between the minimum and the maximum score. The larger the range, the greater the amount of variation of scores in the distribution. The range is calculated by subtracting the lowest score from the highest. In Thea's sample, the range is 40 (95 − 55). The range does not assume equal interval data. It is, like the mean, sensitive to deviant values, because it depends on only the two extreme scores.

We could have a group of four scores ranging from 10 to 20: 10, 14, 19, and 20, for example. The range of this sample would be 10 (20 − 10). If one additional score that was substantially different from the first set of four scores was included, this would change the range dramatically. In this example, if a fifth score of 45 was added, the range of the sample would become 35 (45 − 10), a number that would suggest quite a different picture of the variability of the scores.

Standard Deviation

The standard deviation is the most well-used indicator of dispersion. It provides a picture of how the scores distribute themselves around the mean. Used in combination with the mean, the standard deviation provides a great deal of information about the sample or population, without our ever needing to see the raw scores. In a normal distribution of scores, described previously, there are six standard deviations: three below the mean and three above, as is shown in Figure 14.3.

In this perfect model we always know that 34.13 percent of the scores of the sample fall within one standard deviation above the mean, and another 34.13 percent fall within one standard deviation below the mean. Thus, a total of 68.26 percent, or about two-thirds of the scores, is between +1 standard deviation and −1 standard deviation from the mean.

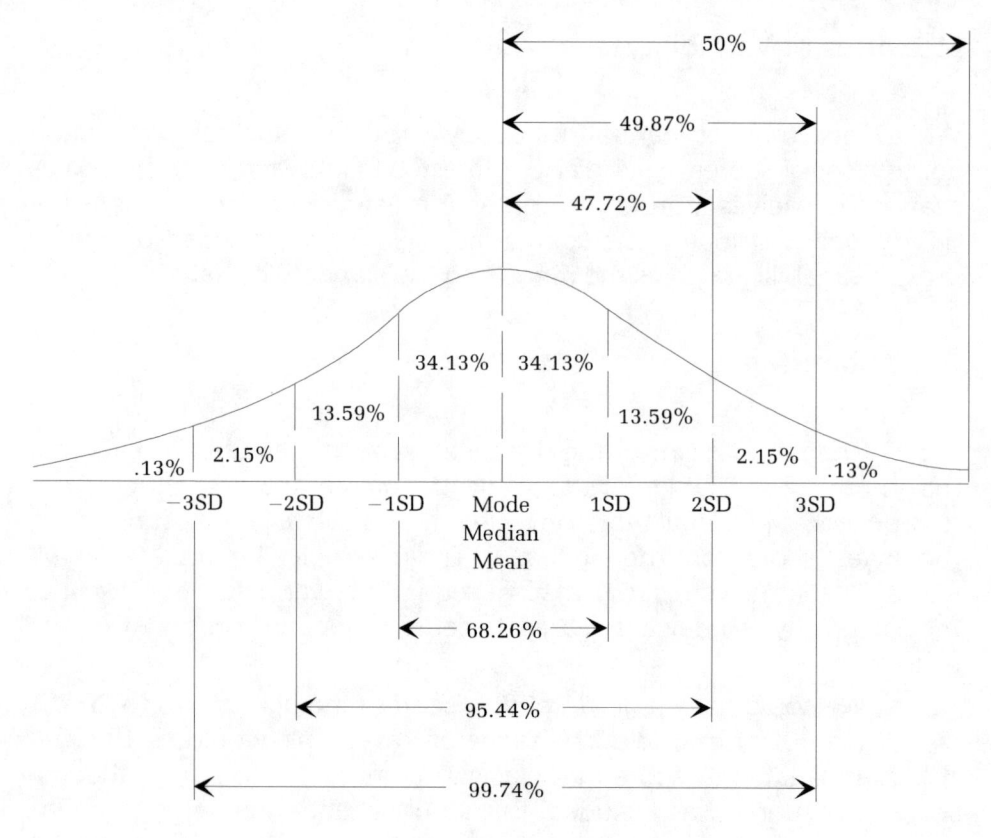

FIGURE 14.3
Proportions of the Normal Curve

This leaves almost one-third of the scores to fall farther away from the mean, with 15.87 percent (50% − 34.13%) above +1 standard deviation, and 15.87 percent (50% − 34.13%) below 1 standard deviation.

In total, when looking at the proportion of scores that fall between +2 and −2 standard deviations, 95.44 percent of scores can be expected to be found within these parameters. Furthermore, 99.74 percent of the scores fall between +3 standard deviations and −3 standard deviations about the mean. Thus, finding scores that fall beyond 3 standard deviations above and below the mean should be a rare occurrence.

The standard deviation has the advantage, like the mean, of taking all values into consideration in its computation. Also similar to the mean, it is utilized with interval or ratio levels of measurement and assumes a normal distribution of scores.

Several different samples of scores could have the same mean, but the variation around the mean, as provided by the standard deviation, could be quite different, as is shown in Figure 14.4*a*.

Two different distributions could have unequal means and equal standard deviations, as in Figure 14.4*b*, or unequal means and unequal standard deviations, as in Figure 14.4*c*.

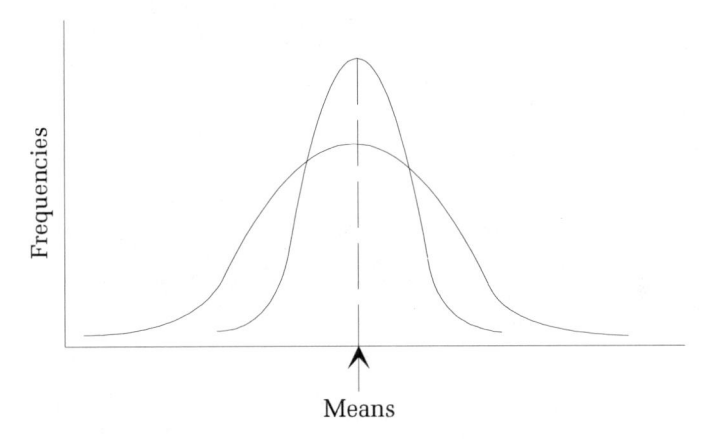

Chapter 14: Analyzing Quantitative Data

(a) Equal means, unequal standard deviations

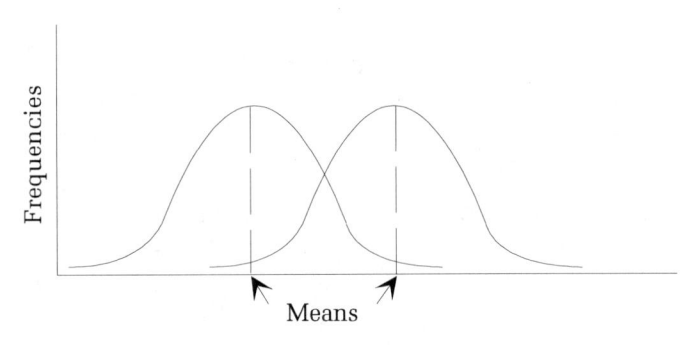

(b) Unequal means, equal standard deviations

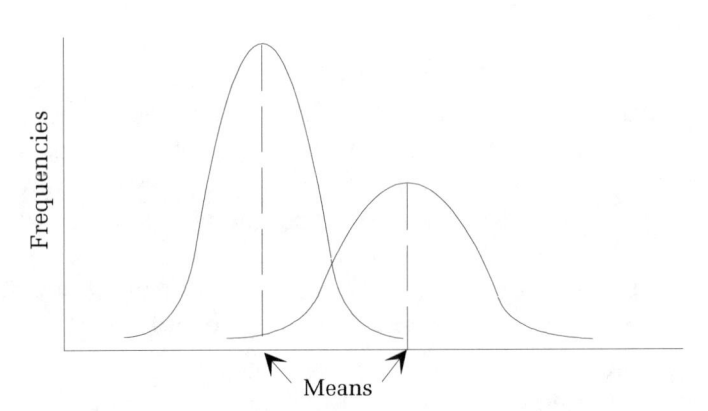

(c) Unequal means, unequal standard deviations

FIGURE 14.4
Variations in Normal Distributions

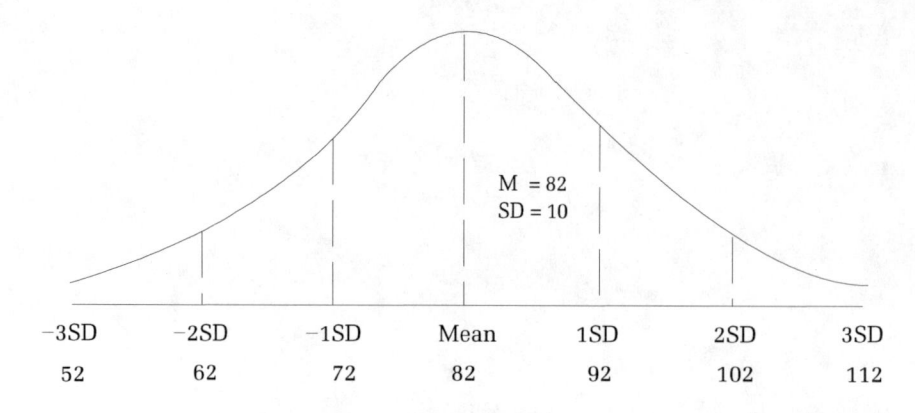

```
                    M = 82
                    SD = 10

   -3SD    -2SD    -1SD    Mean    1SD     2SD     3SD
    52      62      72      82      92     102     112
```

FIGURE 14.5
Distribution of Parental Skill Scores

The standard deviation of the scores of Thea's foster care parents was calculated to be 10. Again, assuming that the variable of knowledge about parenting skills is normally distributed in the population of foster care parents, the results of the *PSS* scores from the sample of parents about whom we are making inferences can be shown in a distribution like Figure 14.5.

As can also be seen in Figure 14.5, the score that would include 2 standard deviations, 102, is beyond the total possible score of 100 on the test. This is because the distribution of the scores in Thea's sample of parents does not entirely fit a normal distribution. The one extremely low score of 55 (see Table 14.1) obtained by one foster care parent would have affected the mean, as well as the standard deviation.

INFERENTIAL STATISTICS

The goal of inferential statistical tests is to rule out chance as the explanation for finding either associations between variables or differences between variables in our samples. Since we are rarely able to study an entire population, we are almost always dealing with samples drawn from that population. The danger is that we might make conclusions about a particular population based on a sample that is uncharacteristic of the population it is supposed to represent.

For example, perhaps the group of foster parents in Thea's training session happened to have an unusually high level of knowledge of parenting skills. If she assumed that all the rest of the foster parents that she might train in the future were as knowledgeable, she would be overestimating their knowledge, a factor that could have a negative impact on the way she conducts her training program.

To counteract the possibility that the sample is uncharacteristic of the general population, statistical tests take a conservative position as to whether or not we can conclude that there are relationships between the variables within our sample. The guidelines to indicate the likelihood that we have, indeed, found a relationship or difference that fits the population of interest are called *probability levels*.

The convention in most social science research is that variables are significantly associated or groups are significantly different if we are relatively certain that in 19 samples out of 20 (or 95 times out of 100) from a particular population, we would find the same relationship. This corresponds to a probability level of .05, written as ($p < .05$).

Probability levels are usually provided along with the results of the statistical test to demonstrate how confident we are that the results actually indicate statistically significant differences. If a probability level is greater than .05 (e.g., .06, .10), this indicates that we did not find a statistically significant difference.

Statistics That Determine Associations

There are many statistics that can determine if there is an association between two variables. We will briefly discuss two: chi-square and correlation.

Chi-Square

The *chi-square test* requires measurements of variables at only the nominal or ordinal level. Thus, it is very useful since much data in social work are gathered at these two levels of measurement. In general, the chi-square test looks at whether specific values of one variable tend to be associated with specific values of another.

In short, we use it to determine if two variables are related. It cannot be used to determine if one variable *caused* another, however. In thinking about the foster care parents who were in her training program, Thea was aware that women are more typically responsible for caring for their own children than men. Even if they are not mothers themselves, they are often in professions such as teaching and social work where they are caretakers. Thus, she wondered whether there might be a relationship between having had previous training in parenting skills and gender, such that women were less likely to have taken such training since they already felt confident in their knowledge of parenting skills.

As a result, her one-tailed hypothesis was that fewer women than men would have previously taken parenting skills training courses. Thea could examine this possibility with her 20 foster care parents using a chi-square test. In terms of gender, Thea had data from the nine (45%) men and 11 (55%) women. Of the total group, four (20%) had previous training in foster care training, while 16 (80%) had not.

TABLE 14.5
Frequencies (and Percentages) of Gender
by Previous Training (From Table 14.1)

Gender	Previous Training Yes	No	Totals
Male	1 (11)	8 (89)	9
Female	3 (27)	8 (73)	11
Totals	4 (20)	16 (80)	20

As shown in Table 14.5, the first task was for Thea to count the number of men and women who had previous training and the number of men and women who did not have previous training. She put these data in one of the four categories in Table 14.5. The actual numbers are called *observed frequencies*. It is helpful to transform these raw data into percentages, making comparisons between categories much easier.

We can, however, still not tell simply by looking at the observed frequencies whether there is a statistically significant relationship between gender (male or female) and previous training (yes or no). To do this, the next step is to look at how much the observed frequencies differ from what we would expect to see if, in fact, if there was no relationship. These are called *expected frequencies*. Without going through all the calculations, the chi-square table would now look like Table 14.6 for Thea's data set.

Because the probability level of the obtained chi-square value in Table 14.6 is greater than .05, Thea did not find any statistical relationship between gender and previous training in parenting skills. Thus, statistically speaking, men were no more likely than women to have received previous training in parenting skills; her research hypothesis was not supported by the data.

Correlation

Tests of correlation investigate the strength of the relationship between two variables. As with the chi-square test, correlation cannot be used to imply causation, only association. Correlation is applicable to data at the interval and ratio levels of measurement. Correlational values are always decimalized numbers, never exceeding ±1.00.

The size of the obtained correlation value indicates the strength of the association, or relationship, between the two variables. The closer a correlation is to zero, the less likely it is that a relationship ex-

TABLE 14.6
Chi-Square Table for Gender by Previous Training
(from Table 14.5)

Gender	Previous Training	No Previous Training
Male	O = 1 E = 1.8	O = 8 E = 7.2
Female	O = 3 E = 2.2	O = 8 E = 8.8

χ^2 = .8, df = 1, p > .05
O = observed frequencies (from Table 14.5)
E = expected frequencies

ists between the two variables. The plus and minus signs indicate the direction of the relationship. Both high positive (close to +1.00) or high negative numbers (close −1.00) signify strong relationships.

In positive correlations, though, the scores vary similarly, either increasing or decreasing. Thus, as parenting skills increase, so does self-esteem, for example. A negative correlation, in contrast, simply means that as one variable increases the other decreases. An example would be that, as parenting skills increase, the stresses experienced by foster parents decrease.

Thea may wonder whether there is a relationship between the foster parents' years of education and score on the *PSS* knowledge test. She might reason that the more years of education completed, the more likely the parents would have greater knowledge about parenting skills. To investigate the one-tailed hypothesis that years of education is positively related to knowledge of parenting skills, Thea can correlate the *PSS* scores with each person's number of years of formal education using one of the most common correlational tests, Pearson's *r*.

The obtained correlation between *PSS* score and years of education in this example is r = −.10 (p > .05). It was in the opposite direction of what she predicted. This negative correlation is close to zero, and its probability level is greater than .05. Thus, in Thea's sample, the parents' *PSS* scores are not related to their educational levels.

Group A:
    Mean = 82
    SD = 10

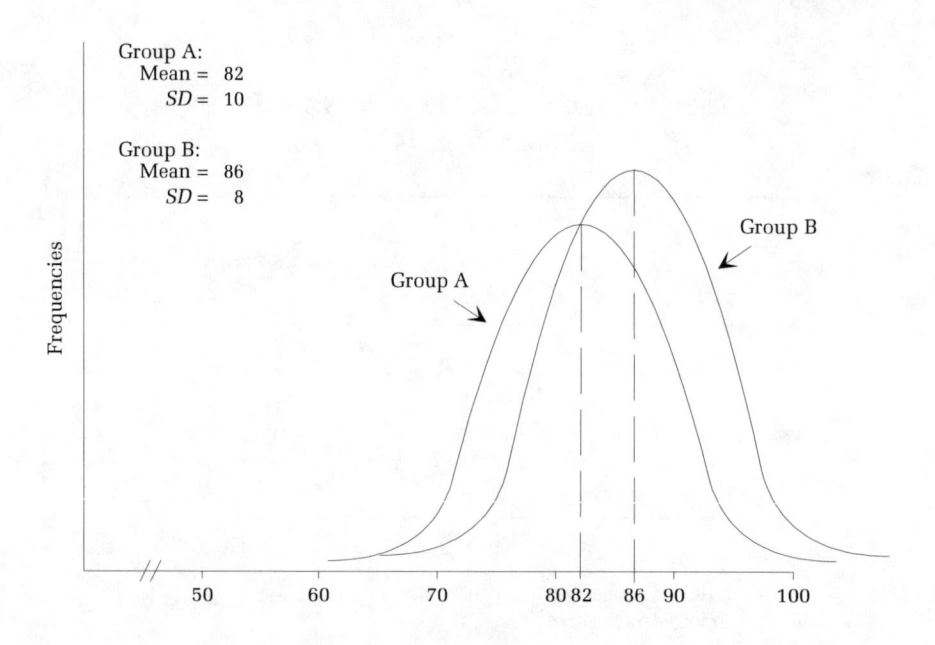

Group B:
    Mean = 86
    SD = 8

FIGURE 14.6
Frequency Distributions of PSS Scores
From Two Groups of Foster Care Providers

If the resulting correlation coefficient (r) had been positive and statistically significant ($p < .05$), it would have indicated that as the knowledge levels of the parents increased so would their years of formal education. If the correlation coefficient had been statistically significant but negative, this would be interpreted as showing that as years of formal education increased, knowledge scores decreased.

If a correlational analysis is misinterpreted, it is likely to be the case that the researcher implied causation rather than simply identifying an association between the two variables. If Thea were to have found a statistically significant positive correlation between knowledge and education levels and had explained this to mean than the high knowledge scores were a result of higher education levels, she would have interpreted the statistic incorrectly.

Statistics That Determine Differences

Two commonly used statistical procedures, t-tests and analysis of variance (ANOVA), examine the means and variances of two or more separate groups of scores to determine if they are statistically different from one another. T-tests are used with only two groups of scores, whereas ANOVA is used when there are more than two groups. Both are characterized by having a dependent variable at the interval or ratio level of measurement, and an independent, or grouping, variable at either the

Chapter 14: Analyzing Quantitative Data

nominal or ordinal level of measurement. Several assumptions underlie the use of both *t*-tests and ANOVA.

First, it is assumed that the dependent variable is normally distributed in the population from which the samples were drawn.

Second, it is assumed that the variance of the scores of the dependent variable in the different groups is roughly the same. This assumption is called *homogeneity of variance*. Third, it is assumed that the samples are randomly drawn from the population.

Nevertheless, as mentioned in Chapter 10 on group research designs, it is a common occurrence in social work that we can neither randomly select nor randomly assign individuals to either the experimental or the control group. In many cases this is because we are dealing with already preformed groups, such as Thea's foster care parents.

Breaking the assumption of randomization, however, presents a serious drawback to the interpretation of the research findings that must be noted in the limitations and the interpretations section of the final research report. One possible difficulty that might result from nonrandomization is that the sample may be uncharacteristic of the larger population in some manner. It is important, therefore, that the results not be used inferentially; that is, the findings must not be generalized to the general population. The design of the research study is, thus, reduced to an exploratory or descriptive level, being relevant to only those individuals included in the sample.

Dependent *t*-Tests

Dependent *t*-tests are used to compare two groups of scores from the same individuals. The most frequent example in social work research is looking at how a group of individuals change from before they receive a social work intervention (pre) to afterwards (post). Thea may have decided that, while knowing the knowledge levels of the foster care parents before receiving training was interesting, it did not give her any idea whether her program helped the parents to improve their skill levels.

In other words her research question became: "After being involved in the program, did parents know more about parenting skills than before they started?" Her hypothesis was that knowledge of parenting skills would improve after participation in her training program.

Thea managed to contact all of the foster care parents in the original group (Group A) one week after they had graduated from the program and asked them to fill out the *PSS* knowledge questionnaire once again. Since it was the same group of people who were responding twice to the same questionnaire, the dependent *t*-test was appropriate. Using the same set of scores collected by Thea previously as the pretest, the mean *PSS* was 82, with a standard deviation of 10. The mean score of the foster care parents after they completed the program was calculated as 86, with a standard deviation of 8.

A *t*-value of 3.9 was obtained, significant at the .05 level, indicating that the levels of parenting skills significantly increased after the foster care parents participated in the skill training program.

The results suggest that the average parenting skills of this particular group of foster care parents significantly improved (from 82 to 86) after they had participated in Thea's program.

Independent *t*-Tests

Independent *t*-tests are used for two groups of scores that have no relationship to each other. If Thea had *PSS* scores from one group of foster care parents and then collected more *PSS* scores from a second group of foster care parents, for example, these two groups would be considered independent, and the independent *t*-test would be the appropriate statistical analysis to determine if there was a statistically significant difference between the means of the two groups' *PSS* scores.

Thea decided to compare the average *PSS* score for the first group of foster care parents (Group A) to the average *PSS* score of parents in her next training program (Group B). This would allow her to see if the first group (Group A) had been unusually talented, or conversely, were less well-versed in parenting skills than the second group (Group B). Her hypothesis was that there would be no differences in the levels of knowledge of parenting skills between the two groups.

Since Thea had *PSS* scores from two different groups of participants (Groups A & B), the correct statistical test to identify if there are any statistical differences between the means of the two groups is the independent *t*-test. Let us use the same set of numbers that we previously used in the example of the dependent *t*-test in this analysis, this time considering the posttest *PSS* scores as the scores of the second group of foster care parents.

As can be seen from Figure 14.6, the mean *PSS* of Group A was 82 and the standard deviation was 10. Group B scored an average of 86 on the *PSS*, with a standard deviation of 8. Although the means of the two groups are four points apart, the standard deviations in the distribution of each are fairly large, so that there is considerable overlap between the two groups. This would suggest that statistically significant differences will not be found.

The obtained *t*-value to establish whether this four-point difference (86 − 82) between the means for two groups was statistically significant was calculated to be $t = 1.6$ with a $p > .05$. The two groups were, thus, not statistically different from one another and Thea's hypothesis was supported.

Note, however, that Thea's foster care parents were not randomly assigned to each group, thus breaking one of the assumptions of the *t*-test. As discussed earlier, this is a serious limitation to the interpretation of the study's results. We must be especially careful not to generalize the findings beyond the groups included in the study.

Also note that in the previous example, when using the same set of numbers but a dependent *t*-test, we found a statistically significant difference. This is because the dependent *t*-test analysis is more robust than the independent *t*-test, since having the same participant fill out the questionnaire twice, under two different conditions, controls for many extraneous variables, such as individual differences, that could negatively influence an analysis of independent samples.

One-Way Analysis of Variance

A one-way ANOVA is the extension of an independent *t*-test that uses three or more groups. Each set of scores is from a different group of participants. For example, Thea might use the scores on the *PSS* test from the first group of foster care parents from whom she collected data before they participated in her program, but she might also collect data from a second and a third group of parents before they received the training. The test for significance of an ANOVA is called an *F*-test.

We could actually use an ANOVA procedure on only two groups and the result would be identical to the *t*-test. Unlike the *t*-test, however, obtaining a significant *F*-value in a one-way ANOVA does not complete the analysis. Because ANOVA looks at differences between three or more groups, a significant *F*-value only tells us that there is a statistically significant difference among the groups: It does not tell us between which ones. To identify this, we need to do a *post-hoc* test. A variety are available, such as Duncan's multiple range, Tukey's Honestly Significant Difference test, and Newman-Keuls, and are provided automatically by most computer statistical programs.

But one caution applies: A post-hoc test should be used *only after finding a significant F-*value, because some of the post-hoc tests are more sensitive than the *F* test and so might find significance when the *F*-test does not. Generally, we should use the most conservative test first, in this case the *F*-test.

In the example of Thea's program, let us say that she collected data on a total of three different groups of foster care parents. The first group of foster care parents scored an average of 82 on the *PSS* (standard deviation 10). The second group scored an average of 86 (standard deviation 8), while the mean score of the third group was 88 with a standard deviation of 7.

The obtained F-value for the one-way ANOVA is 2.63, with a $p > .05$. Thus, we must conclude that there are no statistically significant differences between the means of the groups (i.e., 82, 86, 88). Since the F-value was not significant, we would not conduct any post-hoc tests. This finding would be interesting to Thea, since it suggests that all three groups of foster care parents started out with approximately the same knowledge levels, on the average, before receiving training.

SUMMARY

This chapter provided a beginning look at the rationale behind some of the most commonly used statistical procedures, both those that describe samples and those that analyze data from a sample in order to make inferences about the larger population. The level of measurement of the data is key to the kind of statistical procedures that can be utilized. Descriptive statistics are utilized with data from all levels of measurement. The mode is the most appropriate measure of central tendency for measurements of this level. It is only when we have data from interval and ratio levels that we can utilize inferential statistics—those that extend the statistical conclusions made about a sample by applying them to the larger population.

Descriptive measures of central tendency, such as the mode, median, and mean of a sample or population, all provide different kinds of information, each of which is applicable only to some levels of measurement. In addition to knowing the middle or average of a distribution of scores as provided by measures of central tendency, it is useful to know the value of the standard deviation that shows us how far away from the mean the scores are distributed. It is assumed that many variables studied in social work can be found in a normal distribution in the total population. Consequently many descriptive and inferential statistics assume such a distribution for their tests to be valid.

Chi-square and correlation are both statistical tests that determine whether variables are associated, although they do not show causation. In contrast, t-tests and analysis of variance (ANOVA) are statistical procedures for determining whether the mean and variance in one group (often a treatment group) is significantly different from those in another (often a comparison or control group).

REFERENCES AND FURTHER READING

Coolidge, F.L. (2006). *Statistics: A gentle introduction* (2nd ed.). Thousand Oaks, CA: Sage.

Elliott, A.C., Woodward, W.A. (2007). *Statistical analysis quick reference guidebook.* Thousand Oaks, CA: Sage.

Fielding, J., & Gilbert, N. (2006). *Understanding social statistics* (2nd ed.). Thousand Oaks, CA: Sage.

Gaur, A.S., & Gaur, S.S. (2006). *Statistical methods for practice and research: A guide to data analysis using SPSS.* Thousand Oaks, CA: Sage.

Keller, D.K. (2006). *The Tao of statistics: A path to understanding (with no math).* Thousand Oaks, CA: Sage.

Salkind, N.J. (2007). *Statistics for people who (think they) hate statistics.* Thousand Oaks, CA: Sage.

Weinbach, R.W., & Grinnell, R.M., Jr. (2010). *Statistics for social workers* (8th ed.). Boston: Allyn & Bacon.

Williams, M., Tutty, L.M., & Grinnell, R.M., Jr. (1995). *Research in social work: An introduction* (2nd ed.). Itasca, IL: F.E. Peacock.

Williams, M., Tutty, L.M., & Grinnell, R.M., Jr. (2008). Analyzing quantitative data. In R.M. Grinnell, Jr., & Y.A. Unrau (Eds.), *Social work research and evaluation: Foundations of evidence-based practice* (8th ed., pp. 369–386). New York: Oxford University Press.

Check out our Website for useful links and chapters (PDF) on:
- introductory statistical concepts
- Web-based links to various quantitative statistical tutorials

Go to: www.pairbondpublications.com

Click on: Student Resources
        Chapter-by-Chapter Resources
        Chapter 14

Chapter 15

Analyzing Qualitative Data

PLANNING THE ANALYSIS
    Transcribing the Data
        Deciding What Computer Program to Use, If Any
        Deciding Who Will Transcribe the Data
    Establishing General Rules for the Analysis
        Previewing the Data
        Keeping a Journal
DOING THE ANALYSIS
    First-Level Coding
        Identifying Meaning Units
        Creating Categories
        Assigning Codes to Categories
        Refining and Reorganizing Categories
    Second-Level Coding
        Comparing Categories
LOOKING FOR MEANING
    Interpreting Data and Building Theory
        Developing Conceptual Classifications Systems
        Presenting Themes or Theory
    Assessing the Trustworthiness of the Results
        Establishing Our Own Credibility
        Establishing the Dependability of the Data
        Establishing Our Control of Biases and Preconceptions
SUMMARY
REFERENCES AND FURTHER READING

15

IN THE LAST CHAPTER, WE LOOKED AT METHODS of analyzing quantitative data that were derived from positivist research studies. In this chapter, we turn to the analysis of qualitative data that are derived from interpretive research studies; that is, data collected in the form of words (i.e., text data), most often through interviews, open-ended items on questionnaires, or personal logs. Unlike *numbers*, which are used in quantitative analyses, *words* give us descriptions or opinions from the unique point of view of the person who spoke or wrote the words.

Text data have both disadvantages and advantages. The disadvantages are that words tend to be open to different interpretations and cannot be collected from large numbers of people at a time because the process of collection and analysis is much more time-consuming than is the case for numerical data. The major advantage is that the material is very rich, containing multiple facets, and may provide us with a deeper understanding of the underlying meaning than would be possible if we just collected numbers.

Usually, we need the deeper understanding provided by qualitative data when we know very little about the problem or situation we are investigating and are not yet in a position to formulate theories about it for testing. In other words, we are at the exploratory end of the knowledge continuum, seeking to form patterns out of individual experiences so that we can develop more general theories.

The primary purpose of a qualitative data analysis is to sift and sort the masses of words we have collected from our research participants in such a way that we can derive patterns related to our research question—to identify the similarities and differences presented by individuals and the possible links between them.

Suppose for a moment, that we are investigating postnatal depression among women. We may be interested in what the symptoms are, how new mothers experience them, and how they feel their depression affects their relationships with their newborns. We may also be interested in whether the women who experience postnatal depression are similar in any way with respect to various characteristics, such as age, ethnic background, desire for the baby, partner or family support, socioeconomic status, medical history, and so on.

Indeed, we may have structured our interviews to collect specific data related to these kinds of variables. If we did, then we have already theorized that these characteristics or variables are related to postnatal depression, and we may, even subconsciously, look for patterns in the data that will confirm our theories. There is nothing evil about this. Researchers are human beings with a normal human tendency to make connections between events on the frailest of evidence, and it is a rare researcher who starts on a study with no preconceived notions at all.

The important thing is to be aware of our human frailties with respect to drawing unwarranted conclusions, and to organize our data collection and analysis so that these frailties are minimized as far as possible. A look at some of the assumptions underlying interpretive research might help us to accomplish this:

- We are assuming that the goal of interpretive research is to reach an in-depth understanding of each of our research participants, with respect to the research question, including experiences that are unique to them. We will not achieve this goal if we do not allow participants to express their uniqueness during data collection. Neither will we achieve it if we ignore the uniqueness during data analysis because we were hoping to uncover patterns and this is an anomaly that does not quite fit. It is also easy to ignore uniqueness if it does not fit with the findings of other researchers. An exploratory topic tends to reveal little in the way of previous research, but there is usually some, and it is tempting to disregard a unique experience if it seems to contravene what others have found.

- Information is always provided in some context. In the case of a research study, the context is the way in which the information was elicited; for example, the phrasing of a question. In an interview, the context is the relationship between the interviewer and the interviewee. Preconceptions on the part of the interviewer tend to elicit responses that fit with the preconceptions. Thus, it is important in analysis not to look at just the response but at the emotional atmosphere surrounding the response and the question that was responded to.

There are three major phases involved in a qualitative analysis:

- First, we must plan how we will do the analysis: how we will transcribe spoken or written data into a usable form and what rules we will use to fit the pieces of data together in a meaningful way.

- Second, we must do the analysis, following the general rules we set out at the beginning and perhaps revising these rules along the way if we come across some data to which our rules cannot be sensibly applied. It is important to note though that whatever rules we finally decide on must be applied to all our data.

Changing our minds about rules will mean going back over the material we have worked on already; and, indeed, qualitative analysis is usually a back-and-forth sort of process, involving many re-readings and re-workings as new insights appear and we begin to question our initial theories or assumptions.
- Third, we need to draw meaning from our analysis; that is, to identify relationships between the major themes that have emerged and to build theories around these for testing in the future.

PLANNING THE ANALYSIS

There are two steps involved in planning a qualitative analysis: (1) transcribing the data, and (2) establishing general rules for the analysis.

Transcribing the Data

Transcribing our data itself involves two tasks: (1) deciding what computer program to use, if any; and (2) deciding who will transcribe the data.

Deciding What Computer Program to Use, If Any

If responses are written, as in open-ended items on a questionnaire or personal logs, transcription may be a matter of typing the responses, either just for easier reading or with the aim of using a computer program to assist with the analysis. A few researchers, distrustful of computers, prefer to use a traditional "cut-and-paste" method, physically cutting the manuscript and grouping the cut sections together with other related sections.

A computer need not be used for transcription in this case. A typewriter would do, or even legible handwriting. Some researchers trust the computer just sufficiently to allow it to move selected passages together with other selected passages to electronically form a group of related data.

The majority of word-processing programs can accomplish this. An increasing number of researchers, however, use computer programs that have been developed specifically to assist with the analysis of qualitative data. A few familiar names are ETHNOGRAPH, HYPERQUAL, ATLAS.ti, NUD*IST, and NVivo.

New programs are always coming on to the market and it is wise to consult colleagues or computer companies about which programs might be most helpful for a particular project. It is important to note that no computer program can do the analysis for us. The most it can do is free up time for us to spend on considering the meaning of our data.

Deciding Who Will Transcribe the Data

Some researchers are fortunate enough to have a research or administrative assistant to help them in the transcription process. If this is the case, it is necessary to lay down guidelines right at the beginning about how the material should be transcribed. If an interview has been audio- or videotaped, for example, the questions should be included in the transcript as well as the answers. Nonverbal communications such as pauses, laughing or crying, and voice tone should be included in brackets so that the emotional context of the interview is captured as far as possible in the transcript. Those fortunate researchers with assistants are nevertheless well advised to transcribe at least the first few interviews themselves so that assistants can see what ought to be included.

Another concern is how to format the transcript so that it is easy to read and analyze. It is a good idea to leave a margin of at least two inches along the right-hand side so that we can write notes and codes alongside the corresponding text. It is also a good idea to number each line of the transcript so that we can readily identify each segment of data.

Computer programs designed to assist in qualitative analysis will automatically do this. For example, suppose we worked in a foster care agency and we were asking foster parents who had resigned from the agency in the past year why they had resigned. A few lines from one of our interviews might be transcribed like this:

1. *Sue (angrily):* His behavior was just too much and nobody from the agency told
2. us that he'd set fires before and was probably going to burn our house down. I
3. suppose they thought that if they'd told us that we wouldn't have taken him but
4. I do feel that we were set up from the beginning (sounding very upset). And when
5. we called the agency, there was only an answering machine and it was a whole
6. day before the social worker called us back.
7. *Interviewer:* That's dreadful.

Reading these lines after the transcript is completed might immediately set us thinking about how foster parents' reasons for resigning could be separated into categories. One category might be the foster child's behavior (much worse than the foster parents had been led to expect).

Or this might be two categories: (1) the child's behavior and (2) the discrepancy between the actual behavior and the expected behavior. Another category might be negative feelings toward the agency, with two subcategories: the feeling of having been set up, and the perceived lack of support in a time of crisis. It might not take long at all for these tentative categories to harden into certainties.

Having read six lines from one interview, we now feel we know why foster parents resign and it only remains to confirm the reasons we have found by picking out similar sentiments from our interviews with other foster parents. Job completed!

Actually, we have barely begun. Since we cannot—and do not wish to—stop our minds from jumping ahead in this fashion, we must find some way to organize our thoughts such that our intuitive leaps do not blind us to different and perhaps contradictory insights yielded by other interviews. There are two techniques that might help us to do this: previewing the data, and keeping a journal. Both these techniques will also help us to establish general rules for our analysis—the second and last step in the planning stage.

Establishing General Rules for the Analysis

Establishing general rules for the analysis ensures that our efforts are systematic and the same rules are applied to all our data. We use rules to decide how we could fit together pieces of data in a meaningful way and how these groups of data could be categorized and coded (to be discussed shortly). For example, what criteria or rule do we use to decide whether a child's worsened behavior after a visit with the biological parents should be categorized under "child's behaviors" or under "relationships with biological parents"? Although we clarify and refine the rules throughout the study, by the time we have finished we should have a set of rules that have been consistently applied to every piece of data. We start to think about what rules might apply during the previewing task.

Previewing the Data

The process of transcription might be ongoing throughout the study, with each interview transcribed as soon as it is completed, or transcription might not begin until all the data have been assembled. Whichever method is used, it is important to read all of the transcripts before beginning to formally identify categories.

It is also important to give all of the transcripts and all parts of the transcripts the same amount of attention. We may be in peak form at the beginning of the day while reading the first few pages of the first transcript, for example, but by the end of the day and the end of the third transcript, this initial peak has waned to weary impatience. We would be better advised to read only for as long as we remain interested in the material.

This will obviously mean that the process of previewing the data extends over a longer period, but qualitative data analysis takes time. It is a lengthy process of discovery, whose pace cannot be forced. If we are rereading material, it sometimes helps to read the last third of an interview at moments of high energy instead of the first third. That way, we will not lose valuable insights from later sections of the interview transcript.

Keeping a Journal

Some people love journals and others hate them, but the interpretive researcher cannot afford to be without one. The journal should have been started at the same time as the study was started. It should include notes on the planned method and any changes in the plan, with dates and reasons. For example, perhaps the plan was to interview all of the foster parents who had resigned from the agency during the last year, but some of them would not agree to be interviewed.

We may believe that those who agreed differed in some important respects from those who refused. They were more satisfied with the agency perhaps. If this were the case, the data from our more satisfied sample might lead us to faulty conclusions, possibly causing us to place more emphasis on personal reasons for resignation, such as failing health or family circumstances, and less on agency-related reasons, such as poor information-sharing or support. Whatever the difficulties we encounter and the assumptions we make, our journal should keep an accurate record of them as we move along in the data analysis. Since the work we do must be open to scrutiny by others, it is essential to keep a record of all our activities and the feelings and reasonings behind them.

When the data-collection stage begins, the journal can be used to record personal reactions to the interview situation. For example, we might feel more personal empathy with one foster parent than with another and be tempted to give more weight to the remarks of the parent with whom we sympathized. An unconscious overreliance on one research participant or one subset of research participants will hopefully reveal itself as we read through our journal entries later during the course of the analysis.

When we begin to categorize and code the data, we can use the journal to keep notes about the process, writing down the general rules, revisions to the rules, and questions or comments with respect to how particular pieces of data might be categorized. Let us now turn our attention to actually analyzing qualitative data.

DOING THE ANALYSIS

Once all the data have been previewed, we can start on coding. There are two levels of coding: (1) first-level coding, which deals with the concrete ideas evident in the transcript and (2) second-level coding, which looks for and interprets the more abstract meanings underlying these concrete ideas.

First-Level Coding

There are four tasks in first-level coding: (1) identifying meaning units, (2) creating categories, (3) assigning codes to categories, and (4) refining and reorganizing categories.

Identifying Meaning Units

A meaning unit is a piece of data, which we consider to be meaningful by itself. It might be a word, a partial or complete sentence, or a paragraph or more. For example, let us look once again at the interview segment presented earlier:

1. <u>Sue (angrily): His behavior was just too much and nobody from the agency told</u>
2. *us that he'd set fires before and was probably going to burn our house down. I*
3. *suppose they thought that if they'd told us that we wouldn't have taken him but*
4. **I do feel that we were set up from the beginning (sounding very upset). And when**
5. ***we called the agency, there was only an answering machine and it was a whole***
6. ***day before the social worker called us back.***
7. *Interviewer:* That's dreadful.

In this segment, we might identify four meaning units. The first unit (<u>underlined</u>, line 1) relates to the child's behavior. The second unit (*italics*, lines 2 and 3) relates to lack of information provided by the agency. The third (**bold,** line 4) relates to feeling set up by the agency. The fourth unit (***bold* italics**, lines 5 and 6) relates to poor support on the part of the agency. Of course, different researchers might identify different meaning units or label the same units differently.

For example, lines 4, 5 and 6 might be identified as relating to agency response style rather than poor support and might involve two distinct meaning units, "response method" and "response time." Similarly, the partial sentence in line 2 "he'd set fires before" might be viewed as a separate meaning unit relating to the child's past rather than present behavior.

The first run-through to identify meaning units will always be somewhat tentative and subject to change. If we are not sure whether to break a large meaning unit into smaller ones, it may be preferable to leave it as a whole. We can always break it down later in the analysis and breaking down large units tends to be easier than combining smaller ones, especially once second-level coding begins.

Creating Categories

Once we have identified meaning units in the transcript, our next task is to consider which of them fit together into categories. Perhaps we should have a category labeled "child's behavior" into which we put all meaning units related to the child's behavior, including the second meaning unit identified above "nobody from the agency told us that he'd set fires before and was probably going to burn our house down."

Or perhaps we should have two categories, "child's present behavior" and "child's past behavior," in which case the second meaning unit might belong in the latter category. Or perhaps we feel that the vital words are "nobody told us" and this second meaning unit really belongs in a different category labeled "provision of information by agency." All other meaning units to do with foster parents being given information by the agency would then belong in this same category even though they had nothing to do with the child's behavior.

Since these kinds of decisions are often difficult to make, it is a good idea to note in our journal how we made the decisions we did and what alternatives we considered at the time. What rules did we use to decide whether a particular meaning unit was similar to or different from another meaning unit? How did we define our categories in order to decide whether a group of similar meaning units should be placed in one category or in another?

As we continue to examine new meaning units, we will use these rules to decide whether each new unit is similar to existing units and belongs in an existing category or whether it is different from existing units and needs a new separate category. The number of categories will therefore expand every time we identify meaning units that are different in important ways from those we have already categorized.

Since too many categories will make the final analysis very difficult, we should try to keep the number within manageable limits. This may mean revising our initial rules about how categories are defined and what criteria are used to decide whether meaning units are similar enough to be grouped together. Of course, any change in rules should be noted in our journal, together with the rationale for the change.

The complexity of our categorization scheme also needs to be considered. One meaning unit may, in fact, fall into more than one category, or a group of meaning units may overlap with another group. Large, inclusive categories may consist of a number of smaller, more exclusive subcategories. For example, as we saw, the meaning unit *"nobody from the agency told us that he'd set fires before and was probably going to burn our house down"* has to do both with lack of information provided by the agency and with the child's past behavior.

Sometimes, meaning units cannot be clearly placed into any category and fall into the category of "miscellaneous." When we are tired, most everything may seem to be "miscellaneous," but miscellaneous units should make up no more than ten percent of the total data set. More than that suggests that there is a problem with the original categorization scheme.

The real purpose of a "miscellaneous" category is to prevent our throwing out meaning units which, at first glance, appear to be irrelevant. Such throwing out is risky because at some point we may decide that our whole categorization scheme needs massive revision and we must start the whole process again from scratch.

Assigning Codes to Categories

Codes are simply a shorthand form of the category name. They typically take the form of strings of letters and/or symbols. Codes used in *The Ethnograph*, for example, may be up to ten letters long and can also include symbols. Codes are usually displayed in the margins (often the right margin) of the transcribed text.

If we had a category labeled "child's behavior" we might simply code this CB where the C stands for the foster child and the B stands for behavior. If we want to distinguish between past and present behavior, we might use the codes CPASTB and CPRESB respectively. If there is a category relating to the behavior of the foster parents' own children (perhaps this has worsened since the foster child moved in), we might use the codes FPCPASTB and FPCPRESB respectively, where the FPC stands for the foster parents' child.

In fact, we might make it a rule that codes starting with C, FP, FPC, and A stand for things to do with the foster child, the foster parents, the foster parents' children, and the agency respectively. Then AINF>FP might mean information provided by the agency to the foster parents and AINF<FP might mean information provided by the foster parents to the agency.

It is a good idea to keep the codes as short as possible in the beginning since they tend to become longer as the analysis grows more complex. However, there are many different ways to assign codes and, so long as the code is clearly related to the category, it does not really matter which system is utilized.

Refining and Reorganizing Categories

Before moving on from first-level coding, we need to make a final sweep through the data to ensure that our analysis reflects what our research participants have said. We should consider the logic underlying the rules we made for grouping meaning units and defining categories. We may, for example, be confused about why we created some categories, or feel uncertain about why a particular meaning unit was put into a particular category. We might find that some categories are too complex and need to be split into smaller categories; or some categories are poorly defined; or some of the categories that we expected to emerge from the data are missing altogether.

We might, for example, have expected that some foster parents would resign because of poor health, but there is no category coded FPHEA. Investigation reveals that foster parents did indeed mention their poor health but they always ascribed it to the strain of dealing with the foster child or the foster child's biological parents or the agency.

Hence, the meaning units including poor health have been categorized under "foster child's present behavior (CPRESB)" or "relationships with agency (AREL)" or "relationships with biological parents (BPREL)" and have not been broken down finely enough to isolate poor health as a separate category. We may wish to create such a category or we may prefer to note its absence in our journal, together with other categories that are incomplete or in some way unsatisfactory.

This is a good time to ask a colleague to analyze one or two of our interviews using the rules we have devised. In this way, we can check that the categories themselves and the rules that define them make sense. If our colleague organizes meaning units in a significantly different way, our categorization scheme may need to be substantially revised.

It is probably time to stop first-level coding when all our meaning units fit easily into our current categorization scheme and there are no more units that require the creation of new categories. If interviews are continuing during first-level coding, we will probably find the data becoming repetitive, yielding no new piece of information that cannot be fitted into the present scheme.

Second-Level Coding

The next major step in the data analysis process is second-level coding. As noted earlier, this is more abstract and involves interpreting what the first-level categories mean. During first-level coding, we derived meaning units from interviews with individuals, and we derived categories by comparing the meaning units to see which were similar enough to be grouped together. During second-level coding we will compare the categories themselves in order to uncover possible relationships between them. The point of doing this is to identify themes based on patterns that repeatedly occur among our categories.

Comparing Categories

A comparison of categories in any interpretive study will probably yield many different types of relationships. Coleman and Unrau (2008) have suggested that the following three types of relationships are among those most commonly found:

- *A temporal relationship.* One category may be found to often precede another. For example, foster children whose own parents are abusive might initially behave well because they are happy to be living away from the abuse. Later, however, they may push for a return to their families of origin, and their response to the foster parents may become less positive.
- *A causal relationship.* One category may be perceived to be the cause of another. For example, foster parents may believe that the child's bad behavior after every visit with the biological parent was the cause of their own negative attitudes toward the biological parent. However, it is always risky to assume that one category caused another when, in fact, the opposite may be true. Perhaps the foster parents started off with a negative attitude toward the child's biological parents, the child was more than usually angry about this after visits home, and the child demonstrated rage by behaving badly.
- *One category may be contained within another category.* At this stage of the analysis, we may decide that two categories that we had thought to be separate are in fact linked. For example, foster parents may have said that they were never introduced to the child's biological parents by the agency, and we have categorized this separately from their statement that the agency was not entirely truthful with them about the child's past behavior. Now, we realize that these are both examples of lack of information provided to the foster parent by the agency.

Furthermore, foster parents complained that they were not invited to agency meetings in which the child's progress was reviewed. Lack of information and nonattendance at meetings might combine to form a theme related to the agency's attitude toward the foster parents. The theme might be that the agency does not appear to accept the foster parents as equal partners in the task of helping the child.

When we have identified themes based on patterns among our categories, we code these themes in the same way as we coded our categories. If one of our themes is that the agency does not view foster parents as equal partners, for example, we might code this as A<FP-PART. Once themes have been identified and coded, the process of second-level coding is complete.

LOOKING FOR MEANING

Drawing meaning from our data is perhaps the most rewarding step of a qualitative data analysis. It involves two important steps: (1) interpreting data and building theory, and (2) assessing the trustworthiness of the results.

Interpreting Data and Building Theory

This step involves two tasks: (1) developing conceptual classifications systems, and (2) presenting themes or theory.

Developing Conceptual Classifications Systems

The ultimate goal of an interpretive research study is to identify any relationships between the major themes that emerge from the data set. During first-level coding, we used meaning units to form categories. During second-level coding, we used categories to form themes.

Now we will use themes to build theories. In order to do this, we must understand the interconnections between themes and categories. There are several strategies that might be useful in helping us to identify these connections. Miles and Huberman (1994) have suggested the following strategies for extracting meaning from a qualitative data set:

- *Draw a cluster diagram.* This form of diagram helps us to think about how themes and categories may or may not be related to one another. Draw and label circles for each theme and arrange them in relation to each other. Some of the circles will overlap, others will stand alone. The circles of the themes of more importance will be larger, in comparison to themes and categories that are not as relevant to our conclusions. The process of thinking about what weight to give the themes, how they interact, and how important they will be in the final scheme will be valuable in helping us to think about the meaning of our research study.

- *Make a matrix.* Matrix displays may be helpful for noting relations between categories or themes. Designing a two-dimensional matrix involves writing a list of categories along the left side of a piece of paper and then another list of categories across the top.

  For example, along the side, we might write categories related to the theme of the degree of partnership between the agency and the foster parents. One such category might be whether the foster parents were invited to agency meetings held to discuss the child's progress. Then, along the top we might write categories related to the theme of foster parents' attitudes toward the agency.

  One category here might be whether foster parents felt their opinions about the child were listened to by the agency. Where two categories intersect on the matrix, we could note with a plus sign (+), those indicators of partnership or lack of partnership that positively affect parents' attitudes. Conversely, we would mark with a minus sign (−) those that seem to have a negative affect. Such a matrix gives us a sense of to what degree and in what ways foster parents' attitudes toward the agency are molded by the agency's view of foster parents as partners.

- *Count the number of times a meaning unit or category appears.* Although numbers are typically associated with quantitative studies, it is acceptable to use numbers in qualitative work in order to document how many of the participants expressed a particular theme. We might be interested, for example, in finding out how many participants experienced specific problems related to lack of agency support. We would write the code names for the foster parents interviewed down the left side of a piece of paper and the list of problems across the top.

  To fill in the chart, we would simply place a check mark beside each foster parent's code name if she or he experienced that particular problem. Numbers will help protect our analysis against bias that occurs when intense but rare examples of problems are presented. For example, many foster parents may have felt that they were not given sufficient information about the child's past behavior, but only one may have felt that the agency deliberately set her up. Although, we will certainly not discount this foster parent's experience, we might prefer to view it as an extreme example of the results of poor information sharing.

- *Create a metaphor.* Developing metaphors that convey the essence of our findings is another mechanism for extracting meaning. One example of a metaphor concerning battered women is "the cycle of violence," which effectively describes the tension building between a couple until the husband beats his wife, followed by a calm, loving phase until the tension builds again.

- *Look for missing links.* If two categories or themes seem to be related but not directly so, it may be that a third variable connects the two.

- *Note contradictory evidence.* It is only natural to want to focus on evidence that supports our ideas. Because of this human tendency, it is particularly important to also identify themes and categories that raise questions about our conclusions. All contradictory evidence must be accounted for when we come to derive theories pertaining to our study area: for example, theories about why foster parents resign.

Presenting Themes or Theory

Sometimes it is sufficient to conclude an interpretive study merely by describing the major themes that emerged from the data. For example, we may have used the experiences of individual foster parents to derive reasons for resignation common to the majority of foster parents we studied. When categorized, such reasons might relate to: inadequate training; poor support from the agency; failure on the agency's part to treat the foster parents as partners; negative attitudes on the part of the foster parents' friends and extended family, unrealistic expectations on the part of the foster parents with respect to the foster child's progress and behavior, poor relations between the biological and foster parents, the perceived negative influence of the foster child on the foster parents' own children, marital discord attributed to stress, failing health attributed to stress, and so on. We might think it sufficient in our conclusions to present and describe these categories, together with recommendations for improvement.

On the other hand, we might wish to formulate questions to be answered in future studies. What would change if agencies were to view foster parents as partners rather than clients? If we think we know what would change we might want to formulate more specific questions; for example:

> Are agencies that view foster parents as partners more likely to provide foster parents with full information regarding the child's background than agencies that do not view foster parents as partners?

Or, we might want to reword this question to form a hypothesis for testing:

> Agencies that view foster parents as partners are more likely to provide foster parents with full information regarding the child's background than agencies that do not view parents as partners.

In order to arrive at this hypothesis, we have essentially formulated a theory about how two of our concepts are related. We could carry this further by adding other concepts to the chain of relationships and formulating additional hypotheses:

> Foster parents who have full information about the child's background are less likely to have unrealistic expectations about the child's behavior and progress than foster parents who do not have full information about the child's background.

And,

> Foster parents who have unrealistic expectations about the child's behavior and progress are more likely to experience marital discord (due to the fostering process) than foster parents who do not have unrealistic expectations.

Indeed, we might weave all the various woes our study has uncovered into an elaborate pattern of threads, beginning with the agency's reluctance to view foster parents as partners and ending with the foster parents' resignations. In our excitement, we might come to believe that we have solved the entire problem of resigning foster parents.

If agencies would only change their attitudes with respect to foster parents' status and behave in accordance with this change in attitude, then all foster parents would continue to be foster parents until removed by death. It is at this point that we need to focus on the second stage of our search for meaning: assessing the trustworthiness of our results.

Assessing the Trustworthiness of the Results

There are three major reasons why disgruntled agencies, as well as other actors, may not agree that our results are as trustworthy as we believe: (1) they may doubt our personal credibility as researchers, (2) they may doubt the dependability of our data, and (3) they may think that we have been led astray by our own biases and preconceptions.

Establishing Our Own Credibility

Since an interpretive study depends so much on human judgment, it is necessary to demonstrate that our own personal judgment is to be trusted. Part of this relates to our training and experience. Another important part is the record we made in our journal detailing the procedures we followed, the decisions we made and why we made them, and the thought processes that led to our conclusions. If we can demonstrate that we were qualified to undertake this study and we carried it out meticulously, others are far more likely to take account of our conclusions.

Establishing the Dependability of the Data

If we have been consistent in such things as interview procedures and developing rules for coding, and if we have obtained dependable data through a rigorous, recorded process, then another researcher should be able to follow the same process, make the same decisions, and arrive at essentially the same conclusions. Also, if we ourselves redo part of the analysis at a later date, the outcome should be very similar to that produced in the original analysis.

In order to ensure that we or others could duplicate our work, Coleman and Unrau (2008) have suggested we need to pay attention to the following standard generic issues:

- *The context of the interviews.* Some data-collection situations yield more credible data than others, and we may choose to weight our interviews accordingly. Some authors claim, for example, that data collected later in the study are more dependable than data collected at the beginning because our interviewing style is likely to be more relaxed and less intrusive. In addition, data obtained first-hand are considered to be more dependable than second-hand data, which are obtained through a third party.

  Similarly, data offered voluntarily are thought to be stronger than data obtained through intensive questioning and data obtained from research participants while in their natural environments (e.g., home or neighborhood coffee shop) are to be more trusted than data provided by research participants in a foreign and sterile environment (e.g., researcher's office or interviewing room).
- *Triangulation.* Triangulation is commonly used to establish the trustworthiness of qualitative data. There are several different kinds of triangulation, but the essence of the method

lies in a comparison of several perspectives. For example, we might collect data from the agency about what information was provided to foster parents in order to compare the agency's perspective with what foster parents said.

With respect to data analysis, we might ask a colleague to use our rules to see if he or she makes the same decisions about meaning units, categories, and themes. The hope is that different perspectives will confirm each other, adding weight to the credibility of our analysis.

- *Member checking.* Obtaining feedback from research participants is a credibility technique that is unique to interpretive studies. While such feedback should be an ongoing part of the study (interview transcripts may be presented to research participants for comment, for example), it is particularly useful when our analysis has been completed, our interpretations made, and our conclusions drawn.

Research participants may not agree with our interpretation or conclusions, and may differ among each other. If this is the case, we need to decide whether to exclude the interpretations to which they object, or whether to leave them in, merely recording the dissenting opinions and our position in relation to them.

Establishing Our Control of Biases and Preconceptions

Since all human beings inevitably have biases and preconceptions about most everything, one way to demonstrate that we are in control of ours is to list them. Such a list is useful to both ourselves and others when we want to check that our conclusions have emerged from the data rather than from our established beliefs. Also useful is our journal where we documented our attempts to keep ourselves open to what our research participants had to say. As Coleman and Unrau (2008) point out, we may want to consider the following points in relation to bias:

- Our personal belief systems may affect our interpretation of the data. Member checking has already been mentioned as a way to safeguard against this.
- We might draw conclusions before the data are analyzed or before we have decided about the trustworthiness of the data collected. Such conclusions tend to set to a cement-like hardness and are likely to persist despite all future indications that they were wrong.

- We might censor, ignore, or dismiss certain parts of the data. This may occur as a result of data overload or because the data contradict an established way of thinking.

- We might make unwarranted or unsupported causal statements based on impressions rather than on solid analysis. Even when our statements are based on solid analysis, it is a good idea to actively hunt for any evidence that might contradict them. If we can demonstrate that we have genuinely searched for (but failed to find) negative evidence, looking for outliers and using extreme cases, our conclusions will be more credible.

- We might be too opinionated and reduce our conclusions to a limited number of choices or alternatives.

- We might unthinkingly give certain people or events more credibility than others. Perhaps we have relied too much on information that was easily accessible or we have given too much weight to research participants we personally liked. If we detect such a bias, we can interview more people, deliberately searching for atypical events and research participants who differ markedly from those we have already interviewed. Hopefully, these new data will provide a balance with the data originally collected.

Once we have assessed the trustworthiness of our results, it only remains to write the report describing our methods and findings.

SUMMARY

This chapter has presented a systematic and purposeful approach to data analysis in an interpretive research study. The major steps of a data analysis include transcript preparation, planning the analysis, first-level coding, second-level coding, interpretation and theory building, and assessing the trustworthiness of the results. Although these steps have been presented in a linear fashion, any real qualitative analysis will involve moving back and forth between them in order to produce rich and meaningful findings.

REFERENCES AND FURTHER READING

Bazeley, P. (2007). *Qualitative data analysis with NVivo*. Thousand Oaks, CA: Sage.

Brown, A., & Gibson, W. (2009). *Qualitative data analysis*. Thousand Oaks, CA: Sage.

Coleman, H., & Unrau, Y. (2008). Qualitative data analysis. In R.M. Grinnell, Jr., & Y.A. Unrau (Eds.), *Social work research and evaluation: Foundations of evidence-based practice* (8th ed., pp. 387–405). New York: Oxford University Press.

Denzin, N.K., & Lincoln, Y.S. (Eds.). (2007). *Collecting and interpreting qualitative materials* (3rd ed.). Thousand Oaks, CA: Sage.

Denzin, N.K., & Lincoln, Y.S. (Eds.). (2007). *The landscape of qualitative research* (3rd ed.). Thousand Oaks, CA: Sage.

Denzin, N.K., & Lincoln, Y.S. (Eds.). (2007). *Strategies of qualitative inquiry* (3rd ed.). Thousand Oaks, CA: Sage.

Flick, U. (Ed.). (2007). *The SAGE qualitative research kit.* Thousand Oaks, CA: Sage.

Golden-Biddle, K., & Locke, K. (2007). *Composing qualitative research: Crafting theoretical points from qualitative data* (2nd ed.). Thousand Oaks, CA: Sage.

Grbich, C. (2007). *Qualitative data analysis: An introduction.* Thousand Oaks, CA: Sage.

Gubrium, J.F., & Holstein, J.A. (2008). *Analyzing narrative reality.* Thousand Oaks, CA: Sage.

Knowles, J.G., & Cole, A.L. (Eds.). (2007). *Handbook of the arts in qualitative research: Perspectives, methodologies, Examples and Issues.* Thousand Oaks, CA: Sage.

Lewis, A., & Silver, C. (2007). *Using software in qualitative research: A step-by-step guide.* Thousand Oaks, CA: Sage.

Maxwell, J. A. (2005). *Qualitative research design: An interactive approach* (2nd ed.). Thousand Oaks, CA: Sage.

Miles, M.B., & Weitzman, E. (1999). *Computer programs for qualitative data analysis* (2nd ed.). Thousand Oaks, CA: Sage.

Nagy Hesse-Biber, & Leavy, P. (2006). *The practice of qualitative research: A primer.* Thousand Oaks, CA: Sage.

Richards, L. (2005). *Handling qualitative data: A practical guide.* Thousand Oaks, CA: Sage.

Saldana, J. (2008). *The coding manual for qualitative researchers.* Thousand Oaks, CA: Sage.

Silverman, D. (2006). *Interpreting qualitative data: Methods for analyzing talk, text, and interaction* (3rd ed.). Thousand Oaks, CA: Sage.

Silverman, D., & Marvasti, A. (2008). *Doing qualitative research: A comprehensive guide.* Thousand Oaks, CA: Sage.

Tutty, L.M., Rothery, M.L., & Grinnell, R.M., Jr. (Eds.). (1996). *Qualitative research for social workers: Phases, steps, and tasks.* Boston: Allyn & Bacon.

Check out our Website for useful links and chapters (PDF) on:
- introductory qualitative data-analysis concepts
- Web-based links to various qualitative software packages
- Web-based links to various qualitative data-analysis tutorials

Go to: www.pairbondpublications.com
Click on: Student Resources
　　　　　Chapter-by-Chapter Resources
　　　　　Chapter 15

Check Out Our Website

PairBondPublications.com

√ *Student Workbook Exercises*
√ *Chapter Power Point Slides*
√ *On-line Glossaries*
√ *General Links*
√ *Specific Chapter Links*

Part VII

Research Proposals and Reports

Chapter 16: Quantitative Proposals and Reports   332

Chapter 17: Qualitative Proposals and Reports   354

Chapter 16

Quantitative Proposals and Reports

WRITING QUANTITATIVE RESEARCH PROPOSALS
    Part 1: Research Topic
    Part 2: Literature Review
    Part 3: Conceptual Framework
    Part 4: Questions and Hypotheses
    Part 5: Operational Definitions
        Variable 1: Children
        Variable 2: Domestic Violence
        Variable 3: Children Witnessing Domestic Violence
        Variable 4: Children's Social Interaction Skills
    Part 6: Research Design
    Part 7: Population and Sample
    Part 8: Data Collection
    Part 9: Data Analysis
    Part 10: Limitations
    Part 11: Administration
WRITING QUANTITATIVE RESEARCH REPORTS
    Part 1: Problem
    Part 2: Method
    Part 3: Findings
    Part 4: Discussion
SUMMARY
REFERENCES AND FURTHER READING

16

IN THIS CHAPTER, WE DISCUSS BOTH HOW TO WRITE a quantitative research proposal, which is done *before* the study begins, and how to write a quantitative research report, which is done *after* the study is completed. They will be presented together because a quantitative research proposal describes what is *proposed* to be done, while a research report describes what *has been* done. As will be emphasized throughout the chapter, there is so much overlap between the two that a majority of the material written for a quantitative research proposal can be used in writing the final research report.

This chapter incorporates most of the contents of the preceding ones, so it is really a summary of the entire quantitative research process (and this book) up to report writing. We use an example of Lula Wilson, a social work practitioner and researcher who wants to do a quantitative research study on children who come to her women's emergency shelter with their mothers. She has been working at the shelter for the past two years. The shelter is located in a large urban city.

WRITING QUANTITATIVE RESEARCH PROPOSALS

When writing any research proposal, we must always keep in mind the purposes for its development and be aware of politically sensitive research topics. These are, primarily, to get permission to do the study and, second, perhaps to obtain some funds with which to do it.

There is a third purpose—to persuade the people who will review the proposal that its author, Lula, is competent to carry out the intended study. Finally, the fourth purpose of a proposal is to force Lula to write down exactly what is going to be studied, why it is going to studied, and how it is going to studied.

In doing this, she may think of aspects of the study that had not occurred before. For example, she may look at the first draft of her proposal and realize that some essential detail was forgotten—for instance, that the research participants (in her case, children) who are going to fill out self-report standardized measuring instruments must be able to read.

The intended readers of the proposal determine how it will be written. It is important to remember that the reviewers will probably have many proposals to evaluate at once. Some proposals will need to be turned down because there will not be sufficient funds, space, or staff time, to accept all of them. Thus, proposal reviewers are faced with some difficult decisions on which to accept and which ones to reject. People who review research proposals often do so on a voluntary basis.

With the above in mind, the proposal should be written so that it is easy to read, easy to follow, easy to understand, clearly organized, and brief. It must not ramble or go off into anecdotes about how Lula became interested in the subject in the first place. Rather, Lula's proposal must describe her proposed research study simply, clearly, and concisely.

Now that we know the underlying rationale for the proposal, the next step is to consider what content it should include. This depends to some extent on who will be reviewing it. If the proposal is submitted to an academic committee, for example, it will often include more of a literature review and more details of the study's research design than if it were submitted to a funding organization. Some funding bodies specify exactly what they want included and in what order; others leave it to the author's discretion.

The simplest and most logical way to write the proposal is in the same order that a quantitative research study is conducted. For example, when a quantitative research study is done, a general topic area is decided upon as presented in Chapter 2. This is followed by a literature review in an attempt to narrow the broad research area into more specific research questions or hypotheses. We will now go back and look at how each step of a quantitative research study leads to the writing of a parallel section in a research proposal. Let us turn to the first task of proposal writing, specifying the research topic.

Part 1: Research Topic

The first step in beginning any research study is to decide what the study will be about. The first procedure in writing a proposal, therefore, is to describe, in general terms, what it is that is going to be studied. Lula may describe, for example, her proposed study's general problem area as:

General Problem Area:
Problems experienced by children who witness domestic violence.

The first task is to convince the proposal reviewers that the general problem area is a good one to study. This task is accomplished by outlining the significance of Lula's proposed study in three specific social work areas: its practical significance, its theoretical significance, and its social policy significance. Depending on to whom the proposal is submitted, Lula may go into detail about these three areas or describe them briefly. It may be known, for example, that the funding organization that will review the proposal is mostly interested in improving the effectiveness of individual social workers in their day-to-day practice activities. If this is the case, the reviewers will more likely be interested in the practical significance of Lula's proposed study than its theoretical and/or policy significance.

Therefore, Lula's proposal would neither go into detail about how her study might generate new social work theory, nor elaborate on the changes in social policy that might follow directly from the study's results. Since Lula is going to submit the proposal to the women's emergency shelter where she works, she would be smart in obtaining informal input from the agency's executive director at this stage in writing the proposal. Informal advice at an early stage is astronomically important to proposal writers. In sum, Part 1 of the proposal describes *what* is going to be studied and *why* it should be studied.

Part 2: Literature Review

The second part of a proposal contains the literature review. This is not simply a list of all the books and journal articles that vaguely touch on the general problem area mentioned in Part 1. When a quantitative research study is done, it is basically trying to add another piece to a jigsaw puzzle already begun by other researchers. The purpose of a literature review, then, is to show how *Lula's* study fits into the whole.

The trouble is that it might be a very big whole. There may be literally hundreds of articles and books filled with previous research studies on the study's general topic area. If Lula tries to list every one of these, the reviewers of the proposal, probably her colleagues who work with her at the shelter, will lose both interest and patience somewhere in the middle of Part 2.

The literature review has to be selective—listing enough material to show that Lula is thoroughly familiar with her topic area, but not enough to induce stupor in the reviewer. This is a delicate and sensitive balance. She should include findings from recent research studies along with any classical ones.

On the other side of the coin, another possibility is that previous research studies on Lula's general topic area may be limited. In this case, all available material is included. However, her proposal can also branch out into material that is partially related or describes a parallel topic area. Lula might find a research article, for example, that claims that children whose parents are contemplating divorce have low social interaction skills. This does not bear directly on the matter of problems children have who witnessed domestic violence (the general problem area mentioned above). However, since marital separation can be a result of domestic violence, it might be indirectly relevant.

A literature review serves a number of purposes. First, it shows the reviewers that Lula understands the most current and central issues related to the general topic area that she proposes to study. Second, it points out in what ways her proposed study is the same as, or different from, other similar studies. Third, it describes how the results of her proposed study will contribute to solving the puzzle. Fourth, it introduces and conceptually defines all the variables that will be examined throughout the study.

At this stage, Lula does not operationally define her study's variables—that is, in such a way that allows their measurement. They are only abstractly defined. For example, if Lula is going to study the social interaction skills of children who witness domestic violence, her proposal so far introduces only the concepts of children, domestic violence, children witnessing domestic violence, and children's social interaction skills.

Part 3: Conceptual Framework

A conceptual framework takes the variables that have been mentioned in Part 2, illustrates their possible relationship to one another, and discusses why the relationship exists the way it is proposed and not in some other, equally possible way. The author's suppositions might be based on past professional experience.

For example, Lula has observed numerous children who accompanied their mothers to women's emergency shelters. She has made subjective observations of these children over the past two years and finally wishes to test out two hunches objectively.

First, Lula believes that children who have witnessed domestic violence *seem to* have lower social interaction skills than children who have not witnessed domestic violence. Second, Lula believes that of the children who have been a witness to domestic violence, boys *seem to* have lower social interaction skills than girls. However, these two hunches are based on only two-year subjective observations, which need to be objectively tested—the purpose of her quantitative study.

As we know, ideally, Lula's hunches should be integrated with existing theory or findings derived from previous research studies. In any case, Lula should discuss these assumptions and the reasons for believing them as the basis for the variables that are included in her proposed study.

In sum, Lula wants to see if children who have witnessed domestic violence have lower social interaction skills than those children who have not witnessed it. And of the children who have witnessed domestic violence, she wants to determine whether boys have lower social interaction skills than girls. It must be remembered that the two areas her study proposes to explore have been delineated out of her past experiences and have not been formulated on existing theory or previous research findings.

Part 4: Questions and Hypotheses

As previously discussed at the beginning of this book, when little is known about a topic, only general research questions are asked. Many general research questions relating to Lula's general problem area can be asked. One of the many could be:

> *General Research Question:*
> Is there a relationship between children witnessing domestic violence and their social interaction skills?

On the other hand, when a lot of previous research studies have been previously done, a specific hypothesis can be formulated. A specific hypothesis derived from the above general research question might be:

> *Specific Research Hypothesis:*
> Children who have witnessed domestic violence will have lower social interaction skills than children who have not witnessed such abuse.

Part 5: Operational Definitions

As mentioned, variables are abstractly and conceptually defined in the conceptual framework part of the proposal (Part 3). Part 5 provides operational definitions of them; that is, they must be defined in ways in which they can be measurable.

Let us take Lula's simple research hypothesis previously mentioned. In this hypothesis there are four main variables that must be operationalized before Lula's study can begin: children, domestic violence, children witnessing domestic violence, and children's social interaction skills. Each must be described in such a way that there is no ambiguity as to what they mean.

Variable 1: Children

For example, what constitutes a child? How old must the child be? Does the child have to be in a certain age range, for example between the ages of 5–10? Does the child have to be a biological product of either the mother or the father? Can the child be a stepchild? Can the child be adopted? Does the child have to live full-time at home?

Since Lula's study is at the descriptive level, she may wish to define a child operationally in such a way that permits the largest number of children to be included in the study. She would, however, go to the literature and find out how other researchers have operationally defined "children," and she would use this operational definition if it made sense to her.

However, in a simple study such as this one, a child could be operationally defined as "any person who is considered to be a child as determined by the mother." This is a very vague operational definition, at best, but it is more practical than constructing one such as, "a person between the ages of 5–17 who has resided full-time with the biological mother for the last 12-month period."

If such a complex operational definition was utilized, Lula would have to provide answers to questions as: Why the ages of 5–17, why not 4–18? What is the specific reason for this age range? Why must the child live at home full-time, why not part-time? Why must the mother be the biological mother, why not a non-biological mother? What about biological fathers? Why must the child have had to live at home for the past 12 months, why not two years, four years?

In short, Lula's operational definition of a child must make sense and be based on a rational or theoretical basis. For now, Lula is going to make matters simple: A child in her study will be operationally defined as any child whose mother validates their relationship. This simple operational definition makes the most practical sense to Lula.

Variable 2: Domestic Violence

Let us now turn to Lula's second variable—domestic violence. What is it? Does the male partner have to shove, push, or threaten his partner? How would a child, as operationally defined above, know when it occurs? Does a husband yelling at his partner imply domestic violence? If so, does it have to last a long time? If it does, what is a long time? Is Lula interested in the frequency, duration, or magnitude of yelling—or all three?

A specific operational definition of domestic violence has to be established in order for the study to be of any value. Like most variables, there are as many operational definitions of domestic violence as there are people willing to define it.

For now, Lula is going to continue to make her descriptive study simple by operationally defining domestic violence as, "women who say they have been physically abused by their partners." Lula can simply ask each woman who enters the shelter if she believes she has been physically abused by her partner.

The data provided by the women will be "yes" or "no." Lula could have looked at the frequency, duration, or magnitude of such abuse, but for this study, the variable is a dichotomous one: Either domestic violence occurred, or it did not occur—as reported by the women. Questions regarding its frequency (how many times it occurred), its duration (how long each episode lasted), and its magnitude (the intensity of each episode) are not asked.

Variable 3: Children Witnessing Domestic Violence

The third variable in Lula's hypothesis is the child (or children) who witness domestic violence. Now that operational definitions of a "child" and "domestic violence" have been formulated, how will she know that a child has witnessed such an abuse? Each child could be asked directly, or a standardized checklist of possible verbal and physical abuses that a child might have witnessed can be given to the child, who is then asked how many times such abuse has been observed.

Obviously, the child would have to know what constitutes domestic violence to recognize it. In addition, the child would have to be old enough to respond to such requests, and the operational definition that is used for domestic violence must be consistent with the age of the child. For example, the child must be able to communicate to someone that domestic violence has in fact occurred—not to mention the question of whether the child could recognize it in the first place.

In Lula's continuing struggle to keep her study as simple as possible, she operationally defines "a child witnessing domestic violence" by asking the mothers who come to the women's emergency shelter if their child(ren) witnessed the physical abuse. She is interested only in the women who come to the shelter as a result of being physically abused by their partners. Women who come to the shelter for other reasons are not included in her study. It must be kept in mind that Lula's study is focusing only upon physical abuse and not emotional or mental abuse.

So far, Lula's study is rather simple in terms of operational definitions. Up to this point she is studying mothers who bring their child(ren) with them to one women's emergency shelter. She simply asks the mother if the person(s) with her is her child(ren), which operationally defines "child."

The mother is asked if she believes her partner physically abused her, which defines 'domestic violence." The mother is also asked if the child(ren) who is accompanying her to the shelter saw the physical abuse occur, which operationally defines "children witnessing domestic violence."

Variable 4: Children's Social Interaction Skills

Let us now turn to Lula's fourth and final variable in her hypothesis—the children's social interaction skills. How will they be measured? What constitutes the social skills of a child? They could be measured in a variety of ways through direct observations of parents, social work practitioners, social work researchers, social work practicum students, teachers, neighbors, and even members from the children's peer group. They could also be measured by a standardized measuring instrument such as the ones discussed in Chapters 6 and 7. Lula decides to use one of the many standardized measuring instruments that measure social interaction skills of children, named the Social Interaction Skills of Children Assessment Instrument (*SISOCAI*).

All in all, Part 5 of a proposal provides operational definitions of all important variables that were abstractly defined in Part 3. It should be noted that the four variables that have been operationally defined should be defined from the available literature, if appropriate. (This procedure makes a study's results generalizable from one research situation to another.)

However, there may be times when this is not possible. The proposal must specify what data gathering instruments are going to be used, including their validity and reliability, as presented in Chapters 6 and 7.

In summary, let us review how Lula intends to operationally define her four key variables: child, domestic violence, child witnessing domestic violence, and children's social interaction skills:

1. *Child.* Any person who the mother claims is her child.
2. *Domestic Violence.* Asking the mother if her partner physically abused her.
3. *Child Witnessing Domestic Violence.* Asking the mother if the child(ren) who accompanied her to the shelter witnessed the abuse.
4. *Children's Social Interaction Skills.* The *SISOCAI* score for each child in the study.

The first three operational definitions are rather rudimentary, at best. There are many more sophisticated ways of operationally defining them. However, alternative definitions will not be explored, because Lula wants to keep her quantitative research proposal as uncomplicated as possible as she knows that the shelter does not want a study that would intrude too heavily into its day-to-day operations. On a very general level, the more complex the operational definitions of variables used in a quantitative research study, the more the study will intrude on the research participants' lives—in addition to the agency's day-to-day operations.

Part 6: Research Design

This part of a quantitative research proposal presents the study's research design. Suppose Lula formulated two related research hypotheses from her general problem area mentioned above:

Research Hypothesis 1:
Children who have witnessed domestic violence will have lower social interaction skills than children who have not witnessed such abuse.

Research Hypothesis 2:
Of those children who have witnessed domestic violence, boys will have lower social interaction skills than girls.

In relation to Research Hypothesis 1, Lula's study would use the children who accompanied their mothers to the women's emergency shelter. These children would then be broken down into two groups, (1) those children who witnessed domestic violence, and (2) those children who did not witness it (as determined by the mother).

As presented in Chapter 10, a very simple two-group research design could be used to test Research Hypothesis 1. The average social interaction skill score, via the *SISOCAI*, between the two groups can then be compared.

In relation to Research Hypothesis 2, within the group of children who have witnessed domestic violence, the children's social interaction skills, via the *SISOCAI*, between the boys and girls can be compared. This simple procedure would test Research Hypothesis 2. Once again, as presented in Chapter 10, a simple two-group research design could be used to test Research Hypothesis 2.

In Lula's study, there are two separate mini-research studies running at the same time—Research Hypotheses 1 and 2. All Lula wants to do is to see if there is an association between the social interaction skills of children who have and have not witnessed domestic violence (Research Hypothesis 1). In addition, for those children who have witnessed domestic violence, she wants to see if boys have lower social interaction skills than girls (Research Hypothesis 2).

In Part 6 of a quantitative research proposal, information should be included about what data will be collected, how these data will be collected, and who will be the research participants.

Next, Lula must now describe the data to be collected. She will have her research assistant complete the *SISOCAI* for each child during a half-hour interview. Finally, the conditions under which the data will be gathered are discussed; that is, the research assistant will complete the *SISOCAI* for each selected child one day after the mother entered the shelter. Of necessity, recording all of this will involve some repetition.

For example, Part 5, when discussing operational definitions, discussed what data would be collected, and how. Part 7, the next part, will discuss the study's sample and population (i.e., who will be studied) in much more detail. It is repeated in Part 6, both to give an overview of the whole positivist research process, and to form links between Parts 5 and 7 so that the entire proposal flows smoothly.

Part 7: Population and Sample

This part of the proposal presents a detailed description of who will be studied. Lula's quantitative research study will use the children who accompanied their mothers to one women's emergency shelter who wish to voluntarily participate, and whose mothers agree that they can be included in the study. The children will then each go into one of two distinct groups: those who have witnessed domestic violence, and those who have not (according to the mothers, that is).

Lula's study could have used a comparison group of children from the same local community who have never witnessed domestic violence and have never been to a women's emergency shelter. However, Lula chose to use only those children who accompanied their mothers to the shelter where she works. There is no question of random selection from some population, and it is not possible to generalize the study's findings to any general population of children who have and have not witnessed domestic violence. The results of Lula's study will apply only to the children who participated in it.

Part 8: Data Collection

This part of a quantitative research proposal presents a detailed account of how the data are going to be collected—that is, the specific data collection method(s) that will be used. As we know, data can be collected using interviews (individual or group), surveys (mail or telephone), direct observations, participant observations, secondary analyses, and content analyses. Lula is going to collect data on the dependent variable by having her research assistant complete the *SISOCAI* for each identified child during a one-half hour interview one day after the mother enters the shelter with her child(ren). Those mothers who do not bring their children with them will not be included in Lula's study.

Chapter 16: Quantitative Proposals and Reports

In addition to the children, their mothers are also going to be interviewed to some small degree. Each mother will be asked by Lula if she believes her partner physically assaulted her. These responses will then be used to operationally define "domestic violence." Each mother will also provide data on whether or not her child(ren) who accompanied her to the shelter saw the abuse occur. The mothers' responses will then operationally define whether or not the child(ren) witnessed domestic violence.

Finally, this section should discuss ethical issues involved in data collection. Chapters 11–13 in this book present various data collection methods that can be used in research studies. These chapters should be reread thoroughly before writing Part 8 of a quantitative research proposal.

Part 9: Data Analysis

This part of a quantitative research proposal describes the way the data will be analyzed, including the statistical procedures to be used, if any. Having clearly specified the research design in Part 6, this part specifies exactly what statistical test(s) will be used to answer the research questions or hypotheses.

The *SISOCAI* produces interval-level data, and a child's social interaction skill score on this particular instrument can range from 0–100, where higher scores mean higher (better) social skills than lower scores. Since there are two groups of children that are being used to test both research hypotheses, and the dependent variable (*SISOCAI*) is at the interval-level of measurement, an independent *t*-test would be used to test both research hypotheses.

Part 10: Limitations

There are limitations in every quantitative research study, often due to problems that cannot be eliminated entirely—even though steps can be taken to reduce them. Lula's study is certainly no exception. Limitations inherent in a study might relate to the validity and reliability of the measuring instruments, or to the generalizability of the study's results. Sometimes the data that were needed could not be collected for some reason. In addition, this part should mention all extraneous variables that have not been controlled for.

For example, Lula may not have been able to control for all the factors that affect the children's social interaction skills. Although she believes that having witnessed domestic violence leads to lower social skills for boys as compared to girls, it may not be possible to collect reliable and valid data about whether the children saw or did not see an abuse occur. In Lula's study, she is going to simply ask the mothers, so in this case, she has to take the mothers' word for it. She could ask the children, however.

This would produce another set of limitations in and of itself. For example, it would be difficult for Lula to ascertain whether a child did or did not see a form of domestic violence as perceived by the child. It may be hard for a child to tell what type of abuse occurred. Also the frequency, duration, and magnitude of a particular form of domestic violence may be hard for the child to recall. All of these limitations, and a host of others, must be delineated in this part of the proposal. In addition, asking a child if he or she saw the abuse occur might prove to be a traumatic experience for the child.

Some limitations will not be discovered until the study is well underway. However, many problems can be anticipated and these should be included in the proposal, together with the specific steps that are intended to minimize them.

Part 11: Administration

The final part of a quantitative research proposal contains the organization and resources necessary to carry out the study. First, Lula has to find a base of operations (e.g., a desk and telephone). She has to think about who is going to take on the overall administrative responsibility for the study.

How many individuals will be needed? What should their qualifications be? What will be their responsibilities? To whom are they responsible? What is the chain of command? Finally, Lula has to think about things such as a computer, stationery, telephone, travel, and parking expenses. When all of the details have been put together, an organizational chart can be produced that shows what will be done, by whom, where, and in what order. The next step is to develop a time frame. By what date should each anticipated step be completed?

Optimism about completion dates should be avoided, particularly when it comes to allowing time to analyze the data and writing the final report (to be discussed shortly). Both of these activities always take far longer than anticipated, and it is important that they be properly done—which takes more time than originally planned. When the organizational chart and time frame have been established, the final step is to prepare a budget. Lula has to figure out how much each aspect of the study—such as office space, the research assistant's time, staff time, and participants' time, if any—will cost.

We have now examined eleven parts that should be included when writing a quantitative research proposal. Not all proposals are organized in precisely this way; sometimes different headings are used or information is put in a different part of the proposal. For example, in some proposals, previous studies are discussed in the conceptual framework section rather than in the literature review. Much depends on for whom the proposal is being written and on the author's personal writing style.

Much of the content that has been used to write the eleven parts of a quantitative research proposal can be used to write the final quantitative research report. Let us now turn to that discussion.

WRITING QUANTITATIVE RESEARCH REPORTS

A quantitative research report is a way of describing the completed study to other people. The findings can be reported by way of an oral presentation, a book, or a published paper. The report may be addressed solely to colleagues at work or to a worldwide audience. It may be written simply, so that everyone can understand it, or it may be so highly technical that it can only be understood by a few.

The most common way of sharing a study's findings with other professionals is to publish a report in a professional journal. Most journal reports run about twenty-five double-spaced, typewritten pages, including figures, tables, and references. As we know, quantitative proposals are written with the proposal reviewers in mind. Similarly, a quantitative research report is written with its readers in mind.

However, some of the readers who read research reports will want to know the technical details of how the study's results were achieved, others will only want to know how the study's results are relevant to social work practice, without the inclusion of the technical details.

There are a number of ways to deal with this situation. First, a technical report can be written for those who can understand it, without worrying too much about those who cannot. In addition, a second report can be written that skims over the technical aspects of the study and concentrates mostly on the practical application of the study's findings. Thus, two versions of the same study can be written; a technical one, and a nontechnical one.

The thought of writing two reports where one would suffice will not appeal to very many of us, however. Usually, we try to compensate for this by including those technical aspects of the study that are necessary to an understanding of the study's findings. This is essential, because readers will not be able to evaluate the study's findings without knowing how they were arrived at.

However, life can be made easier for nontechnical audiences by including some explanation of the technical aspects and, in addition, paying close attention to the practical application of the study's results. In this way, we will probably succeed in addressing the needs of both audiences—those who want all the technical details and those who want none.

A quantitative report can be organized in many different ways depending on the intended audience and the author's personal style. Often, however, the same common-sense sequence is followed when the basic problem-solving method was discussed at the beginning of this book. In order to solve a problem, the problem must be specified, ways of solving it must be explored, a solution to solve the problem must be tried, and an evaluation must take place to see if the solution worked.

In general, this is the way to solve practice and research problems. It is also the order in which a quantitative research report is written. First, a research problem is defined. Then, the method used to solve it is discussed. Next, the findings are presented. Finally, the significance of the findings to the social work profession are discussed.

Part 1: Problem

Probably the best way to begin a quantitative research report is to explain simply what the research problem is. Lula might say, for example, that the study's purpose was to ascertain if children who have witnessed domestic violence have lower social interaction skills than children who have not witnessed such abuse. In addition, the study wanted to find out, of those children who have witnessed domestic violence, if boys have lower social interaction skills than girls. But why would anyone want to know about that? How would the knowledge derived from Lula's study help social workers?

Thinking back to Part 1 of her proposal, this question was asked and answered once before. In the first part of her proposal, when the research topic was set out, the significance of the study was discussed in the areas of practice, theory, and social policy. This material can be used, suitably paraphrased, in Part 1 of the final report.

One thing that should be remembered, though, is that a quantitative research report written for a journal is not, relatively speaking, very long. A lot of information must be included in less than twenty-five pages, and the author cannot afford to use too much space explaining to readers why this particular study was chosen. Sometimes, the significance of the study will be apparent and there is no room to belabor what is already obvious.

In Part 2 of the proposal, a literature review, was done in which Lula's proposed study was compared to other similar studies, highlighting similarities and differences. Also, key variables were conceptually defined. In her final report, she can use both the literature review and her conceptually defined variables that she presented in her proposal. The literature review might have to be cut back if space is at a premium, but the abstract and conceptual definitions of all key variables must be included. In Part 3 of her proposal, she presented a conceptual framework. This can be used in Part 1 of the final report, where Lula must state the relationships between the variables she is studying.

In Part 4 of the quantitative research proposal, a research question or hypothesis to be answered or tested was stated. In the final report, we started out with that, so now we have come full circle. By using the first four parts of the proposal for the first part of the quantitative research report, we have managed to considerably cut down writing time. In fact, Part 1 of a quantitative research report is nothing more than a cut-and-paste job of the first four parts of the quantitative research proposal. Actually, if the first four parts of the quantitative research proposal were done correctly, there should be very little original writing within Part 1 of a quantitative research report.

One of the most important things to remember when writing Part 1 of a quantitative report is that the study's findings have to have some form of utilization potential for social workers, or the report would not be worth writing in the first place. More specifically, the report must have some practical, theoretical, or policy significance. Part 1 of a quantitative research report tells why the study's findings would be useful to the social work profession. This is mentioned briefly but is picked up later in Part 4.

Part 2: Method

Part 2 of a quantitative research report contains the method(s) used to answer the research problem. This section usually includes descriptions of the study's research design, a description of the research participants who were a part of the study (study's sample), and a detailed description of the data gathering procedures (who, what, when, how), and presents the operational definition of all variables.

Once again, sections of the original quantitative research proposal can be used. For example, in Part 5 of the proposal, key variables were operationally defined; that is, they were defined in a way that would allow them to be measured. When and how the measurements would occur were also presented. This material can be used again in the final report.

Part 6 of the proposal described the study's research design. This section of the proposal was used—about half way through—to link the parts of the study together into a whole. Since a research design encompasses the entire quantitative research process from conceptualizing the problem to disseminating the findings, Lula could take this opportunity to give a brief picture of the entire process. This part presents who would be studied (the research participants, or sample), what data would be gathered, how the data would be gathered, when the data would be gathered, and what would be done with the data once obtained (analysis).

In the final report, there is not a lot of space to provide this information in detail. Instead, a clear description of how the data were obtained from the measuring instruments must be presented. For example, Lula could state in this part of the report, that "a research assistant rated each child on the *SISOCAI* during a one-half hour interview one day after the mother entered the shelter."

Part 3: Findings

Part 3 of a report presents the study's findings. Unfortunately, Lula's original proposal will not be of much help here because she did not know what she would find when it was written—only what she hoped to find.

One way to begin Part 3 of a report is to prepare whatever figures, tables, or other data displays that are going to used. For now, let us take Lula's Research Hypothesis 1 as an example of how to write up a study's findings. Suppose that there were 80 children who accompanied their mothers to the shelter. All of the mothers claimed they were physically abused by their partners. Lula's research assistant rated the 80 children's social skills, via the *SISOCAI*, one day after they accompanied their mothers to the shelter. Thus, there are 80 *SISOCAI* scores. What is she going to do with all that data?

The goal of tables and figures is to organize the data in such a way that the reader takes them in at a glance and says, "Ah! Well, it's obvious that the children who witnessed domestic violence had lower *SISOCAI* scores than the children who did not see such abuse."

As can be seen from Table 16.1, the average *SISOCAI* score for all of the 80 children is 60. These 80 children would then be broken down into two subgroups: those who witnessed domestic violence, and those who did not—according to their mothers, that is. For the sake of simplicity let us say there were 40 children in each subgroup. In the first subgroup, the average *SISOCAI* score for the 40 children is 45; in the second subgroup, the average *SISOCAI* score for the 40 children is 75. Table 16.1 allows the reader to quickly compare the average *SISOCAI* score for each subgroup.

The reader can see, at a glance, that there is a 30-point difference between the two average *SISOCAI* scores (75 − 45 = 30). The children who witnessed domestic violence scored 30 points lower, on the average, on the *SISOCAI* than those children who did not witnesses domestic violence. Thus, by glancing at Table 16.1, Lula's Research Hypothesis 1 is supported in that children who witness domestic violence had lower social interaction skills than children who did not witness it.

However, it is still not known from Table 16.1 whether the 30-point difference between the two average *SISOCAI* scores is large enough to be statistically significant. The appropriate statistical proce-

TABLE 16.1
Means and Standard Deviations of Social Interaction Skills of Children Who Did and Did Not Witness Domestic Violence

Witness?	Mean	Standard Deviation	n
Yes	45	11	40
No	75	9	40
Average......	60		

dure for this design is the independent t-test, as described in Chapter 14. The results of the t-test could also be included under Table 16.1, or they could be described in the findings section:

> The result of an independent t-test between the *SISOCAI* scores of children who witnessed domestic violence as compared to those children who did not witness it was statistically significant ($t = 3.56$, $df = 78$, $p < .05$). Thus, children who witnessed domestic violence had statistically significant lower social interaction skills, on the average, than children who did not witness such abuse.

Table 16.2 presents the study's findings for Lula's second research hypothesis. This table uses the data from the 40 children who witnessed domestic violence (from Table 16.1). As can be seen from Tables 16.1 and 16.2, the average social skill score for the 40 children who witnessed domestic violence is 45.

Table 16.2 further breaks down these 40 children into two subgroups: boys and girls. Out of the 40 children who witnessed domestic violence, 20 were boys and 20 were girls. As can be seen from Table 16.2, boys had an average social skill score of 35 as compared with the average score for girls of 55. Thus, the boys scored, on the average, 20 points lower than the girls. So far, Lula's second research hypothesis is supported.

However, it is still not known from Table 16.2 whether the 20-point difference between the two average *SISOCAI* scores is large enough to be statistically significant. The appropriate statistical procedure for this design is the independent t-test, as described in Chapter 14. The results of the t-test could be included under Table 16.2, or they could be described in the findings section as follows:

TABLE 16.2
Means and Standard Deviations of Social Interaction Skills of Boys and Girls Who Witnessed Domestic Violence (from Table 16.1)

Gender	Mean	Standard Deviation	n
Boys	35	12	20
Girls	55	13	20
Average.......	45		

The result of an independent *t*-test between the *SISOCAI* scores for boys and girls who witnessed domestic violence was statistically significant (*t* = 2.56, *df* = 38, *p* < .05). Thus, boys had statistically significant lower social interaction skills, on the average, than girls.

Once a table (or figure) is constructed, the next thing that has to be done is to describe in words what it means. Data displays should be self-explanatory if done correctly. It is a waste of precious space to repeat in the text something that is perfectly apparent from a table or figure.

At this point, Lula has to decide whether she is going to go into a lengthy discussion of her findings in this part of the report or whether she is going to reserve the discussion for the next part. Which option is chosen often depends on what there is to discuss. Sometimes it is more sensible to combine the findings with the discussion, pointing out the significance of what has been found as she goes along.

Part 4: Discussion

The final part of a quantitative research report presents a discussion of the study's findings. Care should be taken not to merely repeat the study's findings that were already presented in Part 3. It can be tempting to repeat one finding in order to remind the reader about it preliminary to a discussion, and then another finding, and then a third . . . and, before we know it, we have written the whole of the findings section all over again and called it a discussion. What is needed here is control and judgment—a delicate balance between not reminding the reader at all and saying everything twice.

On the other hand, Lula might be tempted to ignore her findings altogether, particularly if she did not find what she expected. If the findings did not support her hypothesis, she may have a strong urge to express her viewpoint anyway, using persuasive prose to make up for the lack of quantitative objective evidence. This temptation must be resisted at all costs. The term "discussion" relates to what she found, not to what she thinks she ought to have found, or to what she might have found under slightly different circumstances.

Perhaps she did manage to find a relationship between the variables in both of her hypotheses. However, to her dismay, the relationship was the opposite of what she predicted. For example, suppose her data indicated that children who witnessed domestic violence had higher social interaction skills than children who did not witness it (this is the opposite of what she predicted).

This unexpected result must be discussed, shedding whatever light on the surprising finding. Any relationship between two variables is worthy of discussion, particularly if they seem atypical or if they are not quite what was anticipated.

A common limitation in social work research has to do with not being able to randomly sample research participants from a population. Whenever we cannot randomly select research participants the sample cannot be considered to be truly representative of the population in question, and we cannot generalize the study's results back to the population of children who witnessed or did not witness domestic violence in the community. The simplest way to deal with this limitation is to state it directly.

Another major limitation in this study is that we will never know the social skills of children who did not accompany their mothers to the shelter. The social skills of children who stay home may somehow be quite different from those children who accompanied their mothers. In fact, there are a host of other limitations in this simple study, including the simple fact that, in reference to Research Hypothesis 2, boys who did not see domestic violence may also have lower social interaction skills than girls—this was never tested in Lula's study. Nevertheless, we should also bear in mind the fact that few social work studies are based on truly representative random samples. In Lula's study, however, she still managed to collect some interesting data.

All social work researchers would like to be able to generalize their findings beyond the specific research setting and sample. From a quantitative research perspective (not a practice perspective), Lula is not really interested in the specific children in this particular study. She is more interested in children who witness domestic violence in general. Technically, the results of her study cannot be generalized to other populations of children who witness domestic violence, but she can suggest that she might find similar results with other children who accompany their mothers to similar women's shelters. She can imply and can recommend further studies into the topic area.

Sometimes we can find support for our suggestions in the results of previous studies that were not conclusive either, but that also managed to produce recommendations. It might even be a good idea to extract these studies from the literature review section in Part 1 of the report and resurrect them in the discussion section.

On occasion, the results of a study will not agree with the results of previous studies. In this case, we should give whatever explanations seem reasonable for the disagreement and make some suggestions whereby the discrepancy can be resolved. Perhaps another research study should be undertaken that would examine the same or different variables.

Perhaps, next time, a different research design should be used or the research hypothesis should be reformulated. Perhaps other operational definitions could be used. Suggestions for future studies should always be specific, not just a vague statement to the effect that more research studies need to be done in this area.

In some cases, recommendations can be made for changes in social work programs, interventions, or policies based on the results of a study. These recommendations are usually contained in reports addressed to people who have the power to make the suggested changes. When changes are suggested, the author has to display some knowledge about the policy or program and the possible consequences of implementing the suggested changes.

Finally, a report is concluded with a summary of the study's findings. This is particularly important in longer reports or when a study's findings and discussion sections are lengthy or complex. Sometimes, indeed, people reading a long report read only the summary and a few sections of the study that interests them.

SUMMARY

The purpose of writing a quantitative research proposal is fourfold. A research proposal is necessary, first, to obtain permission to carry out the study and, second, to secure the funds with which to do it. Third, the researcher needs to convince the proposal reviewers that he or she is competent enough to do the study. Fourth, we need to think over precisely what we want to study, why we want to study it, what methods we should use, and what difficulties we are likely to encounter.

A quantitative proposal should be well organized and easy to read so that reviewers have a clear picture of each step of the study. The information included in most proposals can be set out under eleven general headings, or parts: research topic, literature review, conceptual framework, questions and hypotheses, operational definitions, research design, population and sample, data collection, data analysis, limitations, and administration.

Using a majority of the material from the quantitative research proposal, a quantitative research report is written that can be broken down into four general headings, or parts: problem, method, findings, and discussion. The four parts of a quantitative research report parallel the eleven parts of the quantitative research proposal. This is not surprising, since both the report and the proposal are describing the same study.

REFERENCES AND FURTHER READING

Fink, A. (2005). *Conducting research literature reviews: From the Internet to paper* (2nd ed.). Thousand Oaks, CA: Sage.

Holosko, M. (2008). Evaluating quantitative research studies. In R.M. Grinnell, Jr., & Y.A. Unrau (Eds.), *Social work research and evaluation: Foundations of evidence-based practice* (8th ed., pp. 423–461). New York: Oxford University Press.

Lock, L.F., Wyrick Spirdusom, W., & Silverman, S.J. (2007). *Proposals that work* (5th ed.). Thousand Oaks, CA: Sage.

Reid, W.J. (2008). Writing reports from research studies. In R.M. Grinnell, Jr., & Y.A. Unrau (Eds.), *Social work research and evaluation: Foundations of evidence-based practice* (8th ed., pp. 409–421). New York: Oxford University Press.

Thody, A. (2006). *Writing and presenting research*. Thousand Oaks, CA: Sage.

Thyer. B. (2008). *Preparing research articles*. New York: Oxford University Press.

Chapter 17

Qualitative Proposals and Reports

WRITING QUALITATIVE RESEARCH PROPOSALS
    Purpose of Writing a Proposal
    Intended Audience
    Content and Writing Style
    Organizing the Proposal
        Part 1: Research Topic
        Part 2: Literature Review
        Part 3: Conceptual Framework
        Part 4: Questions and Hypotheses
        Part 5: Operational Definitions
        Part 6: Research Design
        Part 7: Population and Sample
        Part 8: Data Collection
        Part 9: Data Analysis
        Part 10: Limitations
        Part 11: Administration
WRITING QUALITATIVE RESEARCH REPORTS
    Abstract
    Introduction
    Method
    Analysis and Findings
    Discussion
    References
SUMMARY
REFERENCES AND FURTHER READING

17

LIKE QUANTITATIVE RESEARCH PROPOSALS and reports presented in the previous chapter, qualitative research proposals and their corresponding reports are also similar to one another. If we explain clearly what we intend to do when we write our proposal, for example, and we actually carry out our study as we originally planned, then writing our research report is largely a matter of changing "we will do" (as in the proposal) to "we did" (as in the report). This is true for everything but describing our study's findings in the research report.

As we will see, there is a good deal of similarity between quantitative and qualitative research proposals and reports. After all, the research process follows a logical progression whether it is quantitative or qualitative. We need to know from the beginning of our research study what we want to study, why we want to study it, what methods we will use to study it, how long our research study will take, and how much it will cost. In addition, we need to know what data will be collected, from whom, in what way, and how they will be analyzed.

For the sake of continuity, this chapter—on qualitative proposals and reports—uses roughly the same format that was used in the last chapter that described quantitative proposals and reports. We will also use the same example: Lula Wilson, a social work practitioner who wants to do a qualitative research study on children and their mothers in the women's emergency shelter where she works.

WRITING QUALITATIVE RESEARCH PROPOSALS

Before we begin to write the very first word of a research proposal—whether quantitative or qualitative—we need to know why we want to write it and who will read it. Knowing the purpose and our intended audience helps us to make important decisions about what we should include, in what order, and what writing style should be used.

Purpose of Writing a Proposal

There are three general purposes for writing a research proposal, no matter whether the study being proposed is quantitative or qualitative: (1) We need to obtain permission to do the study, (2) We need to obtain funding for the study, and (3) We need to write down exactly what we intend to study, why, and how.

As we know from Chapter 3, obtaining permission to do our study is often a matter of resolving ethical and informed consent issues to the satisfaction of various ethics committees. Most universities and colleges have ethics committees, which decide whether our proposed study is designed in such a way that the interests of its research participants are ethically addressed. Many social services agencies have their own ethics committees, which vet all proposed research endeavors that involve their clients and staff.

If Lula were associated in any way with a university, for example, and if her women's emergency shelter had its own ethics committee, she would have to obtain permission from both ethics committees before she could begin her study. Even if no ethics committees are involved, Lula would have to discuss her proposed study with her supervisor, who would probably have to obtain official permission from the shelter's Board of Directors.

All research studies require some level of funding. Even if Lula is prepared to do all the work on her own time using her own clients, there will still be direct and indirect costs such as photocopying, travel, phone, fax, and postage. If Lula wants her shelter to cover these costs, she must include a budget in her proposal and get the budget approved before she begins her study. If it is a larger study, necessitating money from a funding body, then Lula must tailor her research proposal to meet the requirements of the particular funding body to which she is applying.

Most funding bodies have application forms that ask the applicant to supply the study's details under specific headings. Usually, funding bodies also want to know how qualified the particular applicant is to undertake the proposed study. In other words, Lula will have to convince the funding body that she, personally, has the experience and educational qualifications necessary to obtain meaningful and trustworthy results from her proposed study.

After permission and funding, the third purpose of writing a proposal is to force Lula to clarify her thoughts a bit more. In the process of describing her proposed study in sufficient detail in an attempt to convince others of its importance, Lula may think of aspects of her study that she has not thought of before. She may realize, for example, that she has little experience with interviewing children, and someone who has more experience with interviewing children may be in a better position to interview them.

Intended Audience

Most research proposals are reviewed by busy people who have a great deal of other work and probably a number of proposals to review. Lula's proposal, therefore should be as short as she can possibly make it. It should concisely describe her proposed study, its budget, and time frame, in a way that is easy to read, to follow, and to understand.

Many proposals have to be rejected because there is insufficient funding or facilities to support them all, and those that are rejected are not necessarily the least worthy in terms of their importance. They are, however, often the least worthy in terms of how well they were organized and written. Lula will therefore be well advised to keep her proposal simple, clear, and brief.

Content and Writing Style

A proposal's content and writing style will largely depend on who is going to review it. As already noted, some funding bodies stipulate on their application forms what, and how much, they want included, and in what format and order. If there is no such stipulation, it is simplest and most logical to write the research proposal in the order that the study will be conducted; that is, the order followed in this chapter. How much to include under what heading depends on the intended audience: A research proposal submitted to an academic committee, for example, often requires more of a literature review than a proposal submitted to a funding organization.

Style similarly depends on the recipient. In most cases, it is safest to write formally, using the third person. As we know, however, qualitative research studies are often more subjective than quantitative ones, their terminology is different, their underlying assumptions are different, and the researcher's own thoughts and feelings are an important component. It may therefore be appropriate to acknowledge the qualitative nature of the study by using a more personal writing style. As will be the case in writing the final research report, the style used depends on the proposal's intended audience and the author's personal judgment.

Organizing the Proposal

As previously noted, if the proposal's recipients have provided no guidance as to how its contents should be organized, it is simplest to present the proposed study in the order in which it would be conducted. That is, the order that follows.

Part 1: Research Topic

This first section of a research proposal does nothing more than introduce the study to its readers. It examines the nature of the research question being explored and its significance, both to social work in general and to the recipient of the proposal in particular. As with quantitative studies, a qualitative study should have practical significance, theoretical significance, or significance with respect to social policy; or it may touch on all three areas. The author's task is to explain what research question is being asked and why the answer to this question will be of benefit, paying particular attention to the interests of the proposal's reviewers. Lula may write, for example, about the general topic area in her study as follows:

General Topic Area:
The problems experienced by children who witness their mothers being physically abused by their fathers.

The results of such a study—knowing what these problems are—might generate new social work theory or it might lead to changes in social policy. If Lula is going to submit her proposal to the women's emergency shelter where she works, however, her fellow social workers are more likely to be interested in how an understanding of the children's problems might help them to address the children's needs on a very practical level. Lula will therefore emphasize the practical significance of her study in the first part of her proposal.

Part 2: Literature Review

As with quantitative research studies, there are four purposes in carrying out a literature review for qualitative studies:

1. To assure the reviewers that Lula understands the current issues related to her research topic.
2. To point out ways in which her study is similar to, or different from, other studies that have been previously conducted. Since many qualitative studies deal with topics about which little is known, Lula may not find very many studies that have explored children's experiences with respect to their witnessing domestic violence. Such a paucity of information will support Lula's contention that her study needs to be conducted.
3. To fit Lula's study into the jigsaw puzzle of present knowledge. Even if there is little knowledge in the area, there will still be some, and Lula's task is to explain how her study will fit with what is known already and will help to fill the knowledge gaps.
4. To introduce and conceptualize the variables that will be used throughout the study. Lulu's proposal, for example, will in-

Chapter 17: Qualitative Proposals and Reports

clude such concepts as children, domestic violence (or partner abuse, marital abuse, wife abuse, whichever term is preferred) and children witnessing domestic violence.

Part 3: Conceptual Framework

As we know, in quantitative research studies, the conceptual framework identifies the possible relationships between and among concepts to one another. Identifying the ways that concepts might be connected lays the groundwork for developing a research question or research hypothesis. In the last chapter, for example, Lula formulated the research hypothesis as follows:

Research Hypothesis:
Children who have witnessed domestic violence will have lower social interaction skills than children who have not witnessed such violence.

That is, her conceptual framework included the idea that a particular concept—children's social interaction skills—was directly related to another concept—whether the children witnessed the abuse or not.

In qualitative studies, the level of knowledge in the topic area will probably be too low to allow such possible connections between and among concepts to be investigated. Children's poor social interaction skills may indeed be one of the problems experienced by children who witness their mothers being physically abused by their fathers, but Lula does not know that yet. Her simple research question at this stage is simply, "What problems do these children experience?"

Relationships between and among concepts can still be hypothesized, however, even at an exploratory level, and even if the hypothesized relationships will not be tested during the course of the study. People reading a qualitative study, for example, are usually more interested in where the study took place and whether the influence of the clinical setting (i.e., the shelter) was appropriately acknowledged in the data analysis.

Lula must therefore take into account the possibility that the problems experienced by the children in her study may have been due to the study's setting (i.e., the shelter) and not so much from their witnessing the abuse. If she conceptualizes this possibility early, she may decide to interview the children's mothers, asking them not only to identify their children's problems, but to describe each problem before and after coming into the shelter.

Similarly, she may want to explore the possibility that the children's problems may have been related to the children being abused themselves and not just to their witnessing their mothers being physically abused by their fathers.

Part 4: Questions and Hypotheses

A qualitative research study rarely tests a research hypothesis. It is very important, however, that the questions to be answered during the course of a qualitative study be clearly formulated before it begins. Lula could formulate quite specific research questions, such as:

- What types of problems are experienced by children who have witnessed domestic violence?
- Does the type of abuse witnessed (e.g., hitting, yelling) affect the type of problems experienced by the children?
- Does the intensity of the abuse—witnessed by the children—affect the problems they experience?
- Does the frequency (e.g., daily, weekly) of the abuse—witnessed by the children—affect the problems they experience?
- Does the duration (e.g., over months, years) of the abuse—witnessed by the children—affect the problems they experience?
- Does the child's gender affect the types of problems they experience?
- Does the child's age affect the types of problems they experience?
- Do the child's problems, as perceived by the mother, affect the mother's decision to leave the abusive relationship?
- Do the child's problems, as perceived by the mother, affect the mother's decision about whether or not to return to the abusive relationship?

If Lula is going to formulate specific research questions, she will probably need to use a fairly structured interview schedule when she collects interview data from the mothers and their children. On the other hand, she may prefer to formulate just a few, more general research questions, such as:

- What types of problems are experienced by children who have witnessed domestic violence?
- What effects do these problems have on the children and their mothers?

Chapter 17: Qualitative Proposals and Reports **361**

In this case, she would use an unstructured interview schedule, which would allow the mothers and children to guide the interviews themselves, relating what is important to them in their own way. Lula's decision about whether to formulate specific or general research questions depends on the level of knowledge about the study's topic area.

If enough knowledge is available to enable her to formulate specific research questions, she will probably do that. If not, one of the purposes of her study would be to gain enough knowledge to allow specific research questions to be formulated in the future.

Part 5: Operational Definitions

As we know, operationally defining a variable in a quantitative research study means defining the variable in such a way that it can be measured. In the last chapter, for example, Lula operationally defined the level of a child's social interaction skills in terms of the child's score derived from a standardized measuring instrument (*SISOCIA*). The idea behind operationally defining a variable in this way is that both its definition and its measurement are consistent and objective.

Lula did not define "children's social interaction skills" herself (except insofar as she selected the measuring instrument) and she did not ask the children or their mothers what they perceived "children's social interaction skills" to be. Similarly, the measured result for each child (a numerical score) did not depend on anyone's personal perception on how well, or how badly, the child interacted socially with others.

Conversely, in qualitative studies, we are not as interested in objectively defining or measuring our concepts as we are when doing quantitative studies. Indeed, we actively encourage our research participants to provide us with their own, subjective definitions, since we are trying to understand their problems as they perceive them to be.

Similarly, we measure the extent, or effect, of a problem in terms of the research participants' subjective viewpoints. Hence, Lula will not have to worry about how to operationally define "a child" or "a child's problem," and she will not have to decide whether "a child witnessing domestic violence" means seeing it, or hearing it, or merely being aware that it is occurring.

Lula might want to collect data about the ages of the children in her study, whether they are the biological children of their mothers, and whether they live full time at home, but none of these data will be used to exclude any child from the study on the grounds that the child is too old or too young, or otherwise does not fit Lula's operational definition of "a child."

Lula does not have an operational definition of a child. In her study, "a child" is operationally defined as "any person whom the mother considers to be her child." Similarly, "a problem" is whatever the mother and/or child considers to be problematic. "Domestic violence" is defined as whatever the research participants think it is; and children who have "witnessed domestic violence" if they and/or their mothers believe that they have.

It might be as well here to put in a word about measurement. The word "measurement" is often associated with numbers, and hence with quantitative studies. To "measure" something, however, only means to describe it as accurately and completely as possible. If we cannot describe it with numbers, we may still be able to describe it with words, and this qualitative type of measurement is just as valid as a quantitative numerical measurement. Hence, Lula is "measuring" the problems experienced by children when she encourages the mothers and their children to describe those problems as accurately and completely as they can.

In quantitative studies, efforts are made to mitigate the effects of researcher bias through objective measurement. In qualitative studies, however, the use of measurement is to capture the subjective experiences of the research participants. Thus, it is vital for Lula to be aware of the effects of her own feelings upon the research participants she will be interviewing.

Any prior assumptions she has made, and any position she might hold, must be clearly outlined at the beginning of her study, so that the reader of her proposal can evaluate the degree to which her study's potential findings would reflect the research participants' opinions rather than Lula's opinions.

Similarly, it is important to record the interests and possible biases of the organization who is funding the study in addition to the agency where the study actually takes place. Would certain findings be more welcome to the funding body or the agency than other findings? Is the researcher under any pressure to emphasize certain aspects of the study's results to the detriment of other aspects? Again, the reader of a research proposal must be able to evaluate the degree to which the proposed study's auspices would potentially affect the study's findings.

A clear statement of the study's purpose might deflect critics who argue that the proposed study did not fulfill other purposes which the critics, themselves, may perceive as more important. Lula's research study might have a practical purpose, for example, where it would be in tune with staff interests who work within the women's emergency shelter that would provide both funding for the study and access to its clients.

Lula simply wants to know what the children's problems are so that the shelter can better meet the needs of the children and their mothers. She is not overly interested in adding to social work theory or changing social policy, although her study's results may indeed have

Chapter 17: Qualitative Proposals and Reports

implications in both of these areas. She is less likely to be criticized for not placing sufficient emphasis on theory and policy in her discussion if she has clearly stated from the beginning that her proposed study's purpose is to inform day-to-day practice activities within her specific shelter.

Part 6: Research Design

We come now to the *how* of the study. This section includes information about what data will be collected, in what way and from whom, and how they will be analyzed. While writing about these matters, there will be many opportunities to address issues related to the study's *trustworthiness*. Evidence of a study's trustworthiness is provided by paying attention to four major concerns:

- Credibility, or truth value
- Transferability, or applicability
- Dependability, or consistency
- Confirmability, or neutrality

The above are roughly equivalent to the quantitative concepts of internal validity and external validity. The first issue related to trustworthiness, credibility (akin to internal validity), is particularly important and is built on the following aspects of a qualitative study:

- *Triangulation of data sources*—collecting data about the same thing from a number of different data sources; also engagement with research participants over a long period of time.
- *Consulting with colleagues*—consulting with them about ethical and legal matters, and about the methods chosen to select the sample of research participants and to collect and analyze the data.
- *Negative case analysis*—ensuring that information from all data sources is included in the data analysis, even when information from one data source seems to contradict themes or conclusions common to other data sources.
- *Referential adequacy*—keeping a complete and accurate record of all personal interviews and observations, such as videotapes, audiotapes, case notes, and transcriptions.
- *Member checks*—research participants are asked to provide feedback on the information collected from the researcher and the conclusions drawn by the researcher.

The second issue related to trustworthiness, transferability (akin to external validity, or generalizability) is addressed through a rich description of the study's clinical setting and research participants. Findings from a qualitative research study are usually not generalized beyond the setting in which the study took place.

The findings may be applicable, however, to other similar client populations: women and children in similar women's emergency shelters elsewhere, for example. Readers can only judge to what degree a study's findings may be applicable to their own clientele if the researcher provides a detailed description of the study's research participants in addition to their special needs and circumstances.

The third issue, dependability (akin to reliability) relates to efforts to maintain consistency throughout the study. Were all interviews conducted in the same setting, according to the same format, and recorded in the same way? Were all research participants asked to provide feedback on the data collected, or only some? During data analysis, were rules concerning categorization and coding consistently applied? Aspects of the study related to credibility and described above may be used to demonstrate dependability as well: for example, referential adequacy, providing evidence of consistent interviewing procedures, and providing evidence that all research participants were asked for their feedback.

The last issue, confirmability (akin to objectivity), has to do with Lula's awareness of her own role in influencing the data provided by the research participants and the conclusions she drew from the data. All qualitative researchers should keep journals in which they record their own thoughts and feelings about the study's research participants and about their interviews and observations.

Lula should note in her journal, for example, why she made the decisions she did about methodological matters, such as sampling procedures, and data collection and analysis techniques. While conducting the data analysis, she will record decisions and concerns about organizing and interpreting the data she collected.

These journal entries disclose the degree of impartiality she brought to the entire research process; and, where she was not impartial, it discloses her awareness, or lack thereof, about her own assumptions and biases. With respect to dependability (discussed above), it provides a record of how consistent her decision making was and how consistently she conducted her interviews and analyzed her data.

Part 7: Population and Sample

In this part of the proposal, Lula provides only a general description of who her research participants will be, together with a rationale for selecting these and not others. In qualitative studies, there is no at-

Chapter 17: Qualitative Proposals and Reports

tempt to select a random sample. Indeed, the sample often consists of all those persons available to be studied who fit broad criteria. Lula could draw her sample of research participants, for example, from all those women who are residents in her women's emergency shelter at a specific time.

Since Lula's study involves the effects on children who witnessed domestic violence, she will need to exclude from her sample all women who do not have children *and* all women who say that their children did not witness the abuse. Lula may personally believe that no child whose mother is being abused can remain unaware of that abuse, and the definition of "witnessing" for her may include a child's awareness, as well as seeing or hearing. In addition, it would be interesting to explore conflicting perceptions between the mothers and their children, when the children believe that they have witnessed domestic violence and their mothers believe that they have not.

Lula is unlikely, however, to elicit information about the effects of witnessing domestic violence from women who do not believe that their children witnessed it; nor are these women likely to give Lula permission to interview their children on the subject.

Lula may decide to include women who do not have their children with them at the shelter. Whether she does so or not will depend on a number of factors. First, how many women can she interview, given her own and the women's time constraints? This will depend on how long she expects each interview to take, which, in turn, depends on such factors as the structure and depth of the interview.

In addition, she must consider the time involved in transcribing and analyzing each interview in its entirety. If the number of women who have their children at the shelter is equal to, or larger than, the number of women Lula can reasonably interview, then she will exclude women whose children are not present.

If the number is smaller, she may consider including these women, but that decision, as well, will depend on a number of factors. Uppermost in the mind of any qualitative researcher is the notion of the study's trustworthiness.

As discussed earlier, one way of establishing the trustworthiness of data is to collect data about the same thing from a number of different data sources. Data on the problems experienced by children, for example, may be collected from three data sources: (1) the children themselves, (2) their mothers, and (3) shelter staff who have observed the children. Such a triangulation of data sources allows assessment of the trustworthiness of the data obtained from any one given source. If children are not present at the shelter, then data on their problems can be obtained only from their mothers and there will be no way to check on the "accuracy" of the data they provide.

Another way to establish trustworthiness is to ask each research participant to comment on the data gathered and the conclusions that the researcher drew from the data. Lula might want to submit the transcript of each interview to the research participant concerned to make sure that she has adequately captured what the participant was trying to say.

Then, she might want to discuss her findings with the other research participants to see if they believe that she has interpreted what they said correctly and has drawn conclusions that seem reasonable to them as well. None of this will be possible if the research participants have left the emergency shelter and disappeared before Lula has transcribed and analyzed her data.

She might, therefore, want to restrict her sample of women to those who are likely to remain in her shelter for a number of weeks or who will go on to a halfway house or some other traceable address. Of course, if she does this, she will lose data from women whose very transience might affect their children's problems and the way those problems are perceived.

Lula must also consider whether to interview the children and, if so, children in what age groups. She may not be skilled in interviewing young children and may feel that children under school age cannot be meaningfully interviewed at all. If there are enough women in the shelter who have older children present, she may consider restricting her sample of research participants to women whose children are, say, 10 years old or older. She will have to justify selecting age 10 instead of 8 or 12, for example, and she will lose data pertaining to the problems experienced by any excluded children.

With this in mind, she may consider enlisting the assistance of a colleague who is more skilled at eliciting information from younger children—through data collection methods such as drama, art, or play—than she is. But, now, she has to think about how such interview data would be analyzed and how she would integrate them with the data collected through her own personal interviews with the mothers. The child's gender may be another consideration. Perhaps Lula has an idea that girls tend to display more internalizing problem behaviors—such as withdrawal and depression—than boys.

And, she may believe that boys tend to display more externalizing behaviors—such as hostility and aggression—than girls. She might therefore want to ensure that her study contains approximately equal numbers of girls and boys.

If she purposefully drew her sample of research participants in this way, she would have to explain that she expected to find more internalizing behaviors in girls and more externalizing behaviors in boys. This would constitute a research hypothesis, which would need to be included in the Questions and Hypotheses section and justified through the literature review. Or perhaps, Lula would phrase it as a research question, simply asking whether the gender of the child was related to the type of problem behavior he or she exhibited.

Similarly, Lula might have an idea that the types of problem behaviors exhibited by children depend on their ethnic background. If she were able to conduct only a small number of interviews, for example, she might purposefully select women and children from different ethnic backgrounds to make up her sample of research participants (called *purposive sampling*, see Chapter 8). Here again, she would have to justify her choice, including a relevant research question or research hypothesis and addressing the matter in the literature review.

Lula thus has a number of factors to consider in deciding whom to include as research participants in her study. The main consideration, however, is always the willingness of the research participant to take part in the study. Like most social work populations, women in emergency shelters are an extremely vulnerable group, and it is vital to ensure that they feel freely able to refuse to participate in the study, knowing that their refusal will in no way affect the quality of the services they receive.

Similarly, the social workers within Lula's women's emergency shelter must also feel able to refuse, knowing that their refusal will not affect the terms of their employment. It is quite likely that Lula will not have the luxury of selecting her research participants in terms of the age, gender, or ethnic background of the children.

More probably, Lula will just interview those women who agree to be interviewed and who also give permission for her, or a colleague, to interview their children as well. The children will not be in a position to sign an informed consent form, as their mothers and the social workers will do, but it is still extremely important to ensure that they understand their rights with respect to refusing to take part in the study or withdrawing from it at any time.

Part 8: Data Collection

This part of the qualitative research proposal provides a detailed account of how the data are going to be collected, together with a justification for using the data collection method selected rather than some other method. Lula could use focus groups, for example, rather than unstructured interviews to collect data from the women.

She could decide not to interview children, but instead to observe the children's behaviors herself, without involving a colleague or other social workers. If she does involve the shelter's social workers, she might decide just to interview them and ask how they define the children's problem behaviors and what problem behaviors they have observed in the children under study.

On the other hand, she might ask them first to define children's problem behaviors, then to purposefully observe certain children with respect to these behaviors, and finally to report their observations back to her. Whatever she decides, she must first justify her decisions and then clearly describe the methods to be used. She should state, for example, that the abused women, the shelter's social workers, and the children aged 10 or over will be interviewed by herself, if that is what she has decided to do.

She should also specify where these interviews will take place, how long approximately each is expected to last, and to what degree the content will be guided by an interview schedule. She should also specify the time frame within which all the interviews will be completed and how the interviews will be recorded. Videotaping, audiotaping, and taking notes during the interview all have their advantages and disadvantages, which need to be discussed.

If a colleague is to work with the younger children, for example, details of the methods used to elicit interview data from these children must be given. In addition, the colleague's credentials must be included at the beginning of the proposal, since this colleague is now a co-researcher and her experience and qualifications will affect the trustworthiness of the study's findings.

Ethical considerations that were not covered in the discussion about selecting research participants should also be addressed in this section. Should Lula obtain completed consent forms from the mothers and assent forms from their children, for example, before she asks the shelter's social workers to observe the children, or to discuss their behaviors with her?

Should she share the social worker's comments with the mothers and their children concerned and tell the social workers beforehand that this is to be done? Should she share data obtained from the children with their mothers, or make it clear that such data will not be shared? Social workers might not be so honest in their comments if they know that the data will be shared: and neither might the children.

In addition, children who know they are being observed might not behave as they otherwise would. These are old dilemmas that always affect data collection methods, and Lula must specify what dilemmas she may encounter and how she plans to resolve them. It is as well to state how the mothers, children, and social workers are to be approached, and precisely what they are going to be told about her study and their own part in it. Samples of consent forms (for the mothers) and assent forms (for the children) should be included as two appendixes at the end of the proposal.

Lula's journal is also a form of data. While it will include little in the way of data collected from her research participants, it will include Lula's reactions to these data and a chronology of her study's process. Lula might therefore want to state in her proposal that she will keep a journal, recording notes on the decisions she is going to make during every stage of her study, with particular reference to the study's trustworthiness.

Part 9: Data Analysis

This part of the research proposal describes the way the data will be analyzed. There are usually no statistical procedures to be discussed, as there may be in a quantitative study, but there are a number of other matters. As presented in Chapter 15, a decision must be made about whether to use a software computer program to aid in the data analysis and, if so, which one. Then, Lula must decide who will transcribe the interviews and how the transcripts should be formatted. She must establish a plan for her data analysis, including some plan for making journal entries.

She might want to add in her proposal that, after she has analyzed the data, using first- and second-level coding methods, and after she has drawn conclusions, she will assess the trustworthiness of her study's findings. She will do this by documenting what she is going to do to establish credibility, transferability, dependability (consistency), and confirmability (control of biases and preconceptions).

Part 10: Limitations

All research studies have limitations. It might even be suggested that one of the main limitations of a qualitative study is that it is not a quantitative one. This is simply not true. Every study is judged on how well it fulfills its own purpose; and one of the purposes of a qualitative study is usually to understand the experiences of the research participants in-depth, including experiences that are unique to them. The purpose of Lula's study is to gain a better understanding of the problems experienced by children who have witnessed domestic violence, from the different perspectives of the mothers, their children, and the shelter's social workers, so that the needs of these children can be better identified and met.

A discussion of a study's limitations should include only factors that impede the fulfillment of this purpose. Lula's study, for example, is not limited because she did not operationally define the concepts "domestic violence" and "children witnessing domestic violence." Part of her study's purpose is to find out how the mothers and their children, themselves, define "domestic violence;" that is, to find out what it was that the children in her study actually witnessed, and what they and their mothers think that "witnessing" includes.

From an ideal standpoint, Lula's study is limited with respect to its transferability (generalizability, in quantitative terms). It would have been ideal if she could have constructed a sampling frame of all the children in the world who had witnessed domestic violence, taken a random sample, and interviewed all these children and their mothers in-depth.

Quantitative researchers sometimes restrict their studies to manageable samples (research participants) and then generalize from the samples to the populations from which they were drawn. It is a limitation, however, if the samples did not adequately represent their populations, thus restricting their abilities to generalize from their samples to their populations. It is not considered a limitation, however, that the quantitatively-orientated studies did not use larger populations in the first place.

Similarly, Lula does not need to apologize for having chosen to work only with those women and their children who were residents in her particular women's emergency shelter at the time she wanted to conduct her study. On the other hand, some of these women whose children had witnessed domestic violence may have refused to participate, and that would be a limitation to Lula's study, since those women may have felt particularly traumatized by their children's involvement, to the point where they felt unable to discuss it. By losing them, Lula would lose a different and valuable perspective.

Another limitation to Lula's study is that many of the children who have witnessed domestic violence may have been abused themselves. It may be impossible for the mothers and/or children to distinguish between the effects of being abused themselves and the effects of witnessing the abuse. The only way Lula could deal with this is to divide her population of children who have witnessed abuse into two groups: those who have been abused, according to their mothers; and those who have not.

Of course, it might be argued that witnessing domestic violence constitutes emotional abuse. If Lula subscribes to this view, she might wish to ask the mothers specifically if their children have been physically or sexually abused, since all the children in her sample will have been emotionally abused according to her own definition.

Nevertheless, in practical terms, Lula can form her two groups of children merely by including a question about physical or sexual abuse during her interviews with them and with their mothers. If Lula identifies this limitation early on while she is conceptualizing her study, she can include the two groups in her study's research design, mentioning the question about domestic violence in the data collection section, and noting, in the data analysis section, that she will accord each group a separate category.

Thus, her study's limitation will have ceased to be a limitation and will have become an integral part of her study. This is one of the purposes of a research design, of course: to identify and address a study's potential limitations so that they can be eliminated, or at least alleviated, to the greatest possible extent before the study actually starts.

Essentially, what Lula has done in thinking about children who have been abused themselves is to identify a confounding or interven-

Chapter 17: Qualitative Proposals and Reports **371**

ing variable that might interfere with the relationship between their witnessing domestic violence and their experiencing problems, if any, due to witnessing it. Inevitably, there will be a host of other confounding variables since no one can tell for certain whether children's particular problematic behaviors are due to witnessing domestic violence or to some other factor(s). Lula will be able to conclude only that children who witnessed domestic violence experienced certain problems, not that the problems were caused by witnessing the violence in the first place.

Failure to establish causality, however, is only a limitation if the establishment of causality was one of the purposes of the study. In this case, it was not; and indeed the kind of rigorous research design needed to establish causality is usually inappropriate in a qualitative study.

Lula may find that her sample of children is not diverse enough in terms of age, gender, or ethnic background to allow her to draw conclusions about the effects of these variables on their problem behaviors. Again, this is a study limitation only if she has stated her intention to draw such a conclusion. The major limitation that Lula is likely to encounter in her study is related to the issue of credibility or truth value (internal validity, in quantitative terms).

How will she know whether the mothers and their children were truthful in relating their experiences or whether their remarks were geared more toward pleasing her or making themselves appear more socially desirable?

And, if their remarks were based on memories of previous abusive behaviors, how far were those memories reliable? These are common dilemmas in both research and clinical interviews. One way to handle them is through triangulation: obtaining data on the same issue (variable) from more than one data source. Another way is to constantly reflect on the quality of the data being obtained throughout the interview process, and to record the results of these reflections in the researcher's journal. The following are examples of the kinds of questions Lula might ask herself while she is pursuing her reflections:

- Is the interviewee withholding something—and what should I do about it?
- What impact might my race, age, social status, gender, or beliefs have on my interviewee?
- What difference might it make that I work at the women's emergency shelter?
- Did what the interviewee said ring true—or did she want to please me, or look good, or protect someone else, or save herself embarrassment?

- Why am I feeling so stressed after this interview?
- Am I getting the kinds of data that are relevant for my study?

These questions might improve the quality of the data obtained by making Lula more aware of possible sources of error. Even if they do not, Lula will have shown that she has recognized her study's limitations and will take the necessary steps to deal with each limitation.

Part 11: Administration

The final part of a research proposal deals with the organization and resources necessary to carry out the proposed study. Lula might want to separate her role as a researcher from her role as one of the shelter's social workers, for example, by equipping herself with a desk and computer in a room other than that which she usually uses.

If "researcher space" is not a problem, Lula will still need to think about where she should base her operations: where she will write up her notes, analyze her data, and keep the records of her interviews. Then, she has to think about administrative responsibilities. Will she take on the overall responsibility for her study herself or will that fall to her supervisor? What will be the responsibilities of her colleague and the shelter's social workers? To whom will they report? What is the chain of command?

When Lula has put together the details of who does what, in what order, and who is responsible to whom, she will be in a position to consider a time frame. How long will each task take and by what date ought it to be completed? It is very easy to underestimate the amount of time needed to analyze qualitative data and to feed the information back to the research participants for their comments (member checking).

It is also easy to underestimate the time needed to write the final report. Neither of these tasks should be skimped and it is very important to allow adequate time to complete them thoroughly: more time, that is, than the researcher believes will be necessary at the beginning of the qualitative study.

Finally, Lula must consider a budget. If she has to purchase a software computer program to help her analyze her interview data, who will pay for it? Who will cover the costs related to transcribing the data and preparing and disseminating the final report? How much money should she allocate to each of these areas? How much should she ask for overall?

When she has decided on all this, Lula will have completed her research proposal. As discussed, not all proposals are organized in this way, but all essentially contain the information that has been discussed in the preceding sections. This same information can be used to write the final research report, and it is to this that we now turn our attention.

WRITING QUALITATIVE RESEARCH REPORTS

As with a quantitative research report, a qualitative report is a way of describing the research study to other people. How it is written, and to some degree what it contains, depends on the audience it is written for. Lula may want to present her study's findings, for example, only to the board of directors and staff of the women's emergency shelter where she works. In this case, it will be unnecessary to describe the clinical setting (i.e., the shelter) in detail since the audience is already familiar with it. This very familiarity will also mean that Lula must take extra care to protect the identities of her research participants since personal knowledge of the women and children concerned will make it easier for her audience to identify them.

Lula will probably want to submit a written report—particularly if her shelter funded her study—but she may also want to give an oral presentation. As she imagines herself speaking the words she has written, she may find that she wants to organize the material differently or use a different style than she would if she were preparing a written report. Perhaps she will use less formal language, or include more detail about her own thoughts and feelings, or shorten the direct quotes made by the research participants.

Other possible outlets for her work include books, book chapters, journal articles, and presentations at conferences. Again, depending on the audience, she might write quite simply or she might include a wealth of technical detail, perhaps describing at length the methods she used to categorize and code her interview data.

In order to avoid writing a number of reports on the same study aimed at different audiences, she might choose to include in the main body of the report just sufficient technical detail to establish the study's trustworthiness, while putting additional technical material in appendixes for those readers who are interested. Whatever approach she chooses, it is important to remember that qualitative research studies are based on a different set of assumptions than quantitative ones.

As we know by now, the goal of a qualitative study is to understand the experiences of the study's research participants in-depth, and the personal feelings of the researcher cannot be divorced from this understanding. It is therefore often appropriate to report a qualitative study using a more personal style, including both quotes from interviews with the research participants and the researcher's own reflections on the material. The aim is to produce a credible and compelling account that will be taken seriously by the reader.

The material itself can be organized in a number of ways, depending on whether it is to be presented in book form or more concisely, in the form of a journal article. An article usually contains six parts: (1) an abstract; (2) an introduction; (3) a discussion of methodology; (4) a presentation of the analysis and findings; (5) a conclusion, or discussion of the significance of the study's findings; and (6) a list of references.

Abstract

An abstract is a short statement—often about 200 words—that summarizes the purpose of the study, its methodology, its findings, and its conclusions. Journal readers often decide on the basis of the abstract whether they are sufficiently interested in the topic to want to read further.

Thus, the abstract must provide just enough information to enable readers to assess the relevance of the study to their own work. A statement of the study's research question, with enough context to make it meaningful, is usually followed by a brief description of the study's methodology that was used to answer the research question.

Lula might say, for example, that she interviewed eight women and eleven children who were residents in a women's emergency shelter in a small town in Alberta, Canada, plus three of the shelter's social workers. She might go on to identify the problems experienced by the children who had witnessed domestic violence, stating that these problems were derived from analyses of interview data. Finally, she would outline the practical implications from the study for social work practice resulting from a greater understanding of the children's problems.

Introduction

The main body of the report begins with the introduction. It describes the *what* and *why* components of the study, which Lula has already written about in the first five parts of her research proposal. If she goes back to what she wrote before, she will see that she has already identified her research question and put it into the context of previous work through a literature review.

She has discussed why she thinks this question needs to be answered, clarified her own orientation and assumptions, and commented on the interests of the women's emergency shelter or other funding organizations. In addition, she has identified the variables relevant to her study and placed them within an appropriate framework. In short, she has already gathered the material needed for her introduction, and all that remains is to ensure that it is written in an appropriate style.

Chapter 17: Qualitative Proposals and Reports

Method

After the *what* and *why* components of the study comes the *how*. In the methods section, Lula describes how she selected her sample of research participants and how she collected her data. She would provide a justification for why she chose to use the particular sampling and data collection methods.

Again, if she looks back at her research proposal, she will see that she has already written about this in Parts 6, 7, and 8: research design, population and sample, and data collection, respectively. As before, she can use this same material, merely ensuring that it is written in a coherent and appropriate style.

Analysis and Findings

Materials on data analysis and findings are often presented together. Descriptive profiles of research participants and their direct quotes from interviews are used to answer the research question being explored. In her proposal, Lula has already written the part on data analysis in her research proposal, which stated which computer program she was going to use (if any), and the utilization of first- and second-level coding methods. In her research report, however, she would want to identify and provide examples of the meaning units she derived from the first-level coding process. One segment from one of her interviews might have gone as follows:

1. **Pam (sounding upset): The poor kid was never the same after that. The**
2. **first time, you know, it was just a slap on the butt that she might even**
3. **have mistaken for affection, but that second time he slammed me right**
4. **against the wall and he was still hitting my face after I landed. (pause) No**
5. mistaking that one, is there, even for a four-year old? (longer pause) No, well,
6. I guess I'm kidding myself about that first time. She knew all right. *She was*
7. *an outgoing sort of kid before, always out in the yard with friends,* but then
8. she stopped going out, and she'd follow me around, kind of, as if she was
9. afraid to let me out of her sight.

Lula may have identified three meaning units in this data segment. The first (in bold, lines 1–4) relates to what might and might not constitute domestic violence in the mind of a 4-year-old child. The second (in *italics*, lines 6 and 7) relates to the child's behavior prior to witnessing the abuse; and the third (underlined, lines 7–9) relates to the child's behavior after witnessing the abuse. In her report, Lula might want to identify and briefly describe the meaning units she derived from all her interviews, occasionally illustrating a unit with a direct quote to provide context and meaning.

As discussed in Chapter 15, Lula's next task in the analysis is to identify categories, assign each meaning unit to a category, and assign codes to the categories. A description of these categories will also come next in her report. She may have found, for example, that a number of mothers interpreted their child's behavior after witnessing domestic violence as indicative of fear for the mother's safety. Instead of one large category "child's behavior after witnessing domestic violence."

Lula may have chosen instead to create a number of smaller categories reflecting distinct types of behavior. One of these was "after witnessing domestic violence, child demonstrates fear for mother's safety," and Lula coded it as *CAWFMSAF*, where *C* stands for "the child," *AW* stands for "after witnessing domestic violence," and *FMSAF* stands for "fear for mother's safety."

Depending on the number and depth of the interviews conducted, Lula may have a very large number of meaning units, and may have gone through an intricate process of refining and reorganizing in order to come up with appropriate categories. In a book, there will be room to describe all this, together with Lula's own reflections on the process; but in a journal article, running to perhaps 25 pages overall, Lula will have to be selective about what parts of the process she describes and how much detail she provides.

Although meaning units and categories are certainly a major part of Lula's findings, the majority of readers will be more interested in the next part of the analysis: comparing and contrasting the categories to discover the relationships between and among them in order to develop tentative themes or theories. By doing this, Lula may have been able to finally identify the problems most commonly experienced by children who have witnessed domestic violence.

She may even have been able to put the children's problems in an order of importance as perceived by the mothers and their children. In addition, she may have been able to add depth by describing the emotions related to the children's problems: perhaps guilt, on the mother's part, or anger toward the father, or a growing determination not to return to the abusive relationship. These themes will constitute the larger part of Lula's analysis and findings section, and it is to these themes that she will return in her discussion.

Discussion

This part of the research report presents a discussion of the study's findings. Here, Lula will point out the significance of her study's findings as they relate to the original purpose. If the purpose of her study was to inform practice by enabling the shelter's social workers to better understand the needs of children who have witnessed domestic violence, then Lula must provide a link between the children's problems and their needs resulting from those problems. She must also point out exactly how the shelter's social workers' practice might be informed.

If she has found from her study, for example, that children who witnessed domestic violence tend to experience more fear for their mothers' safety than children who had not witnessed the abuse, then a related need might be to keep the mother always within sight. Social workers within the shelter who understand this need might be more willing to tolerate children underfoot in the shelter's kitchen, for example, and might be less likely to tell Mary to "give Mom a moment's peace and go and play with Sue." These kinds of connections should be made for each theme that Lula identified in her study.

The final part of a research report often has to do with suggestions for future research studies. During the process of filling knowledge gaps by summarizing the study's findings, Lula will doubtless find other knowledge gaps that she believes ought to be filled. She might frame new research questions relating to these gaps; or she might even feel that she has sufficient knowledge to enable her to formulate research hypotheses for testing in future research studies.

References

Finally, both quantitative and qualitative researchers are expected to provide a list of references that will enable the reader to locate the materials used for documentation within the report. If the manuscript is accepted for publication, the journal will certainly ask for any revisions it considers appropriate with regard to its style. It is important to note that quotes from a study's research participants do not have to be referenced, and adequate steps should always be taken to conceal their identities.

SUMMARY

The purposes of writing a research proposal are threefold: to obtain permission to do the study; to obtain funding for the study; and to encourage the author to think carefully through what he or she wants to study and what difficulties are likely to be encountered.

The proposal itself should be clear, brief, and easy to read. Although proposals may be differently organized depending on who is to receive them, the information included in most proposals may be logically set out under general headings identified in this chapter. The information contained under most of these headings can also be used to write the research report. Since the proposal outlines *what will be done* and the research report describes *what was done*, the proposal and the report should parallel each other closely, unless the implementation of the study differed widely from what was planned.

REFERENCES AND FURTHER READING

Coley, S.M., & Scheinberg, C.A. (2008). *Proposal writing: Effective grantsmanship* (3rd ed.). Thousand Oaks, CA: Sage.

Fink, A. (2005). *Conducting research literature reviews: From the Internet to paper* (2nd ed.). Thousand Oaks, CA: Sage.

Lock, L.F., Wyrick Spirdusom W., & Silverman, S.J. (2007). *Proposals that work* (5th ed.). Thousand Oaks, CA: Sage.

McClelland, R.W., & Austin, C.D. (1996). Phase four: Writing your report. In L.M. Tutty, M.A. Rothery, & R.M. Grinnell, Jr. (Eds.), *Qualitative research for social workers: Phases, steps, and tasks* (pp. 120–150). Boston: Allyn & Bacon.

Raines, J.C. (2008). Evaluating qualitative research studies. In R.M. Grinnell, Jr., & Y.A. Unrau (Eds.), *Social work research and evaluation: Foundations of evidence-based practice* (8th ed., pp. 445–461). New York: Oxford University Press.

Reid, W.J. (2008). Writing reports from research studies. In R.M. Grinnell, Jr., & Y.A. Unrau (Eds.), *Social work research and evaluation: Foundations of evidence-based practice* (8th ed., pp. 409–421). New York: Oxford University Press.

Thody, A. (2006). *Writing and presenting research.* Thousand Oaks, CA: Sage.

Thyer. B. (2008). *Preparing research articles.* New York: Oxford University Press.

Williams, M., Unrau, Y.A., & Grinnell, R.M., Jr. (2005). Writing qualitative proposals and reports. In R.M. Grinnell, Jr., & Y.A. Unrau (Eds.), *Social work research and evaluation: Quantitative and qualitative approaches* (7th ed., pp. 428–441). New York: Oxford University Press.

Wolcott, H.F. (2001). *Writing up qualitative research* (2nd ed.). Thousand Oaks, CA: Sage.

Check out our Website for useful links and chapters (PDF) on:
- Web-based links to various tutorials on how to write research reports and proposals
  Go to: www.pairbondpublications.com
  Click on: Student Resources
            Chapter-by-Chapter Resources
            Chapter 17

Check Out Our Website

PairBondPublications.com

  √ *Student Workbook Exercises*
  √ *Chapter Power Point Slides*
  √ *On-line Glossaries*
  √ *General Links*
  √ *Specific Chapter Links*

Credits

This edition has been adapted and modified from the previous even editions: Grinnell, R.M., Jr., & Williams, M. (1990). *Research in social work: A primer.* Itasca, IL: F.E. Peacock; Williams, M., Tutty, L.M., & Grinnell, R.M., Jr. (1995). *Research in social work: An introduction* (2nd ed.). Itasca, IL: F.E. Peacock; Williams, M., Unrau, Y.A., & Grinnell, R.M., Jr. (1998). *Introduction to social work research* (3rd ed.). Itasca, IL: F.E. Peacock; Williams, M., Unrau, Y.A., & Grinnell, R.M., Jr. (2003). *Research methods for social workers* (4th ed.). Peosta, IA: Eddie Bowers Publishing; Williams, M., Unrau, Y.A., & Grinnell, R.M., Jr. (2005). *Research methods for social workers* (5th ed.). Peosta, IA: Eddie Bowers Publishing; Grinnell, R.M., Jr., Williams, M., & Unrau, Y.A., (2008). *Research methods for social workers: A generalist approach for BSW students* (6th ed.). Peosta, IA: Eddie Bowers Publishing; and Grinnell, R.M., Jr., Williams, M., & Unrau, Y.A., (2009). *Research methods for BSW students* (7th ed.). Kalamazoo, MI: Pair Bond Publications.

Photographs: All photographs in Part openings by Harvey Finkle Photography, 1524 Sansom Street, Philadelphia, PA 19102. Used with permission.

Chapters 14 & 16; Boxes 3.1, 3.2, 10.1, & 10.2: From: Williams, M., Tutty, L.M., & Grinnell, R.M., Jr. (1995). *Research in social work: An introduction* (2nd ed.). Itasca, IL: F.E. Peacock. Used with permission.

Box 6.1, 6.2, & Figure 6.2: Adapted and modified from: "Measuring Variables," by Nancy S. Kyte and Gerald J. Bostwick, Jr. in Richard M. Grinnell, Jr. (Ed.), *Social work research and evaluation: Quantitative and qualitative approaches* (6th ed.). Copyright © 2001 by F.E. Peacock Publishers. Used with permission.

Figure 6.1: Walter W. Hudson. Copyright © 1993 by WALMAR Publishing Company. Used with permission.

Figure 7.1: From P.N. Reid, and J.H. Gundlach, "A scale for the measurement of consumer satisfaction with social services," *Journal of Social Service Research, 7,* 37-54. Copyright © 1983 by P.N. Reid and J.H. Gundlach. Used with permission.

Figures 9.8, 9.9, & 9.10 (and related text): Adopted from: Richard Polster and Mary Mary Lynch. "Single-Subject Designs." In Richard M. Grinnell, Jr. (Ed.), *Social work research and evaluation* (2nd ed.). Copyright © 1985 by F.E. Peacock Publishers. Used with permission.

Box 11.1: From: "Utilizing existing statistics," by Jackie D. Sieppert, Steven L. McMurtry, and Robert W. McClelland, in Richard M. Grinnell, Jr. (Ed.), *Social work research and evaluation: Quantitative and qualitative approaches* (6th ed.). Copyright © 2001 by F.E. Peacock Publishers. Used with permission.

Index

A design, 170
AB design, 174–175
ABA design, 176–178
ABAB design, 176–178
ABC design, 175
ABCD design, 175
Abstracts, as used within qualitative research reports, 374–375
Accessing information, 44–45
Accountability and research, 3
Accuracy of authority figures, 8–9
Administration:
    as addressed in a qualitative research proposal, 372–373
    as addressed in a quantitative research proposal, 344–345
Alternate-form method, as used within reliability, 125–126
Analysis of covariance, 192
Analysis of findings, as used within qualitative research reports, 375–377
Analysis of variance, one way, 307–308
Analyzing:
    data, 93–94
    qualitative data, 310–330
    quantitative data, 85–87, 282–309
Anonymity, the use of in the informed consent process, 54–55
ANOVA, 307–308
Applicability, as addressed in a qualitative research proposal, 363
Applied research, 28
Approaches to the research method, 28–29
Authority figures, accuracy of, 8–9
Authority, as a way of knowing, 7–9
Availability sampling, 161

Awareness, of values, 14

B design, 171–172
BAB design, 178
BB_1 design, 173
BC design, 174
BCBC design, 178–179
Beliefs and knowledge, 12
Biases, establishing our own in a qualitative data analysis, 327–328
Bribery, in the informed consent process, 55–56

Case-level research designs, 166–185
Categories:
    assignment of codes to, 319–320
    creation of in qualitative data analysis, 318–319
Causality research questions, 37–38
Causality-comparative interaction research questions, 38
Causality-comparative research questions, 38
Central tendency, measures of, 293–296
Checklists, as used in measurement, 140, 145
Chi-square, 301–302
Classical experimental design, 224–225
Classification of research questions, 34–38
Clients, how can we help them, 3
Cluster random sampling, 160
Coercion, use of in research studies, 55–56
Cohort studies, 213–214

381

Collecting data, 92–93:
    within the qualitative research approach, 107, 250–265
    within the quantitative research approach, 107, 230–249
Comparing categories, as used in a qualitative data analysis, 321–322
Comparing research results, 94–95
Comparison group posttest-only research design, 220–221
Comparison group pretest-posttest research design, 221–222
Compensatory equalization, as a threat to internal validity, 202
Compensatory rivalry, as a threat to internal validity, 203
Completeness, of research hypothesis, 82
Composition research questions, 36–37
Computers:
    as used in a qualitative data analysis, 313–314
    as used in a quantitative data analysis, 288–289
Concepts;
    development of, 75
    variables within, 76
Conceptual classification systems, as used in qualitative data analyses, 322–326
Conceptual framework:
    as addressed in a qualitative research proposal, 359–360
    as addressed in a quantitative research proposal, 336–337
Concurrent validity, 128
Confidentiality, in social work research, 50
Conformability, as addressed in a qualitative research proposal, 363
Consent, informed, 50–59
Consistency, as addressed in a qualitative research proposal, 363
Constant errors, 130–131
Consulting with colleagues, as addressed in a qualitative research proposal, 363

Content validity, of measuring instruments, 127–128
Content, as addressed in a qualitative research proposal, 357
Contrast errors, 131
Control groups, use of, 192–193
Correlated variation, 192
Correlation, 302–304
Correspondence, as used within measurement, 117
Council on Social Work Education, 3–4
Creator and disseminator of knowledge, 21–22
Credibility:
    as addressed in a qualitative research proposal, 363
    establishing our own in a qualitative data analysis, 326
Criterion validity, of measuring instruments, 128–129
Cross-sectional group-level survey research design, 209–210

Data:
    definition of, 101
    interpretation of, 94
    secondary, 241–244
    types of, 232–233
Data analysis:
    as addressed in a qualitative research proposal, 369
    as addressed in a quantitative research proposal, 343
    qualitative, 310–330
    quantitative, 282–309
Data collection, 84–85, 92–93
Data collection:
    as addressed in a qualitative research proposal, 368–369
    as addressed in a quantitative research proposal, 342–343
    quantitative data, 230–249
    within the qualitative research approach, 108
Data collection method:
    criteria for selecting a, 272–277
    definition of, 231–232
    evaluation of, 279

implementation of, 278–279
selecting a, 266–280
Data sources, definition of, 231–232
Deception, the use of in the informed consent process, 55–56
Deductive reasoning, 18–19
Demoralization, as a threat to internal validity, 203
Dependability:
    as addressed in a qualitative research proposal, 363
    of qualitative data, 326–327
Dependent t-tests, 305–306
Dependent variables, 79–80
Descriptive case-level designs, 174–175
Descriptive group-level research designs, 215–224
Descriptive research studies, 33
Descriptive statistics, 85–86, 289–300
Descriptive-comparative research questions, 37
Developing:
    concepts, 75
    the research question, 73–83
Diaries, as used in measurement, 143–144
Differential selection of research participants, as a threat to internal validity, 200–201
Diffusion of treatments, as a threat to internal validity, 203
Directional hypothesis, 80–81
Discussion:
    as addressed within qualitative research reports, 377–378
    as addressed within quantitative research reports, 351–352
Disseminating a study's results, 95–96
Dissemination of qualitative research findings, 108
Duplication;
    as used within measurement, 118–119
    striving toward, 71

Educational policy statements, CSWE, 4–5
Error:
    of central tendency, 131
    of leniency, 131
    of severity, 131
Errors:
    constant, 130–131
    measurement, 130–132
    random, 131–132
Ethical acceptability, of research questions, 41
Ethical principles, obeying, 92
Ethics, research, 48–64
Evidence, sources of, 9–10
Existence research questions, 35–36
Existing statistics, use of in research studies, 244–248
Experience, as a way of knowing, 12–13
Experiments, ideal, 188–196
Explanatory:
    case-level research designs, 176–185
    group-level research designs, 224–228
    research studies, 33–34
Exploratory:
    case-level research designs, 166–185
    group-level research designs, 206–215
    research studies, 30–33
External validity, 204–206
Extraneous variables, holding constant, 191–192

Face validity, 129
Feasibility, of research questions, 41
Findings, as addressed in a quantitative research report, 348–350
First-level coding, 317–320
Formats, of measuring instruments, 140–141
Formulating research questions, 38–39
Frequency distributions, 291–293
Fulfilling funding requirements, 62
Funding bodies, 3

Gathering data, 93–94
General research problem area, refinement of, 89–90
Group-level research designs, 186–228

Halo effect, 131
Historical data, as a qualitative data-collection method, 262–264
History, as a threat to internal validity, 197–198
Honesty, when doing research, 15–16
Hudson's *Index of Self-Esteem*, 119–120
Hypotheses:
    as addressed in a qualitative research proposal, 360–361
    as addressed in a quantitative research proposal, 337
    construction of, 80–81
    directional, 81
    evaluation of, 81–83
    nondirectional, 80–81

Ideal experiments, 188–196
Identifying variables within concepts, 76
Independent *t*-tests, 306–307
Independent variable, manipulation of, 79–80, 189–190
Inductive reasoning, 18–19
Inferential statistics, 86–87, 300–308
Information:
    accessing, 44–45
    definition of, 101
Informed consent letter, example of, 53–54
Informed consent, obtaining, 50–55
Initial impressions, of research problem, 88
Instrumentation error, as a threat to internal validity, 199–200
Intended audience, as addressed in a qualitative research proposal, 357

Interaction effects, as a threat to internal validity, 202
Internal validity, 197–203
Interpretation of data, 94
Interpreting data, as used in a qualitative data analysis, 322–326
Interpretive, as a way of thinking, 99–102
Interrupted time-series research design, 222–224
Interval measurement, 286
Interval variables, 137
Introduction, as used within qualitative research reports, 375
Intrusion into the lives of research participants, as a criterion for selecting a data–collection method, 275–276
Intuition, as a way of knowing, 13
Inventories, as used in measurement, 140, 145

Journal, keeping a, 316–317
Journals, as used in measurement, 143–144
Justifying decisions already made, 60–61

Knowing, ways of, 7–20
Knowledge and beliefs, 12
Knowledge level continuum, 29–34

Levels of measurement 283–288
Libraries, use of, 44–45
Limitations:
    as addressed in a qualitative research proposal, 369–371
    as addressed in a quantitative research proposal, 343
    of research studies, 95
Literature review, 88–89:
    as addressed in a qualitative research proposal, 358–359
    as addressed in a quantitative research proposal, 335–336
Logs, as used in measurement, 144–145

Longitudinal case-level study design, 212–213

Matched pairs, as used in random assignment, 195
Maturation, as a threat to internal validity, 198
Mean, as a measure of central tendency, 296
Meaning units, identification of, 317–318
Measurability, striving toward, 67–68
Measurement, 115–133:
    errors, 130–132
    levels, 283–288
Measures of central tendency, 293–296
Measuring instruments, 134–150:
    selection of, 121–129
    standardized, 146–149
    types of, 143–146
    validity of, 127–129
Measures of variability, 297–300
Measuring variables, 90–91
Median, as a measure of central tendency, 296
Member checking, as used in a qualitative data analysis, 327
Member checks, as addressed in a qualitative research proposal, 364
Method:
    as addressed in a qualitative research report, 375
    as addressed in a quantitative research report, 347–348
Misuses of research results, 60–62
Mode, as a measure of central tendency, 296
Mortality, as a threat to internal validity, 201
Multi-group posttest-only research design, 210–211
Multiple baseline designs, 179–185
Multiple realities, 100
Multiple-treatment interference, as a threat to external validity, 205

Narrative interviewing, as a qualitative data-collection method, 252–256
National Association of Social Workers, 3–7
Negative case analysis, as addressed in a qualitative research proposal, 363–364
Neutrality, as addressed in a qualitative research proposal, 363
Nominal measurement, 284
Nominal variables, 135–136
Nondirectional hypothesis, 80–81
Nonprobability sampling, 160–164
Non-reactivity, as used in measuring instruments, 123–124
Normal curve, 295
Normal curves, variations of (figure), 299

Objectivity, striving toward, 69–70
Observations, structured, as a quantitative data-collection method, 237–241
Observer reliability, 126
One-group posttest-only group-level research design, 207–209
One-group pretest-posttest research design, 219–220
One-way analysis of variance, 307–308
Operational definitions:
    as addressed in a qualitative research proposal, 361–363
    as addressed in a quantitative research proposal, 337–341
Ordinal:
    measurement, 284–286
    variables, 136–137
Organization, as addressed in a qualitative research proposal, 357

Panel studies, 214–215
Participant observation, as a qualitative data-collection method, 256–259
Performance appraisals, 62
Phases of scientific inquiry, 16–19
Phases, within the qualitative research approach, 102–108
Politics, research, 59–60
Population:
    as addressed in a qualitative research proposal, 365–367
    as addressed in a quantitative research proposal, 342
    as used in sampling, 154–155
Positivist, as a way of thinking, 67–72
Practice and research, comparison of, 19–20
Predictive validity, 128–129
Presentation of research findings, 87, 108
Pretest-treatment interaction, as a threat to external validity, 204
Previous research findings, as a criterion for selecting a data-collection method, 277
Probability sampling, 155–160
Problem identification, within the qualitative research approach, 104–106
Problem solving process, 19–20
Problem specification, as addressed in a quantitative research report, 346–347
Problems and questions, 26–47
Program participation, as a criterion for selecting a data-collection method, 273–274
Public relations, 61
Pure research, 28
Purpose, as addressed in a qualitative research proposal, 355–356
Purposive sampling, 161–162

Qualitative data analysis, 310–330:
    collection of data within a, 230–265
    planning of a, 313–317
    previewing within a, 315–316
    transcribing within a, 313–315
Qualitative research:
    approach, 28–29, 98–114
    findings, dissemination of, 108
    process (figure), 103
    proposals, 354–379
    reports, 354–379, 373–378
Quantification, as used within measurement, 118
Quantitative:
    data analysis, 85–87, 282–309
    research approach, 28–29, 66–97
    research proposals, 332–353
    research reports, writing of, 345–353
Questionnaires, as a quantitative data-collection method, 233–237
Questions formulation, within the qualitative research approach, 104–106
Questions:
    as addressed in a qualitative research proposal, 360–361
    as addressed in a quantitative research proposal, 337
Quota sampling, 162

Random assignment of research participants, ethics of, 57
Random errors, 131–132
Random numbers table (table), 156
Randomized cross-sectional survey research design, 218–219
Randomized one-group posttest-only research design, 216–217
Randomized posttest-only control group research design, 225–226
Range, as a measure of variability, 297
Ratio measurement, 286–288
Ratio variables, 137
Reactive effects:
    as a threat to external validity, 205
    as a threat to internal validity, 201–202
Realities, multiple, 100

Reality, perceptions of, 109
Reasoning:
  deductive, 18–19
  inductive, 18–19
Reducing uncertainty, striving toward, 70
References, as used within qualitative research reports, 378
Referential adequacy, as addressed in a qualitative research proposal, 364
Refining categories, as used in qualitative data analysis, 320
Reid-Gundlach Social Service Satisfaction Scale, 142
Relations between experimental and control groups, as a threat to internal validity, 202–203
Relationship research questions, 37
Relevance:
  of research hypotheses, 81–82
  of research questions, 40
Reliability, of measuring instruments, 124–125
Research and accountability, 3
Research and practice, comparison of, 19–20
Research approaches:
  comparison of, 108–111
  qualitative, 98–114
  quantitative, 66–97
Research attitude, 28
Research consumer, 21
Research, definition of, 27
Research design, 83–84
Research design:
  as addressed in a qualitative research proposal, 363–365
  as addressed in a quantitative research proposal, 341
  case-level, 166–185
  group-level, 186–228
Research ethics, 48–64
Research findings, presentation of, 87
Research findings, presenting qualitative, 108
Research hypotheses:
  as addressed in a qualitative research proposal, 360–361
  completeness of, 82
  testing of, 83
Research method, 16
Research method, characteristics of, 14
Research participants:
  as used in research studies, 100
  assigning to groups, 194
  ethics with the use of, 49–59
  selection of, 151–165
Research problems and questions, 26–47, 73–83, 87–88
Research proposals, writing of quantitative, 332–353
Research questions:
  as addressed in a qualitative research proposal, 360–361
  causality, 37–38
  causality-comparative interaction, 38
  causality-comparative, 38
  classification of, 34–38
  composition, 36–37
  descriptive-comparative, 37
  existence, 35–36
  factors contribution to good, 40–41
  formulation of, 38–39
  relationship, 37
Research reports, writing of quantitative, 332–353
Research results, misuses of, 60–62
Research roles, 20–23
Research steps, within the quantitative approach, 72–87
Research studies:
  descriptive, 33
  designing in an ethical manner, 56–57
  explanatory, 33–34
  exploratory, 30–32
  identification of previous, 43–44
Research topic:
  as addressed in a qualitative research proposal, 358
  as addressed in a quantitative research proposal, 334–335
Researchability, of research questions, 40
Researcher bias, as a threat to external validity, 205–206

Resources, as a criterion for selecting a data-collection method, 276
Reversal case-level designs, 176–179
Reviewing the literature, 41–44
Rival hypotheses, controlling of, 190–191
Roles, research, 20–23

Sample:
    as addressed in a qualitative research proposal, 365–367
    as addressed in a quantitative research proposal, 342
    deciding on a, 91–92
Sample selection, 91–92
Sample size, determining, 162–163
Sampling, as used with research participants, 151–165
Sampling frame, as used in sampling, 154–155
Scales, as used in measurement, 140
Secondary content data, as a qualitative data-collection method, 260–261
Secondary data, as a quantitative data-collection method, 241–244
Second-level coding, as used in qualitative data analysis, 321–322
Selecting measuring instruments, 121–129
Selecting research participants, 151–165
Selection-treatment interaction, as a threat to external validity, 204
Sensitivity to small changes, within measuring instruments, 122–123
Sharing, research findings, 15
Simple random sampling, 155–156
Size of research study, as a criterion for selecting a data-collection method, 272–273
Skeptical curiosity, of research findings, 15
Snowball sampling, 162
Social service programs, ethics within, 59–62
Social work research, definition of, 27
Sources of evidence, 9–10
Specificity of variables, as a threat to external validity, 204–205
Specificity, of research hypothesis, 82
Split-half method, as used within reliability, 126
Standard deviation, as a measure of variability, 297
Standardization, as used within measurement, 117–118
Standardized measuring instruments, 146–149
Standardized procedures, striving toward, 72
Statistical regression, as a threat to internal validity, 200
Statistics:
    descriptive, 85–86, 289–300
    inferential, 86–87, 300–308
Stratified random sampling, 157–159
Structured observations, as a quantitative data-collection method, 237–241
Study's findings, informing others about, 57–59
Style, as addressed in a qualitative research proposal, 357
Subjects, as used in research studies, 101
Summative measuring instruments, 146
Survey data:
    advantages of, 236–237
    recording of, 236
    reliability and validity of, 236
Survey questionnaires, use of, 233–237
Systematic random sampling, 156–157

Testing, as a threat to internal validity, 199
Testing, research hypothesis, 83
Test-retest method, as used within reliability, 124–125

Theory building, as used in a qualitative data analysis, 322–326
Time order of variables, 188–189
Time, as a criterion for selecting a data-collection method, 276–277
Tradition, as a way of knowing, 10–11
Transcribing qualitative data, 313–315
Transferability, as addressed in a qualitative research proposal, 363
Trend studies, 213
Triangulation of data sources, as addressed in a qualitative research proposal, 363
Triangulation, as used in a qualitative data analysis, 327
Trustworthinesss of qualitative data, 326–328
Truth value, as addressed in a qualitative research proposal, 363
t-tests:
    dependent, 305–306
    independent, 306–307

Utility, of measuring instruments, 121–122

Validity, of measuring instruments, 127–129
Value awareness, 14
Value bases, 109–110
Value labels, 76–79
Values, 102
Variability, measures of, 297–300
Variables:
    description of, 117–121, 135–137
    measurement of, 90–91, 115–133
    relationships between, 190

Ways of knowing, 7–20:
    authority, 7–9
    experience, 12–13
    intuition, 13
    research method, 13–19
    tradition, 10–11
Web, use of, 44–45
Worker cooperation, as a criterion for selecting a data-collection method, 274–275
Writing up a study's results, 95–96